Transportation in a Climate-Constrained World

Transportation in a Climate-Constrained World

Andreas Schäfer, John B. Heywood, Henry D. Jacoby, and Ian A. Waitz

The MIT Press
Cambridge, Massachusetts
London, England

MIT Press books may be purchased at special quantity discounts for business or sales promotional use. For information, please email special_sales@mitpress.mit.edu or write to Special Sales Department, The MIT Press, 55 Hayward Street, Cambridge, MA 02142.

This book was set in Sabon on 3B2 by Asco Typesetters, Hong Kong and was printed and bound in the United States of America.

Library of Congress Cataloging-in-Publication Data

Transportation in a climate-constrained world / Andreas Schäfer . . . [et al.].
p. cm.
Includes bibliographical references and index.
ISBN 978-0-262-01267-6 (hardcover : alk. paper)—ISBN 978-0-262-51234-3 (pbk. : alk. paper) 1. Transportation—Environmental aspects. 2. Combustion gases—Environmental aspects. 3. Greenhouse gas mitigation. 4. Air quality management. 5. Transportation and state. I. Schäfer, Andreas.
TD195.T7T746 2009
363.73′1—dc22 2008039613

10 9 8 7 6 5 4 3 2 1

Contents

Preface

Modern life is enabled by the transportation developments of the past century. While vehicle technology has undergone rapid transformations, our reliance on oil products as a transportation fuel has remained unchanged. Burning gasoline, diesel, or jet fuel in an engine to provide movement produces carbon dioxide, the most abundant anthropogenic greenhouse gas (GHG). Transportation is the fastest-growing source of energy-related GHG emissions, and within this sector, passenger travel is the main component.

Policy makers are confronted with difficult decisions on how to respond to the threat of climate change. The Kyoto Protocol, the first legally binding treaty that mandates the reduction of anthropogenic GHG emissions, is expected to be only the first in a series of commitments that will eventually involve all major GHG-emitting countries. The types of policy measures taken to satisfy these treaty targets will have lasting impacts on the technology we use and could affect the way we live, consume, and travel. Collating and analyzing relevant information about these potential developments is thus critical.

However, someone pursuing an interest in this subject may feel overwhelmed with the amount of information. There are thousands of newspaper articles, newsletters, conference papers, peer-reviewed papers, and other types of publications on the topic. And getting on top of the subject is further complicated by many of the publications being aimed at a scientific audience or focusing on only specific details within this broad area. Furthermore, not all studies lead to similar conclusions.

This book is the first attempt to systematically integrate the various factors affecting GHG emissions for all major modes of passenger transport on a U.S. and global scale. To be comprehensive, it covers a wide range of determinants, including travel demand, consumer and industry behavior, transportation and fuel technology, and climate policy. The book draws heavily from studies in this area that we have conducted at the Massachusetts Institute of Technology during the past decade. It thus integrates results from a range of research programs, research centers, and departments at MIT, including the MIT Joint Program on the Science and Policy of Global Change, the Cooperative Mobility Program at the Center for Technology, Policy & Industrial Development, the Laboratory for Energy and the Environment, and the Departments of Mechanical Engineering and Aeronautics & Astronautics. However, for this book to be inclusive, it also incorporates outputs from numerous other research efforts from around the world.

We address this book to a broad audience. At one end of the spectrum, it is intended to reach the informed reader interested in this subject. However, by keeping the book broad, we hope a scientist entering a career in this area will also discover new information and find it a useful reference. Given the wide audience it is intended to reach, we expect readers to approach it in different ways. Those generally interested in the subject may read all eight chapters. Since each chapter is self-contained, readers interested in specific topics can focus on a specific chapter in isolation of other chapters. The technical reader may find the many technical details and references in the endnotes useful.

While much of the book's material is based on our joint studies, other experts have been involved in research that underlies our analysis. Among them, David Victor has collaborated on global travel demand (chapter 2), and Malcolm Weiss on the life-cycle assessment of alternative fuels (chapter 6). We are also grateful to Jim Hileman and Tom Reynolds for help on aircraft operations and technology (chapters 3 and 5). Lynnette Dray, Tony Evans, Jim Hileman, Tom Reynolds, and Malcolm Weiss provided helpful comments on sections of the manuscript. This book also benefits from hard-working graduate students, in particular Raffi Babikian, Solomon Jamin, Joosung Lee, and Stephen Lukachko. We also wish to thank Bob Zalisk, who carefully edited the first manuscript in view of making it understandable to a broad audience.

Thanks are also due to our institutions, the Massachusetts Institute of Technology and the University of Cambridge, which gave us the time and resources such a book project requires. In particular, we thank the MIT Joint Program on the Science and Policy of Global Change and its sponsors, which include U.S. government agencies, a consortium of corporations, industry associations and foundations, and an anonymous donor. We also thank the industry sponsors of the Cooperative Mobility Program. Last, but all the more, we thank our families, whose patience has been tested at length.

Despite the invaluable support from colleagues and a wide range of sponsors, the ideas expressed in this book and the responsibility for their accuracy rests with the authors.

Abbreviations

ASK	Available seat kilometer
bbl	Barrel (1 bbl = 42 gal = 159 L)
BTL	Biomass-to-liquids
CAFE	Corporate Average Fuel Economy
CBO	Congressional Budget Office
CCSP	Climate Change Science Program
CTL	Coal-to-liquids
CH_4	Methane
CNG	Compressed natural gas
CO_2	Carbon dioxide
DOC	Direct operating costs
E	Energy
EJ	Exajoule (10^{18} joules)
E/PKT	Passenger travel energy intensity
EPPA	The MIT Emissions Prediction and Policy Analysis model
EPPA-Ref	EPPA reference run
ETS	Emissions Trading Scheme
FTP-75	Federal Test Procedure-75
g	Grams
gCO_2-eq	Grams of CO_2 equivalent
gal	U.S. gallon (1 gal = 3.785 L)
GGE	Greenhouse gas emissions
GHG	Greenhouse gas
GE	Gasoline equivalent
GJ	Gigajoule (10^9 joules)
GTL	Gas-to-liquids
GDP	Gross domestic product
GWP	Gross world product

ha	Hectare
hp	Horsepower (1 hp = 0.746 kW)
I	Investments
ICAO	International Civil Aviation Organization
IEA	International Energy Agency
IPCC	Intergovernmental Panel on Climate Change
J	Joule
kg	Kilograms
km	Kilometer (1 km = 0.622 miles)
kW	Kilowatts
kWh	Kilowatt hour
L	Liter (1 L = 0.264 gallon)
lbs	Pounds (1 lb = 0.454 kg)
LDV	Light-duty vehicle (automobiles and light trucks, i.e., minivans, vans, pickup trucks, and sport-utility vehicles)
LCB	Lifetime CO_2 burden
L/D	Lift-to-drag ratio
Li-ion	Lithium ion
LNG	Liquefied natural gas
MBTE	Methyl tertiary-butyl ether
MER	Market exchange rates
MJ	Megajoule (1 million joules)
mpg	Miles per gallon
NextGen	Next Generation Air Transportation System
NiMH	Nickel metal hydride
NO_x	Nitrogen oxides
N_2O	Nitrous oxide
NRC	National Research Council
OPEC	Organization of Petroleum Exporting Countries
PAX	Number of passengers
PJ	Petajoule (10^{15} joules)
pkm	Passenger kilometers
pmi	Passenger mile
PKT	Passenger kilometers traveled
PNGV	Partnership for a New Generation of Vehicles
ppmv	Parts per million by volume
PPP	Purchasing power parity
Q	Net energy content

SESAR	Single European Sky Air Traffic Management Research initiative
SFC	Specific fuel consumption
SRES	IPCC Special Report on Emission Scenarios
UN	United Nations
V	Speed
vkm	Vehicle kilometer
VKT	Vehicle kilometers traveled
VLJ	Very light jet (jet-engine-powered aircraft with a weight below 4.5 tons [10,000 lbs])
vmi	Vehicle mile
W	Weight
Wh	Watt hours
ZEV	Zero-Emission Vehicle

Transportation in a Climate-Constrained World

1

Introduction: From Local Impacts to Global Change

Consider this traffic situation: the roads are highly congested and clogged with broken-down vehicles. Exhaust emissions are a thousand times higher than those of the average vehicle today and leave air quality miserable. Traffic-related fatalities claim one of every two hundred residents each year. The streets are dangerous, noise levels are intolerable, and a noxious smell hangs over all. Beyond the packed streets, the increasing demand for transportation fuel is about to threaten other basic needs of daily life.

A bleak forecast by the sixteenth-century seer Nostradamus? In fact, the passage describes the societal impact of the horse-drawn urban transportation system of the second half of the nineteenth century.

The magnitude of this impact becomes even more impressive when seen in a specific location. In Victorian London, an 1850 traffic count recorded a thousand horse-drawn vehicles per hour passing over London Bridge in the course of a day. Given total horse droppings of about five tons per horse per year, new markets developed to take care of this "exhaust" problem. The function of one new job, crossing sweeping, was to create a passage through the often liquid, ankle-deep manure so that London pedestrians could cross from one side of the street to the other.[1] At the time, it was forecast that within a few decades, current rates of traffic growth would bury London under six feet of horse manure.

Clearly, the urban transportation system of the nineteenth century had reached its limits. Further increases in horse traffic would have been constrained not only by the already extreme levels of congestion, emissions, and noise, but also by the limited amount of land available for growing horse fuel at reasonable cost. According to one estimate, the demand for hay in the United States had already taken up about one-third of the cropland at that time.[2] (As we will see later in this book, this land constraint reappears in today's drive for *modern* biofuels.)

Given these limiting conditions, it was just a question of time as to when a revolution in transportation technology would emerge. Although the first horse*less* carriages were widely regarded with skepticism, their advantages soon became obvious. The early automobiles released hundreds of times lower tailpipe emissions, occupied less than one-half of the road space, and liberated huge cropland areas to feed a growing population. The rest is history.

The Rise of the Automobile

Analysts identify two phases of motor vehicle adoption in the United States. From about 1900 to the early 1920s, the automobile was expensive; it was mainly wealthier horse owners who switched to the motor vehicle.[3] During that period the U.S. automobile fleet increased at almost 40 percent per year. Such rapid growth, sustained over such a long period, has remained unmatched in the subsequent history of transportation. The next phase of motor vehicle adoption allowed a growing portion of the middle class to own an automobile, a development made possible by lasting economic growth, a sharp decline in vehicle purchase costs (mainly because of the advent of mass production), and the introduction of vehicle-financing plans.

During the 1930s the automobile became the single most important mode of transport in the United States, two decades before the construction of the interstate highway system.[4] Given long-term growth in vehicle ownership at rates of about 2 percent per year during the second phase of vehicle adoption, mass motorization—when on average every U.S. household owned one automobile—was achieved by 1952. In combination with generous government housing programs, rising car ownership enabled an increasing portion of the middle class to realize the American dream of affordable housing in a safe suburban environment. In light of increasing car dependency and decreasing urban population densities, the gradual decline of public transportation was unavoidable. In the 1980s, a Chicago transit official declared that mass transit was "no longer relevant to the American way of life."[5]

Although the onset of motorization was earliest in the United States, it has become a global phenomenon. In Western Europe the rapid increase in the number of vehicles per person started only after the Second World

War. Since then, Western Europe has maintained a roughly constant delay of about thirty years behind U.S. motorization levels. This delayed growth resulted in part from lower average household income, but also in part from a different industry approach to marketing. U.S. manufacturers aimed their products at the mass market early on and soon were able to exploit economies of scale. In contrast, European manufacturers initially targeted their products toward the wealthy upper class, necessarily incurring higher costs and reduced sales.

In both the United States and Western Europe (along with other industrialized countries), the vehicle market has essentially become a replacement market, where new vehicles mainly substitute old or outdated ones. In contrast, today's largest growth market is in the developing economies. Given the currently small size of their vehicle fleets, auto ownership is growing at about 30 percent per year, a rate nearly as high as that in the U.S. market a century ago. Since about 80 percent of the world population lives in the developing countries, the largest wave of motorization is yet to come.

Figure 1.1 shows the century-long growth of the world motor vehicle fleet in the United States, other industrialized countries, and the rest

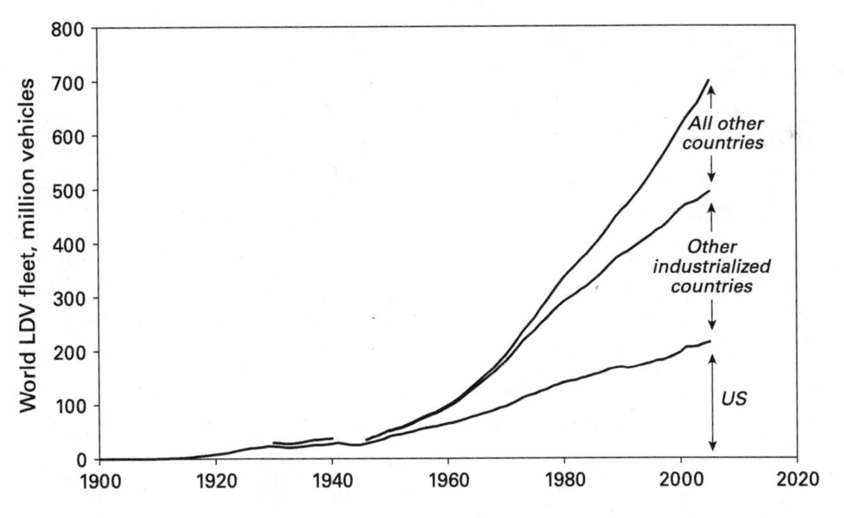

Figure 1.1
Historical growth of the world light-duty vehicle fleet, 1900–2005. Sources: U.S. Department of Transportation, various years. *Highway Statistics*. U.S. Federal Highway Administration, Washington, DC. Motor Vehicle Manufacturers Association, 1996. *World Motor Vehicle Data*, Detroit, MI. Authors' database.

of the world. The motor vehicle fleet includes automobiles and other light-duty vehicles (LDVs; vans, minivans, pickup trucks, and sport-utility vehicles) that are predominantly used for passenger travel.[6] In 2005, the world LDV fleet comprised 700 million vehicles, up from essentially zero in 1900. Over the past five decades alone, the vehicle fleet grew by 5 percent per year, corresponding to a doubling of its size every fifteen years. Should that rate continue to hold into the future, the global fleet of LDVs would increase to about 2 billion in 2030.

The growing motor vehicle fleet has transformed national economies by enabling the integration of distant labor, product, and consumer markets. In addition, the vehicle industry itself has become a pillar of the economy. A 2003 report by the University of Michigan and the Center for Automotive Research concludes that the automobile industry alone provides one out of ten jobs in the United States, either directly or indirectly.[7] Rising automobile ownership has also contributed to a higher quality of life. Compared with a person living in the nineteenth century, a representative of the automobile age has greatly improved access to education and health care, much greater freedom of choice in where to live and work, and where, when, and how to travel. These invaluable opportunities have lead to the perception of the automobile as an icon of personal freedom.

These benefits have not come without challenges, however. Although the transition from the horse to the automobile has provided significant environmental and societal benefits on a transport-system level, many of these improvements have been outpaced by growth in transport demand and associated side effects. At the same time, the original challenge that innovations in transportation technology were expected to resolve has spread. Take traffic congestion as an example, a phenomenon originally observed in city centers. Although automobiles occupy only less than one-half the road space of horse carriages, the enormous growth in automobile ownership has led to congestion levels as severe as in the horse age but at a larger geographic scale. Traffic congestion today typically covers entire metropolitan areas.

A similar increase in geographic scale of transport-related impacts has been observed for urban air pollution, even though significant technical improvements have been achieved. In the horse age, the immediate impact of solid and liquid emissions was confined to urban areas, mainly owing to the limited range of the flies that transmitted infectious diseases. In the automobile age, gaseous vehicle pollutants are transported over hundreds of kilometers and have effects at distant locations. Already during the early 1950s, automobile emissions were shown to contribute to the photochemical smog in large areas of the Los Angeles basin. That mixture of reactive pollutants was soon found in other U.S. cities and abroad.

In response to concerns about deteriorating urban and regional air quality, regulatory measures were adopted. In 1966 California introduced the first tailpipe emission standards for hydrocarbons and carbon monoxide, and shortly thereafter federal legislation regulated so-called criteria pollutants (carbon monoxide, lead, nitrogen dioxide, sulfur dioxide, particulate matter, and ground-level ozone) that affect human health. This and subsequent U.S. legislation led to the development and adoption of the catalytic converter and cleaner transportation fuels, which decoupled the criteria emissions from gasoline use. Between 1970, when the federal clean air legislation was passed, and 2002, U.S. automobile and light truck emissions of carbon monoxide, nitrogen oxides, and unburned hydrocarbons declined by 60 to 70 percent even while gasoline use increased by 60 percent and vehicle kilometers traveled multiplied by a factor of 2.5.[8] Governments from many other countries have followed suit, and similar reductions in criteria emissions are being achieved worldwide. Examples of other vehicle-level improvements that have been outpaced by the strong growth in road traffic, and whose impacts have increased in geographic scale, include noise and traffic accidents.[9]

And how did the vehicle industry respond to traffic congestion? While automobile manufacturers cannot mitigate traffic congestion itself, the industry has started offering onboard satellite-guided navigation devices, which in combination with traffic updates allow drivers to avoid the most congested areas. Vehicle manufacturers also offer more advanced entertainment systems and onboard work opportunities to mitigate the *impact* of traffic congestion on car occupants.

The Growing Competition from Air Travel

Although automobiles began to dominate U.S. travel in the 1930s, for several decades long-distance trips continued to be made predominantly via rail. Among rail options, the streamliners, lightweight diesel-electric trains, offered high-speed intercity connections.[10] Yet, after the Second World War, government investments in interstate highways and airports contributed to the demise of this first-generation high-speed train.

The strong increase in commercial air travel after the Second World War was also enabled by technological innovations like the introduction of the jet engine in the early 1950s, allowing a significant increase in aircraft capacity and speed. By the mid-1950s aircraft had already displaced buses and intercity railways to become the second most important mode of U.S. intercity travel. Since about 1960 the strong growth in air travel has captured market share from automobiles as well. By 2005 commercial aircraft accounted for about half of all passenger kilometers traveled (PKT) on long-distance trips, here defined as travel at trip distances greater than 100 kilometers (62 miles).[11] Aircraft ultimately dominate travel markets at distances greater than 1,000 kilometers; that is, distances that would require automobile drivers to spend at least one night in a hotel. Over the past five decades, the demand for air travel in the United States has grown by nearly 9 percent per year, compared with about 2 percent per year for the LDV fleet.

Similar growth trends can be observed in other parts of the industrialized world, even in countries with an extensive high-speed rail network. In Western Europe also, aircraft provided half of all long-distance PKT in 2005, up from 2 percent in 1950. (Despite the dense network, high-speed rail accounted for only 3 percent of the 2005 total PKT in long-distance travel.) In Japan, the country with the longest continuous history of high-speed rail, air travel supplied one-third of total long-distance PKT, with Shinkansen trains accounting for an additional one-sixth. While the air travel share currently is lower than in rich countries, even stronger growth in air travel is seen in the rapidly developing Asian economies, where rising income, poor-quality surface transport, and large travel distances are contributing to the trend.

Figure 1.2 shows the historical growth in U.S. and world passenger air traffic. Although the era of commercial air travel began after the Second World War, PKT started to grow strongly only in the 1960s, when a critical number of large jet aircraft entered the U.S. fleet. In 2005, U.S. aircraft provided about 1,250 billion passenger kilometers (pkm), an average of 4,200 kilometers (2,610 miles) per person. Total U.S. air traffic volume roughly compares to that of all other industrialized countries, and to that of the remaining countries. All together, in 2005, the commercial world aircraft fleet provided about 4,000 billion pkm, up from

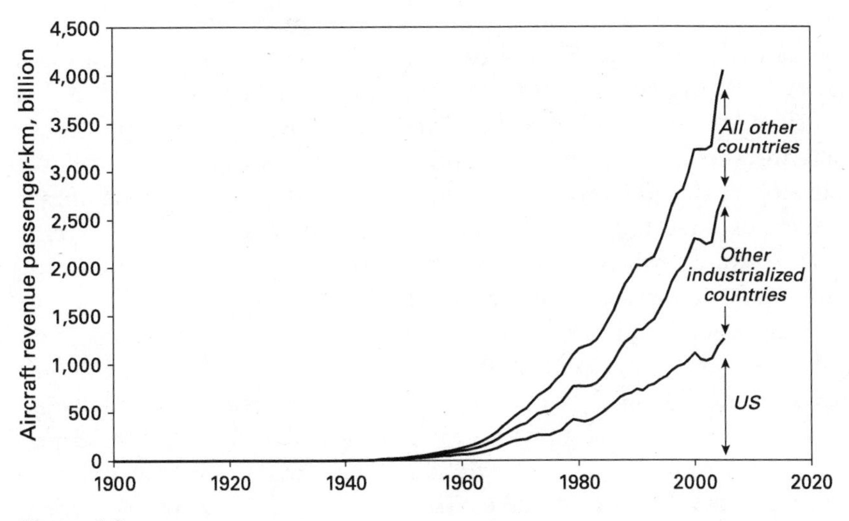

Figure 1.2
Historical growth of world commercial passenger aviation, 1900–2005. Sources: U.S. Census Bureau, 1975. *Historical Statistics of the United States, Colonial Times to 1970*, U.S. Department of Commerce, Washington, DC. Air Transport Association of America, 1950–2007. *Economic Report* (formerly *Air Transport Facts and Figures*), U.S. Department of Commerce, Washington, DC.

virtually zero in 1950, and corresponding to an average growth in excess of 9 percent per year.

In 2005, more than 60 percent of the 4,000 billion pkm supplied by aircraft occurred in international traffic. Not counting the large North American market, in which 70 percent of all aircraft-related PKT occurred in domestic travel, international air traffic accounted for nearly 80 percent of all world aviation-related PKT in 2005.[12] While the defeat of the immediate "distance barrier" has made the automobile an icon of personal freedom, overcoming national and continental boundaries has turned air travel into a symbol of globalization. The very existence of many industries and services has become reliant on the fast and efficient movement of people and goods over long distances.

As with the automobile, strong growth in air travel demand has forced the aircraft industry to try to mitigate its vehicles' environmental and societal impacts. Unfortunately, even impressive improvements at the vehicle level have been outpaced by growth in travel demand, so impacts have increased in geographic scale. Take surface air quality, for example. In 1973, the U.S. Environmental Protection Agency regulated emissions of smoke, unburned hydrocarbons, carbon monoxide, and nitrogen oxides for several classes of subsonic aircraft engines. Due to the focus on surface air quality, such regulations apply to landing–take off cycles, which extend to an altitude of 915 meters (3,000 feet); emissions above that threshold have remained unregulated to date.[13] Over time, the emission standards have become increasingly stringent, and significant reductions have been achieved per aircraft operation. Yet, in contrast to automobiles, where the catalytic converter and cleaner transportation fuels have decoupled local air pollutant emissions from fuel use, the growing demand for air travel has caused most of these emissions to increase. Between 1970 and 2002, the number of departures at U.S. airports from scheduled passenger and freight aircraft increased by 80 percent. Over the same period, emissions increased by a low of 33 percent (sulfur dioxide) to a high of 200 percent (large particulates); carbon monoxide and nitrogen oxide emissions increased by about 60 percent. Only emissions of unburned hydrocarbons declined, by about 60 percent.[14] In addition, the geographic scale of aviation-related air pollutants has increased. Aircraft emissions are increasingly being understood to contribute to regional environmental impacts, even when aircraft operate at cruise altitude.[15]

Another impact that has grown in geographic scale is air traffic congestion. Absent any capacity increase or changes in operational strategies, air traffic congestion would have already become a binding constraint for a significant further rise in air travel. However, the increase in traffic congestion at primary airports has induced the airline industry to adopt new business models aimed at greater use of secondary and tertiary airports. As a result, airport operations have been dispersed to multiple points in many metropolitan areas. Most prominent are the budget airlines, which—as a result of increasing traffic congestion at primary airports along with prospects of cost savings at secondary airports—started a network of airline services parallel to that of the large commercial carriers as early as the 1970s.[16] More recently, very light jets (VLJs) equipped with precision satellite navigation have begun to open up a third, largely parallel network of airline operations, with the prospect of operating routinely among the several thousand U.S. airfields that do not have control towers or radar. If successful, these comparatively affordable single-pilot aircraft, which seat between three and six passengers, would further spread aircraft operations (and air pollution) over a regional scale. The geographic scale of other impacts has been less pronounced. Significant progress has been achieved in aviation safety, but these improvements have also been offset by the strong growth in traffic.[17] And aircraft noise has been reduced, even on an absolute scale.[18]

Although some of the environmental and societal concerns could not be resolved completely yet, continuous improvements in aircraft technology and fuels are likely to result in further reductions in pollutant emissions and noise on a per aircraft and per operation basis. However, most projections suggest that, at least over the next ten to thirty years, these reductions will not be sufficient to compensate for the expected increase in air travel.[19] Fortunately, in the longer run, concepts for future aircraft designs and cleaner fuels could diminish these air travel impacts.

Oil Dependence and Climate Change

While the reduction of many environmental and societal impacts on a vehicle and operation basis have been offset by the strong growth in traffic, there is a growing sense that all can eventually be controlled, at least over the long term. Two concerns, however, remain unresolved. Both are related to the dependence of our transportation system on petroleum products: gasoline, diesel, and jet fuel account for 97 percent of all transportation fuels in the United States and 94 percent on a global average.[20]

One concern is oil dependence. Over the past century, the U.S. demand for petroleum products has grown strongly. However, domestic oil production could not keep pace with that development. After the Second World War, oil imports began to exceed exports, and the share of imported oil has risen ever since, from 36 percent in 1975 to 65 percent in 2005.[21] Since the great bulk of the world's oil reserves is located in politically less stable regions, the rising dependence on oil imports raises national security concerns. The vulnerability of the global fuel supply systems was made clear by the oil supply disruptions during the 1973 Middle East War and the 1979 Iranian Revolution, and oil and gas security remain important concerns of the foreign and military policies of importing regions, importantly including the U.S.

In an effort to mitigate oil dependence, the U.S. Congress enacted the world's first fuel economy regulations. Introduced in 1978 with the intention of doubling the fuel economy of new motor vehicles within a decade, the regulations forced vehicle manufacturers to design products that met increasingly stringent fuel-efficiency targets. In addition to

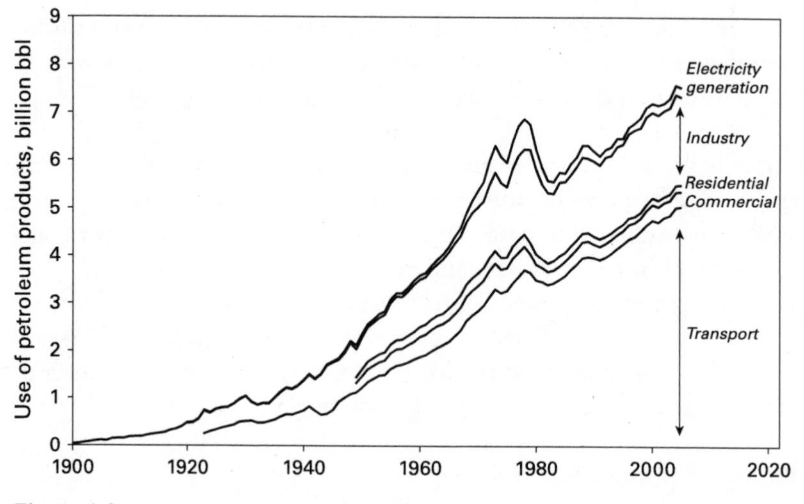

Figure 1.3
Use of petroleum product by sector in the United States, 1900–2005. Sources: U.S. Census Bureau, 1975. *Historical Statistics of the United States, Colonial Times to 1970*, U.S. Department of Commerce, Washington, DC. U.S. Energy Information Administration, 2007. *Annual Energy Review 2006*, U.S. Department of Energy, Washington, DC.

government responses, the expectation of further oil price spikes led to market and other economic adjustments; the amount of oil used in electricity generation, industry, and the residential sector declined in favor of less price-volatile fuels. Because of the stronger responsiveness of non-transportation sectors, transportation accounted for an even larger share of oil consumption after this period of market disruption.

The century-long growth in the consumption of petroleum products and their use across sectors in the United States is reflected in figure 1.3. Between the early 1920s and 2005, consumption of petroleum products rose more than tenfold. During this same period, passenger and freight transport increased their share of oil use from less than half to two-thirds. As a result, U.S. transport-sector oil use increased more than twentyfold. Given the likely continuation of the strong increase in the world motor vehicle fleet (figure 1.1) and of air travel (figure 1.2), the demand for petroleum products, especially for transportation fuels, is likely to continue to grow also on a global scale.

The strong dependence of transport systems on oil products and the projected growth in consumption not only raises energy security concerns. Burning a liter of gasoline, diesel, or jet fuel in an automobile or aircraft engine releases nearly 2.5 kilograms (around 5.5 lbs) of carbon dioxide (CO_2), a major greenhouse gas (GHG), into the atmosphere. Already, the atmospheric concentration of CO_2 has increased from a preindustrial level of 280 parts per million by volume (ppmv) in 1800 to about 380 ppmv in 2005. Given the projected increase in human activity, concentration levels will continue to rise, changing the radiative balance of Earth and affecting the global climate. The projected implications of the anthropogenic (human-influenced) greenhouse effect are significant. An increase in the mean Earth temperature leads to the thermal expansion of oceans, the melting of the ice shelves, and thus to a sea level rise. An increase in the mean Earth temperature also induces an increase in extreme weather events, such as heat and cold waves, droughts, heavy rains, and tropical storms. Some of these ecosystem alterations form the basis for secondary impacts, including the spread of tropical diseases outside their current latitude band, mass migration of people most affected by climate change, and economic losses.

Due to its abundance, CO_2 is the most important contributor to the anthropogenic greenhouse effect. Other GHG emissions, however, can have a stronger warming effect. Among those are methane emissions from agricultural practices, animal farming, landfills, energy-related activities, and other sources. Over a span of one hundred years, methane has a 21 times stronger climate impact than the same mass of CO_2. A still stronger climate impact of 310 times that of a mass unit of CO_2 results from nitrous oxide (N_2O) emissions if measured over the same time horizon.[22] N_2O emissions result from soil cultivation, nitrogen fertilizer use, and animal waste. As we will discuss in more detail, some strategies that aim at reducing CO_2 emissions can result in an increase in these stronger GHG emissions and thus reduce the potential of GHG emission reduction as given by CO_2 alone.

Two other important components of the greenhouse effect are water vapor and clouds. Along with CO_2, water vapor is the other major product of fossil fuel combustion, but water emissions at Earth's surface have no climate effect. If released at cruise altitude by commercial aircraft, however, water vapor emissions can be more significant. The precise climate impact of aircraft water emissions, especially the extent to which they result in persistent contrails, depends on several factors, including the prevailing ambient atmospheric conditions and the amount and types of particles formed in the engine exhaust.

While line-shaped contrails contribute to global warming, they can evolve into larger regions of cirrus clouds, which also tend to warm Earth, although the amount of the warming is subject to great uncertainty. A similar complication is the climate impact of aircraft nitrogen oxide (NO_x) emissions, because they cause changes in the concentration of other greenhouse gases. These changes partly offset each other, and their impacts are regional and thus not strictly additive. Overall, however, the combined global warming impacts of the various consequences of commercial aircraft are estimated to be larger than the impacts of the CO_2 emissions alone.[23] (However, because of the formation of troposphere ozone, automobiles also contribute to climate change to a larger extent than suggested by their CO_2 emissions alone.)

The Intergovernmental Panel on Climate Change (IPCC) summarizes the results of computer models that simulate the dynamics of the biosphere, ocean, and atmosphere. The IPCC concludes that an increase in the atmospheric CO_2-equivalent concentration to 550 ppmv—roughly twice the preindustrial level—would result in a global average temperature increase of about 3°C above the preindustrial equilibrium. The associated global average sea level rise would correspond to 0.6–1.9 meters (2.0–6.2 feet). Limiting the effects to such levels, however, would be very ambitious, because global CO_2 emissions would need to peak before 2030. A later peak in CO_2 emissions would correspondingly result in a larger atmospheric concentration, a stronger temperature increase, and a higher sea level rise.[24]

GHG emissions differ from urban air pollutants in two important ways. Because the atmospheric lifetime of CO_2 is on the order of one hundred years, changes in the composition of the atmosphere are long lasting. Thus, while the impact of urban air pollutants have been limited to a regional level, the long lifetime of CO_2 and other long-lived greenhouse gases means that their impacts expand to global scale. In addition, unlike urban air pollution, there is no practical end-of-pipe technology that could be used to reduce CO_2 emissions from transport systems. Since CO_2 is formed by the oxidation of carbon atoms in the transport fuel, all reduction options must aim at burning less (nonrenewable) carbon-containing fuel. Because of oil dependence and the huge scale of the transportation system today, changing automobile and aircraft technologies and their supporting fuel systems is an enormous task that may require instituting the largest technological transformation since the transition from the horse to the automobile itself.

Table 1.1 indicates GHG emissions by type, country, and source in 2005, ranked according to per capita emission levels. Not counting the CO_2 emissions from land-use change, global GHG emissions amounted to nearly 40 billion tons of CO_2 equivalent. As can be seen, their distribution across countries is inhomogeneous. Nearly one-fifth of the world GHG emissions were emitted by the United States, with more than 7 billion tons of CO_2 equivalent, which include emissions of CO_2, methane, N_2O, and various industrial gases. A close second is China, which in 2007 overtook the United States as the leading emitter of GHGes, albeit with a much larger population. In general, CO_2 accounts for the large majority of total GHG emissions, and nearly all CO_2 emissions (aside from those related to forest destruction) result from activities involving fossil fuels (the "energy-related" CO_2 emissions in table 1.1). In the United States, passenger travel accounts for 22 percent of all energy-related CO_2 emissions and for 18 percent of total national GHG emissions. This emission level is especially significant when comparing to the total GHG emissions from other countries. The CO_2 emissions from U.S. *total* (passenger and freight) *transportation* of 1,920 million tons are larger than the *total national* GHG emissions from each of the other countries with the exception of China and Russia. U.S. *passenger* travel CO_2 emissions alone are larger than the *total national* GHG emissions of Germany and are about twice the national GHG emissions from the United Kingdom or France.

Global passenger travel currently releases a comparatively small share of total GHG emissions. In 2005, world passenger travel accounted for roughly 14 percent of the global energy-related CO_2 emissions. Yet, the relative importance of passenger travel CO_2 emissions is likely to increase in the future mainly because of structural changes in the world economy. Early in the economic development process, agriculture is usually the dominant production sector. With economic growth, agriculture is bypassed first by industry and then by services. Within services, total transport use (industrial and personal) takes an ever-increasing economic role. Since each of these sectors consumes energy to produce goods or services, we observe a similar sector shift for energy use and CO_2 emissions. Indeed, the historical data in figure 1.4 show these shifts in CO_2 emissions, from the residential sector (for heating, cooling, and running appliances), to the industry sector, and finally to the service sector, with transportation being by far the single largest energy consumer of the service sector.[25]

Table 1.1
Greenhouse gas emissions (in million tons of CO_2 equivalent) for selected countries and the world by source in 2005

	GHG emissions		CO_2 emissions					
	Total	Per capita		Energy-related				
					Transportation-related			
						Passenger travel		
	Mt[a] CO_2-eq[b]	tCO_2-eq	$MtCO_2$	$MtCO_2$	$MtCO_2$	$MtCO_2$	% $EnCO_2$[c]	% $TrpCO_2$[d]
Australia	534	26.3	393	384	91	60	16	66
United States	7,300	24.6	6,140	6,090	1,920	1,340	22	70
Russia	2,170	15.0	1,780	1,740	165	86	5	52
Germany	1,020	12.4	894	873	184	132	15	72
United Kingdom	696	11.6	596	558	171	106	19	62
Japan	1,380	10.8	1,320	1,290	281	170	13	61
France	576	9.5	434	417	154	85	20	55
China	6,940	5.3	5,300	5,170	335	135	3	40
India	1,900	1.7	1,270	1,250	110	47	4	43
World	38,400	5.9	28,200	27,900	6,370	3,890	14	61

Sources: GHG and CO_2 emissions were mainly derived from UN Framework Convention on Climate Change (UNFCCC), *Gas Inventory Data—Detailed Data by Party*; http://unfccc.int/ghg_data/ghg_data_unfccc/ghg_profiles/items/3954.php. These figures were complemented with CO_2 emission data from international air traffic, which are not included in the UNFCCC data. The basis for estimates of CO_2 emissions from international air traffic is International Energy Agency, 2007. *Energy Balances of OECD and Non-OECD Countries*, IEA/OECD, Paris. The IEA *Energy Balances* were also the basis for estimating missing CO_2 emissions data for China, India, and the world. Source for land-to-atmosphere CO_2 fluxes: Denman, K.L, G. Brasseur, A. Chidthaisong, P. Ciais, P.M. Cox, R.E. Dickinson, D. Hauglustaine, C. Heinze, El. Holland, D. Jacob, U. Lohmann, S. Ramachandran, P.L. da Silva Dias, S.C. Wofsy, X. Zhang, 2007. Couplings Between Changes in the Climate System and Biogeochemistry, IPCC Fourth Assessment Report, Working Group I Report, *Climate Change 2007—The Physical Science Basis*, Cambridge University Press, Cambridge, UK.

[a] Mt = million metric tons.

[b] CO_2-eq emissions include CO_2 emissions and those of other greenhouse gases, converted to CO_2 emissions on the basis of a 100-year global warming potential. Data exclude land-to-atmosphere CO_2 fluxes of 1,800–9,900 $MtCO_2$.

[c] Percent energy use-related CO_2 emissions.

[d] Percent transportation-related CO_2 emissions.

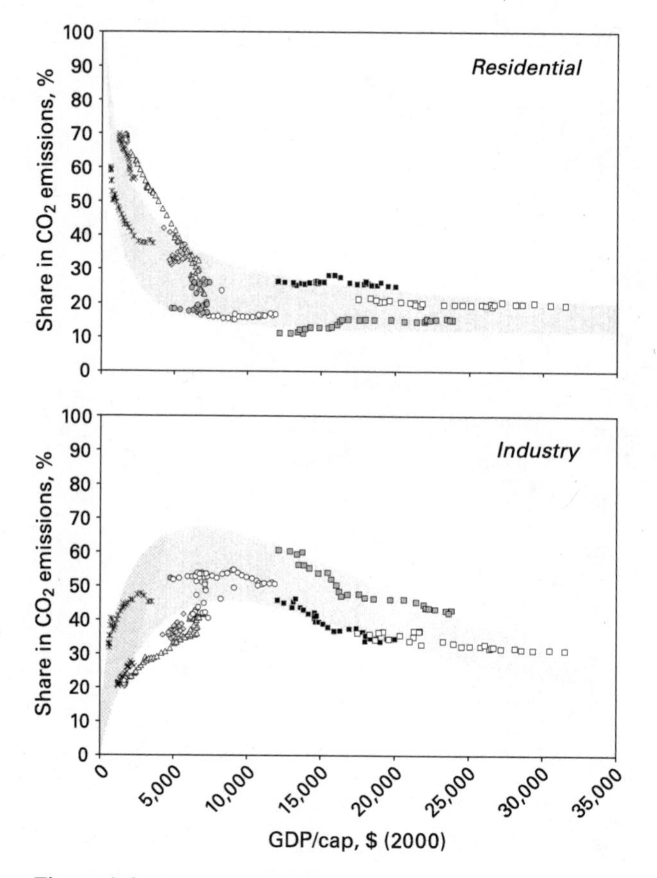

Figure 1.4
Structural change in CO_2 emissions for eleven world regions, 1971–1998. Source: Schäfer, A., 2005. Structural Change in Energy Use, *Energy Policy*, 33(4): 429–437.

While the structural shifts shown in figure 1.4 lead to a continuous increase in transportation's share of energy use and CO_2 emissions, the rising relative importance of passenger travel is also caused by its faster growth in comparison to freight transport. At early stages of industrialization, most transportation energy is used in freight transport, which is vital for setting up the basic urban and industrial infrastructure. Passenger movements are conducted largely by non-motorized transport modes and by motorized two-wheelers, buses, and railways. With subsequent growth, first in LDV motorization and later in air travel, the share of

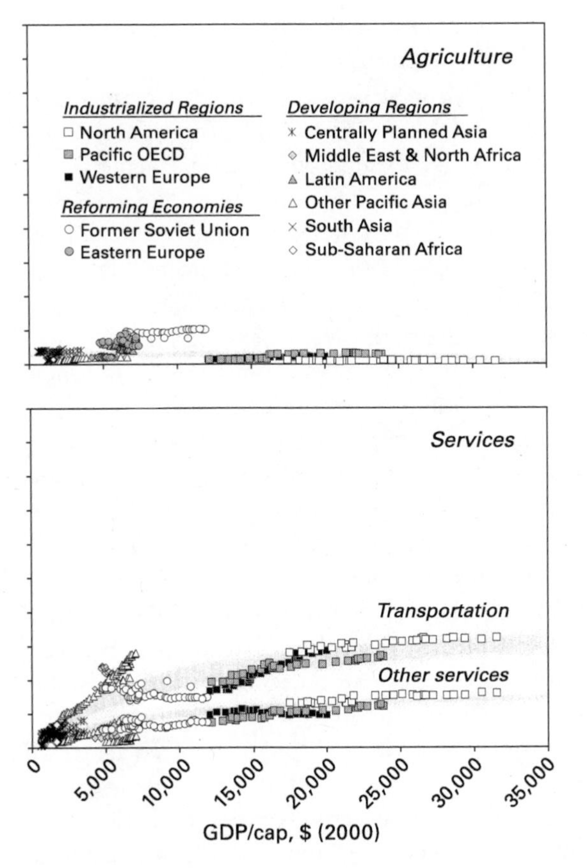

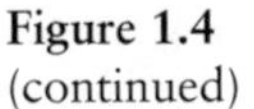
Figure 1.4
(continued)

passenger transport energy use and CO_2 emissions rises strongly, saturating at a roughly 70 percent share of total transport CO_2 emissions, as can be seen from the right column in table 1.1. The combination of these trends, the structural shift toward services and the rising relative importance of passenger travel, argue that passenger travel GHG emissions are likely to continue their growth relative to other sectors.

Why This Book?

Governments have reacted to the threat of climate change in various ways. In the United States, efforts to mitigate GHG emissions have been voluntary to date and have focused on research and technology development rather than controls or price measures applied directly on the emissions. In contrast, the European Union (EU) is imposing policies to meet low-emission targets it accepted with its ratification of the Kyoto Protocol, importantly including a union-wide emissions trading scheme (ETS). This scheme anticipates the inclusion of aviation but excludes surface transport. In surface transport, freight is excluded from any policy mechanism, but a voluntary approach has been tried for automobile transport, with vehicle manufacturers agreeing to achieve increasingly stringent CO_2 tailpipe emission targets for their vehicles over time. However, because the first target has not been met, the EU Commission has adopted a mandatory industry-wide CO_2 tailpipe-emission limit.

The lack of coherent strategy results in part from policy makers receiving mixed signals about their efforts to reduce emissions from this sector. Independent researchers often point to an untapped potential for reducing GHG emissions. Some believe that by "doing it right," a vast improvement in automobile and aircraft fuel efficiency is technologically feasible and economically affordable.[26] In contrast, automobile manufacturers point to the many technological difficulties that need to be overcome before low-GHG-emission vehicles can be produced at reasonable cost. They argue that if their products are forced to meet stringent GHG-emission targets, the rise in consumer costs will lead to reduced sales and factory layoffs.

The intensity of this debate over the potential trade-offs between reduced GHG emissions (and oil dependence) and cost is overshadowed by an apparently insatiable human demand for more travel, at higher speeds, and in greater comfort. To help inform this discussion, and the crucial public policy decisions that are at issue, we attempt to do three things in this book: first, assess the opportunities for transportation technologies and fuels to achieve GHG-emission reductions; second, evaluate the potential limitations on those possibilities; and finally, review the structural challenges that face the implementation of promising technological advances—and the policy approaches that would be required to overcome them.

Since the growth in travel demand has undermined many improvements that were achieved on a vehicle level in the past, we conduct this analysis in the context of a projected plausible future demand for passenger mobility. Together, the projected change on a vehicle level and the anticipated change in travel demand will determine the change in absolute levels. Such projection leads to the first decision to be made. Whenever looking into the future, an immediate issue that arises is the study's time horizon. On the one hand, our interest in observing the long-term effects of policies put into place over the next decades requires a possibly long time horizon. On the other hand, our ability to assess the technology development fades the further we look into the future. Thus, the time horizon is a compromise. We chose the year 2050, since we believe that some of the promising technologies now under development will take about twenty years to enter the market—a period over which we still feel confident to project these new technologies' main characteristics. It would then take another twenty years to displace much of the then-prevailing fleet of air and ground vehicles.

Outline of the Book

The amount of GHG emissions from passenger travel depends on various factors, including how much travel is undertaken, the type and use of transport modes and technologies, and the transport fuel used. These determinants, and the structure of the chapters to follow, can be summarized using the algebraic statement shown on the right-hand side of equation 1.1, which describes the overall identity of passenger travel greenhouse gas emissions (GGE). Among all the influences on this sector's energy (E) use, growth in travel demand—PKT, passenger kilometers traveled—is the most obvious. The relationship is direct: a doubling in global PKT causes a proportional rise in energy use, all other factors being equal. Since understanding travel growth is crucial to making an assessment of the urgency and scale of GHG mitigation—remember the compensation of many improvements on a vehicle level in the past—in chapter 2 we discuss the past and possible future trends in world-regional and global travel demand.

$$GGE = \frac{GGE}{E} \cdot \frac{E}{PKT} \cdot PKT \qquad (1.1)$$

For a given transport mode, energy use is also determined by the technology characteristics of the transport system, that is, the amount of energy use per PKT and how efficiently it is used. "Use efficiency" is determined by the average occupancy rate, the driving cycle of a given vehicle, and other factors. These determinants of energy intensity are analyzed in chapter 3. The product of PKT (projected in chapter 2) and

E/PKT (examined in chapter 3) describes passenger travel energy use. In chapter 3, we also project future levels of passenger travel energy use and GHG emissions in a constant technology scenario.

Chapters 4 to 6 describe the technology and fuel opportunities for reducing (the projected) GHG emissions. Chapter 4 discusses the technology opportunities for reducing the energy intensity (E/PKT) of LDVs, and chapter 5 describes those for reducing the energy intensity of passenger aircraft. Translating energy demand into GHG emissions requires knowledge of the type of transportation fuel in use, characterized by the amount of GHG emissions released per unit of energy consumed (GGE/E). As we will discuss in more detail in chapter 6, each type of transportation fuel and the underlying production process results in a specific value of GGE/E and thus a distinct global warming impact.

In these three chapters, we identify a range of technology and fuel options that could greatly reduce GHG emissions per unit of travel activity. The question of why these technologies have not yet been introduced (on a large scale) is addressed in chapter 7, in which we discuss policies that could help bring low-GHG technologies and fuels into the market and influence total levels of use, PKT. Chapter 8 summarizes our view of future prospects and the challenges facing policy makers.

Limitations of This Study

This book examines the opportunities and limitations of current and future technologies and fuels for mitigating GHG emissions from passenger travel. It also discusses the policies that may be used to speed these technologies into the market and to influence total travel. Such a broad scope necessarily imposes some limitations.

One limitation is national scope. Although we describe the evolution of GHG emissions from passenger travel with respect to the entire world, most of our technology analysis focuses on the United States. This choice has a practical justification: the air and ground vehicle data at the required degree of detail is more easily available for the U.S. transport system than for any other one in the world. However, we do not see this "geographic technology focus" as a drawback. The U.S. is the largest national market in the world and will likely remain so over the next few decades. Its sheer size is overwhelming. As we show in table 1.1, CO_2 emissions from U.S. passenger travel alone are greater than the total GHG emissions of nearly any other country. In addition, as we will argue, motor vehicle technology is similar across the industrialized world and increasingly to that used in developing countries. The similarity applies to aircraft technologies to an even larger extent and is also true of measures to influence total vehicle use, particularly through changes in fuel price.

Another limitation is modal. Although our demand projections include all major modes of motorized travel, our technology assessment focuses on the two major modes, automobiles and aircraft. These two modes already dominate passenger travel in virtually all industrialized countries, and trends we identify in chapter 2 will likely lead to their dominance in the developing world as well.

Finally, this book neither provides revolutionary proposals nor formulates groundbreaking recommendations. Rather, it explores practical means by which lower GHG-emission technologies and fuels could evolve over time. The scale of the problem precludes any revolution in our view: it inherently requires an evolutionary change process (though several of the steps may appear revolutionary to some). Overall, we will explore and discuss changes that have the potential for significant real-world impact within the next few decades. With this focus, we give scant attention to options that would fundamentally transform the transport system. For example, some combination of land-use controls, massive investment in urban public transit, and rapid intercity surface systems could be used to mitigate GHG emissions. However, such policies could affect emissions only on a time horizon beyond the one considered here. Changes in technologies and in fuel price will be the key ingredients of a policy package that can effectively reduce the rate of GHG-emission growth over the next thirty to fifty years.

2

The Global Demand for Passenger Travel

From the beginning of the horse age to the end of the first stage of automobile adoption in the early 1920s, the economic and environmental impacts of passenger mobility on the United States were noticeable mainly on a local scale. However, with the onset of mass motor travel and the rising significance of air traffic in the 1950s, the impacts have become increasingly global. From an economic perspective, they can be seen in the integration of new labor and product markets, through the geographic expansion of industry supply chains, and in the growing network of services that reach more remote areas of the world. From an environmental point of view, transportation activities increasingly contribute to climate change. Because of the long atmospheric lifetime of carbon dioxide (CO_2) emissions (and their resulting global spread), the sources of these gases at all geographic locations contribute equally to the greenhouse effect; CO_2 emissions from a taxicab operating in Mumbai have the same impact on Earth's radiation balance as those released by a sedan cruising on an interstate highway in the United States. The global nature of both the benefits and detriments of transportation requires a broad perspective. Thus, this chapter's focus is the projection of future levels of worldwide passenger mobility.

As appreciated by baseball philosopher Yogi Berra, "prediction is difficult, especially about the future." Following this insight, we only aim to describe an internally consistent future of the world passenger transport system. One way to do this is through understanding the fundamental relationships that have transformed our transportation system in the past. If these forces and constraints remain largely unchanged, or if we can characterize the way they might depart from past behavior, then we can gain an understanding of the likely transport-system evolution into the future.

Following this logic, we assembled a unique database of world passenger travel from 1950 to 2005. Since most of the underlying datasets are incomplete, we made estimates of data where necessary.[1] For easier data management, we aggregated the enormous amount of country data into eleven homogenous world regions; these regions are consistent with those used in an emission scenarios report by the Intergovernmental Panel on Climate Change (IPCC).[2]

With this database in hand, we continued our study by identifying the main determinants of world-regional travel demand to explain the past five decades of passenger travel within the eleven world regions and the world as a whole. We used these same determinants to project world passenger mobility for all major modes of transport over the next five decades. The assumption underlying the projections was that economic motivations and impediments, technological advances, environmental influences (positive and negative), and other drivers of change are largely consistent with those of the past.

Again, our goal in this effort has not been to generate sophisticated predictions of expected future passenger mobility, but rather to understand the implications of several influences on travel demand in order to get the main trends right. We ask, finally, whether the range in projected travel demand can be achieved when several potentially limiting factors are taken into account.

Determinants of Aggregate Travel Demand

Many factors determine how much people travel and what mode of transport they choose. Growth in per capita income and population are the two most important drivers. Over the past fifty-six years, global average per capita income has more than tripled, and world population has more than doubled.[3] As we shall see, the combined eightfold growth of the gross world product (GWP) has led to a nearly proportional increase in passenger mobility, measured here as the aggregate traffic volume in passenger kilometers traveled (PKT). As discussed in chapter 1, the growth in passenger travel has been enabled in part by technological progress, which has led to transport modes that can operate at ever higher speeds. At the same time, technological progress has contributed to a reduction in travel costs of all major modes of transport. The decline in air travel prices by two-thirds contributed to a rapidly rising demand for air travel during the second half of the twentieth century. Finally, the historical growth in travel demand and mode choice has also been determined by consumer behavior. In the past, consumers have spent roughly constant shares of their available money and time on travel. In the following sections, we discuss these money and time budgets, but we begin with the costs of travel.

Costs of Travel

In 1882, the average fare for traveling one kilometer by a U.S. railway was 1.15 cents (1882) or around 20 cents (2000). By 2002, the average fare of railway intercity travel had declined to about 5 cents (2000).[4] During the 120-year period, average income, as measured by gross domestic product (GDP) per capita, increased by a factor of ten; therefore, the affordability of railway travel has increased even further.[5] Similar trends toward increasing affordability can be observed for all modes of transport, despite significant improvements in speed, comfort, and reliability of transport systems over the past century.

Figure 2.1 outlines the evolution of travel costs per PKT in the U.S. since 1950. Among all transport modes, air traffic has experienced the greatest decline, dropping by two-thirds over the past fifty years. In comparison, the 30 percent decline in travel costs from low-speed public transport modes (which include urban, commuter, and low-speed intercity railways and all types of buses) has been comparatively modest. Over the past five decades, only the average costs for owning and operating a light-duty vehicle have risen slightly. The main underlying reason was the declining occupancy rate (vehicle kilometers traveled [VKT] is divided by fewer and fewer occupants, causing driving costs per PKT to increase), a consumer trend that will be discussed in more detail in chapter 3. In fact, on a VKT basis, travel costs have also declined from 32 cents per vehicle-kilometer (c/vkm) in 1950 to 25 c/vkm in 2005 (52 cents per vehicle mile [c/vmi] to 40 c/vmi). Nonetheless, irrespective of the denominator, since average U.S. income has tripled since 1950, affordability has increased significantly for all modes. Similar trends can be observed in all countries for which long-term data are available.

Despite occasional ups and downs, the average travel cost (of all modes of transportation taken together) has remained approximately constant at about $0.12 per passenger kilometer (pkm) or $0.19 per passenger mile (pmi).[6] (Only recently, average travel costs have increased markedly, because of rising oil prices leading to higher automobile travel costs.) Since the costs per PKT have declined for all transport modes

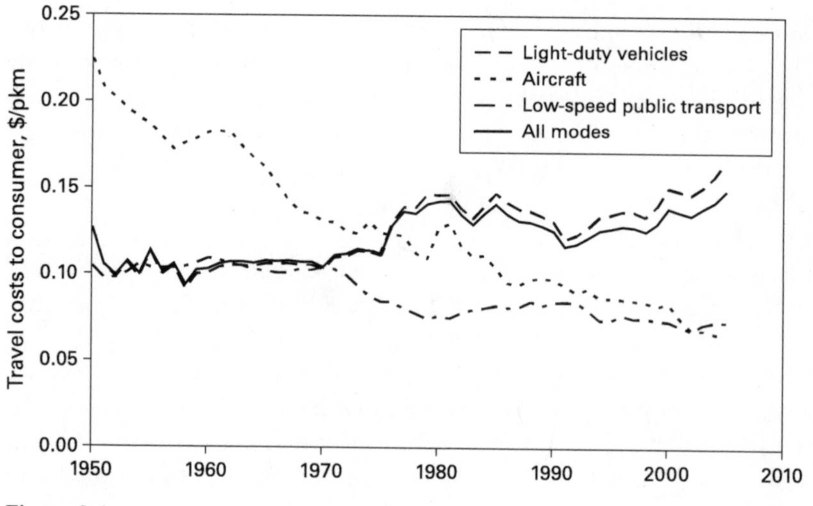

Figure 2.1
Travel costs by mode for the United States, 1950–2005. Sources: Eno Transportation Foundation, various years. *Transportation in America*, Washington, DC. U.S. Census Bureau, 2008. *Statistical Abstract of the United States*, U.S. Department of Commerce, Washington, DC. U.S. Bureau of Economic Analysis, 2008. *National Income and Product Accounts*, Washington, DC; www.bea.gov/. U.S. Bureau of Transportation Statistics, various years. *National Transportation Statistics*, U.S. Department of Transportation, Washington, DC. American Public Transportation Association, various years. *Public Transportation Fact Book*, Washington, DC.

except the automobile, but average travel costs have remained roughly constant, consumers must have changed their mix of travel modes when satisfying their desire or need to travel—a development that will be discussed in detail below.

Figure 2.1 shows that average U.S. travel costs have remained roughly constant since 1950. How did these costs evolve before 1950? Unfortunately, average travel costs for earlier years cannot be estimated with sufficient confidence, because some of the underlying data are not available. However, more recent data from other, less-motorized countries show an initial increase in average transport costs with rising motorization. This increase can be attributed to the higher specific costs (costs per PKT) for owning and operating a light-duty vehicle (LDV) compared with those of public transportation. Travel costs begin to saturate at a motorization level of about three hundred cars per one thousand people, an ownership level at which each household, on average, has acquired an automobile.

(As discussed in chapter 1, the United States reached that level in the early 1950s, which explains the roughly constant average travel costs in figure 2.1). The initial increase and subsequent saturation of average travel costs can be seen from more recent data from countries with lower motorization levels. For example, in Western Europe, costs have increased from $0.09 per pkm ($0.15/pmi) in 1963 (the first year when expenditure data became available for a critical number of countries) until more or less reaching the saturation point at about $0.14 per pkm ($0.23/pmi) in 1980, when the motorization level reached about three hundred LDV per one thousand people.

The rising affordability of all transport modes, especially in light of the significant increase in speed, comfort, and reliability, has been the result of various forces. Among those, technological change has played a major role. An early example is Henry Ford's mass-produced Model T. While early automobiles were manufactured individually by highly skilled craftsmen who hand built commissioned vehicles, Ford's mass production process reduced the assembly time of an appropriately redesigned vehicle by a factor of ten. Using the craftsman approach, the time required to build a Model T chassis was about twelve hours; Ford's moving assembly line reduced that period to only ninety-three minutes.[7] The associated cost reductions were then amplified by using less-skilled and thus cheaper labor. Overall, the price of a Model T declined by about 75 percent between 1908 and the mid-1920s, going from $12,300 to $3,200.[8] Subsequent improvements in production techniques and vehicle design, supported by progress in materials, computer science and simulation, have continuously contributed to a long-lasting substitution of capital for labor, that is, an increasing trend toward automation across all parts of the supply chain. And more recently, these cost reductions have been complemented through restructuring the vehicle and transportation industries. This overhaul has taken place across several dimensions of the industries, including organizing the manufacturing process using "lean production" principles, a technique that strives to yield more, and higher-quality, outputs with fewer inputs.[9] As we discuss in chapter 3, increasing competition among vehicle manufacturers has contributed to a further reduction in vehicle costs through price discounts and other marketing measures.

Advances in technology have also reduced travel costs *directly*. One example is the reduction of the cockpit personnel in commercial aircraft, where the flight engineer was displaced by a computer. And as we discuss at length in chapters 3, 4 and 5, fuel consumption by automobiles and aircraft has declined dramatically. While a Model T had a gasoline consumption of 15 liters (L) per 100 vehicle-kilometers (vkm), progress in vehicle and fuel technology has reduced that consumption by about one-third for today's average new automobile in the United States. (Correspondingly, the Model T's fuel efficiency of 15.7 miles per gallon has roughly doubled over time.) As we discuss in chapters 3 and 4, these fuel-use reductions would have been significantly larger if the safety and comfort characteristics of the Model T had remained unchanged. And as chapter 5 describes, even greater reductions in fuel consumption have been achieved by aircraft over the past four decades.

Reductions in travel costs have also been attained through introducing market principles into formerly regulated transportation services. The deregulation of the U.S. air transport market in 1978, for example, is believed to have contributed to the continuous decline in airfares.[10] More recently, that trend of declining air travel costs has been reinforced by the emergence of low-cost carriers.

Travelers also benefit from subsidies that are paid by society as a whole. These societal payments partly consist of direct subsidies to road, rail, and air transportation. In 2001, total U.S. federal, state, and local government transportation expenditures amounted to $176 billion. About 69 percent of these expenditures were covered by fuel taxes paid by the transportation users and operators.[11] The difference between the expenditures and the tax revenues thus amounted to a subsidy of $54 billion. Dividing this amount by total PKT in that year results in an upper bound of 0.7 cents per pkm (1.1 cents per pmi). (The subsidy to passenger travel, however, is smaller than this estimate, since the transport infrastructure is also used for moving freight.)[12]

In addition to government subsidies, society as a whole bears various other costs associated with transportation. Travel costs are artificially low, because they do not account for the time lost in traffic jams, the health and welfare costs of accidents, and a range of environmental impacts, including climate change. We review these costs in chapter 7 when we discuss technology policies for mitigating greenhouse gas (GHG) emissions.

All these factors, technological and nontechnological, have contributed to the declining travel costs of all modes (except for the costs of U.S. automobile travel between 1950 and 2005, if measured per unit

PKT). This trend toward rising affordability has not only led to growing travel demand (as we will discuss below), but also to a consumer choice for larger and more powerful vehicles as described in chapter 3. With ongoing technological progress, further cost reductions are likely. In combination with growing income, the affordability of travel is likely to continue to increase.

Travel Time Budget

Societies consist of individuals who have widely different preferences, capabilities, and opportunities. This heterogeneity is apparent in many measurable characteristics, such as income level, family size, and available modes of travel. A survey that records the travel time of individuals on a given day would also reflect this heterogeneity and produce a skewed distribution. Some respondents would spend a modest amount of their time traveling (say 10 minutes); the largest number of respondents would spend a moderate amount (say 30 minutes); a lesser number would spend a significant amount traveling (say 2 hours); and a few would travel for many hours. However, while the time particular individuals spend traveling varies, the *average* amount of time spent traveling *per person* is surprisingly stable over a wide range of civilizations and income levels. First postulated by the late Yacov Zahavi in similar form in the 1970s, this (aggregate) "travel time budget" stability is readily apparent in figure 2.2 for a wide range of societies (characterized by differences in GDP per capita).[13] At low income levels, residents in African villages spend slightly more than one hour per day traveling—roughly 5 percent of their time—mainly to fetch water and firewood and for commuting to and from agricultural work. At high income levels, residents in the automobile-dependent societies of Japan, Western Europe, and the United States—where half of all trips are for leisure purposes—spend a similar amount of time just traveling.

Travel time per person has not only remained invariant with income; the average daily travel time per person also has remained largely stable, even when time allocations for other activities have changed. Figure 2.3 shows the time allocations for various activities as they relate to the amount of time dedicated to work in fifteen different settings within the Western Hemisphere in 1965–1966 (empty symbols) and ten European countries in about 2000 (full symbols). Work time, the independent variable shown on the horizontal axis, is the average number of hours worked per day over the course of a whole week (i.e., the weekly work

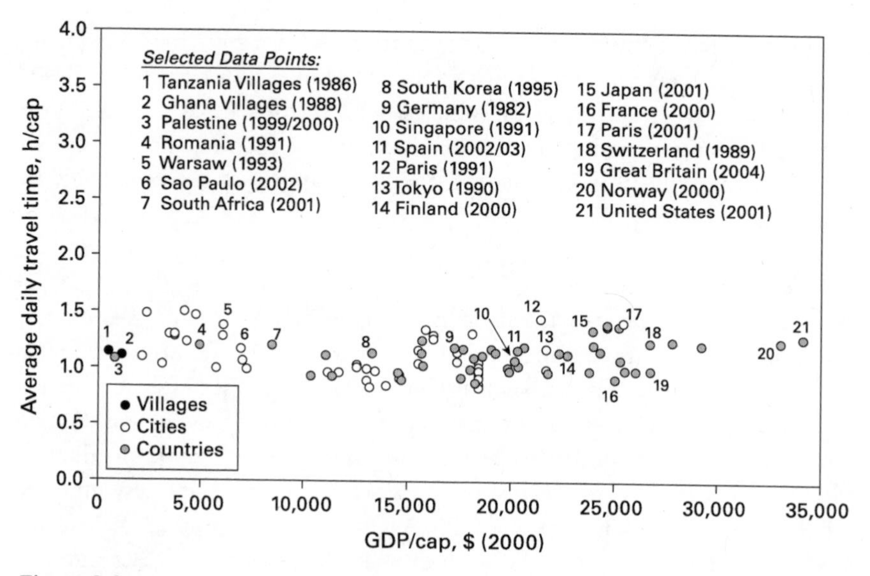

Figure 2.2
Average daily travel time in hours per person as a function of GDP per capita. Source: updated dataset of Schäfer, A., D.G. Victor, 2000. The Future Mobility of the World Population, *Transportation Research A*, 34(3): 171–205.

hours divided by seven days). Historically, the average working hours per week have declined.[14] Both figures show that the decline in work time went along with an increase in the amount of time dedicated to sleep.

Other time allocations as a result of changes in work time are predictable only at a given point in time. For example, beginning in 1965–1966, declining work time led to increases in leisure time—by 2000, the time dedicated to leisure activities had become "saturated" at about six hours per person per day. A similar discontinuity can be observed in the average amount of time devoted to personal care and meals, whereas the amount of time dedicated to household and family declined by about one hour between 1965–1966 and 2000.

While time allocations for all of these activities have changed over time, the only comparatively stable time expenditure has been on travel, on both a cross-sectional and a longitudinal basis. This fact suggests that travel is not exclusively a "disutility" that people generally want to minimize; to some extent people seem to desire the feeling of movement and speed for their own sake. However, when time expenditures for travel

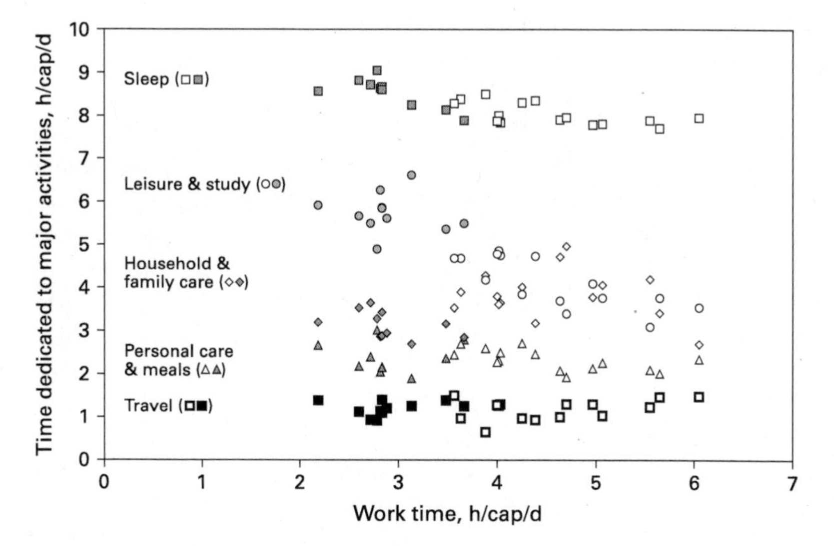

Figure 2.3
Time allocation to various activities as a function of work time in fifteen different settings within the Western Hemisphere in 1965–1966 (empty symbols) and in ten European countries at about 2000 (full symbols). Sources: Szalai, A., P.E. Converse, P. Feldheim, K.E. Scheuch, P.J. Stone, 1972. *The Use of Time: Daily Activities of Urban and Suburban Populations in 12 Countries*, Mouton, The Hague. European Commission, 2003. *Time Use of Different Stages of Life—Results from 13 European Countries*, Office for Official Publications of the European Communities, Luxembourg.

rise, transportation itself can become uncomfortable, although the "annoyance threshold" will be different for different people.[15]

Cesare Marchetti of the International Institute for Applied Systems Analysis suggests that the travel-time budget has likely remained unchanged since the beginning of human civilization. He points out that the average distance of the walls of ancient Persepolis, Rome, Vienna, and other historic cities to the city center was about two to three kilometers. One round trip a day—from the walls to the city center and back—would take slightly more than one hour to walk.[16] With the introduction of faster transportation systems—horse tram, electric tram, motorbuses, and automobiles—city boundaries grew at a rate roughly proportional to the average speed of the fastest mode of transportation available, continuously allowing a return trip from the outskirts to the city center to be taken within the same travel-time budget. Today the city radius of Rome has grown to about 10 kilometers, allowing car drivers to take one trip from the city boundary to the center and back within a travel time of about 1.2 hours under average (congested) travel conditions.[17]

Researchers have tried to explain the surprising invariance of average travel-time expenditures from various perspectives, including the behavioral and the psychological, or they simply interpret the stability as a by-product of time allocations given to primary activities. However, no one perspective provides an ironclad explanation. (As shown in figure 2.3, travel time cannot be simply the time left over from time "given" to other activities, since it shows such stability over a wide range of working hours). It appears that the resulting mean travel time (of about 1.2 hours per day) somehow reflects the average time society as a whole is willing to travel.[18]

While the travel-time budget is essentially stable, figure 2.2 shows that it can vary between one and one and a half hours. The range of this variability in travel time is greatest across cities (shown as empty circles). Since average trip distances are longer and congestion levels higher in urban areas, residents of larger cities typically experience longer travel times. Such differences are partially averaged out when the data are aggregated over larger geographic regions. The range of average travel time is smaller across countries (grey circles) and should be still smaller on a world-regional level, the geographic unit in our study.

The variations are also artifacts of the different survey techniques used to measure travel activities. Since these methods have become more sophisticated over time and better capture incidental travel that typically had been underreported by survey respondents in the past, the time devoted to travel by residents of the same country seems to rise over time and daily travel distance to increase. For example, early U.S. national travel surveys relied completely on respondents' memories in reporting their travel activities, but in more recent surveys, starting in 1995, they have been asked to document all travel in a diary. This methodological improvement alone has "increased" the per person trip rate by 15 percent, and as a second-order effect, has shown travel time to rise.[19]

In addition to inconsistent measurements, however, some of the variability in figure 2.2 may also be the result of changes within society, such as the increasing participation of women in the labor force. According to a West German travel survey conducted in 1989, a travel time of 1.21 hours per active traveler per day was the average of widely different time expenditures by different socioeconomic groups. Within the labor force, state employees had the highest travel time (1.43 hours per traveler), followed by nonstate white-collar employees (1.30 h), the unemployed (1.24 h), and blue-collar workers (1.16 h). By contrast, those not employed but who stayed home to care for a household and/or children spent an average of only 0.99 hours per day traveling.[20] Thus, an increase in the labor force by those who formerly stayed at home would cause average travel time to rise, all other factors being equal.

Travel time can also rise if it is used more productively. Some automobiles have already been converted to "offices on wheels," equipped with telephones, fax machines, and laptop computers. These consumer options may be simply an industry response to a developing trend. According to a 2002 survey by the U.S. Department of Transportation, drivers eat or drink during 30 percent of all trips, use cell phones during 18 percent of all trips, use wireless Internet access during 3 percent of all trips—and even indulge in personal grooming during 8 percent of all trips.[21] (Unfortunately, no time series data are available that would help identify and characterize these behaviors as a trend.) If these habits become more widespread in the future, travel times might increase. Absent any offsetting mechanism, these changing patterns of travel imply that behaviors that have existed over long periods and across societies may change.[22]

Since we cannot dismiss the possibility of an increase in average travel time, we take a pragmatic approach in our projections of travel demand and mode shares. We assume that travel-time budgets will remain stable at 1.2 hours per person per day but conduct sensitivity tests to make apparent the potential impacts of changes in travel-time budgets on the projected future levels of travel demand and mode choice. As we will show, changing the travel-time budget within plausible limits will not significantly change the fundamental dynamics underlying people's demand for passenger mobility.

Travel Money Budget

A similar metamorphosis from variability at disaggregate levels to stability at aggregate levels can be observed with respect to the share of travel-money expenditures, that is, the percentage of income dedicated to travel. Although highly variable on an individual level, averages of the income percentage devoted to travel follow a pattern similar to that of travel costs. Zahavi observed that households that rely exclusively on nonmotorized modes of travel and public transportation spend only

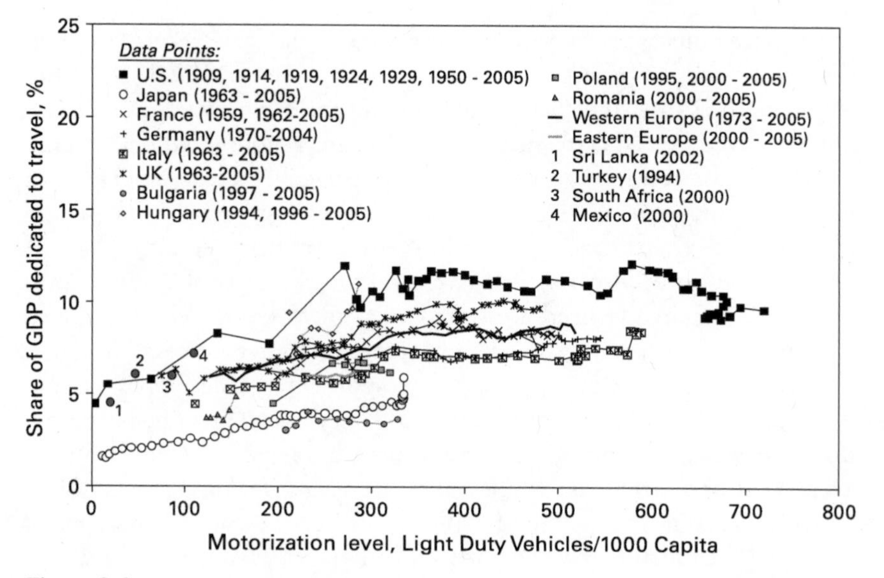

Figure 2.4
Travel expenditures as a fraction of GDP over motorization levels. Source: updated dataset of Schäfer, A., 1998. The Global Demand for Motorized Mobility, *Transportation Research A*, 32(6): 455–477.

about 3–5 percent of their income on travel, while those owning at least one automobile dedicate 10–15 percent of their income to transportation needs.[23]

This relationship also applies to entire countries at the aggregate level. Figure 2.4 presents the initial growth and subsequent saturation of the "travel-money budget" (defined here as total travel expenditures divided by GDP), dependent on the number of LDVs (automobiles and light trucks used for passenger travel) per person, for different countries and regions. The travel-money budget increases from about 5 percent of GDP at motorization levels close to zero cars per capita (nearly all U.S. households in 1909 or Sri Lanka today) to around 10 percent of GDP at about three hundred cars per one thousand capita (the United States in the early 1950s and Western Europe in the 1980s). This per country ownership level corresponds to the threshold at which most individual households own an automobile (one car per household of three to four persons).

Consistent with Zahavi's household-based observation, the share of income spent as travel money has remained stable at higher motorization levels. The different saturation levels in figure 2.4 can be explained in part by relative price differences in national economies. The same figure also demonstrates the impact of aggregation on the "smoothness" of travel money expenditure shares: it shows a high variability at the level of nations but relatively smooth trends at world-regional levels (such as Western Europe).

The oil crises in 1973 and 1979, which raised the cost of automobile travel, are an instructive test of the stability of the travel-money budget. In response to rising fuel prices (a 50 percent jump in 1979), travelers reduced other transportation costs—demanding, for example, smaller, less expensive, and more fuel-efficient new vehicles. Despite the two rapid rises in fuel prices, economic recessions, and fluctuations in new car prices, travel-money expenditure shares remained nearly unchanged between 1970 and 1985, oscillating between 10.3 percent and 12.2 percent of GDP.[24]

The Past Five Decades in World Travel Demand

We now use these determinants of travel demand to explain the past five decades of global passenger mobility. We begin with total travel and then discuss the changes in transport modes. Later in this chapter, we use these same determinants to project future levels of mobility and mode shares.

Historical Trends in Total Mobility

Over the history of transportation, travel has become increasingly affordable. While average costs per PKT have leveled off, consumers have continuously spent a roughly constant fraction of their rising income on travel. Thus, rising GDP has translated into rising PKT. The direct correlation between GDP per capita and per capita PKT is shown in figure 2.5 for eleven world regions and the world as a whole, spanning over fifty-six years of historical data. Because of the large regional differences in per person GDP and PKT, figure 2.5 is presented on a double-logarithmic scale. As can be seen, each additional unit of GDP has generated a corresponding rise in PKT (through increased use of automobiles, buses, railways, and aircraft)—independent of the stage of economic development, the political system, or cultural values. Given this uniform growth, economists attribute an "income elasticity" of 1 to travel demand. However, the causal arrow does not exclusively point from GDP to PKT;

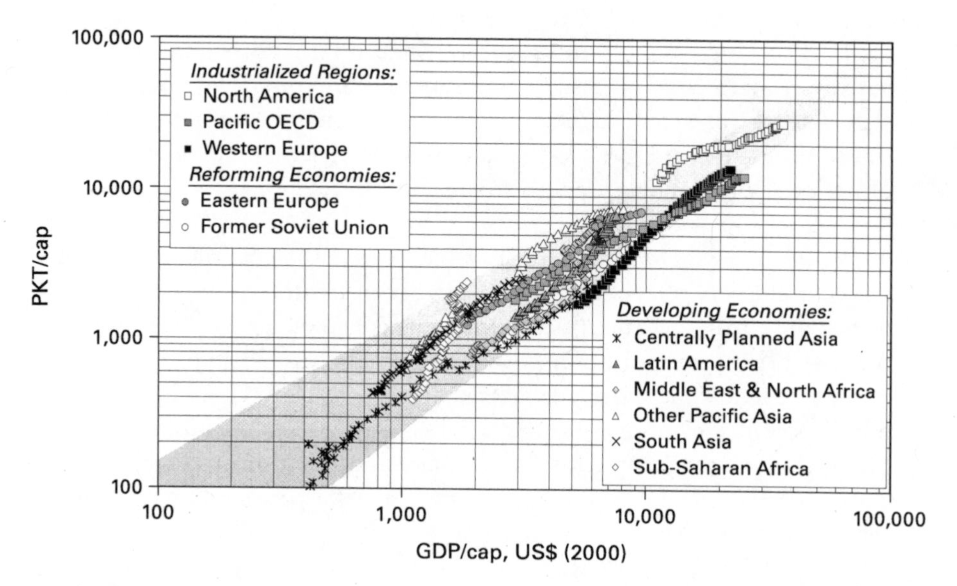

Figure 2.5
Passenger kilometers traveled (PKT) per capita over per capita GDP (in purchasing power parity) for eleven world regions and the world between 1950 and 2005. Source: updated dataset of Schäfer, A., 1998. The Global Demand for Motorized Mobility, *Transportation Research A*, 32(6): 455–477.

wealth and mobility are codependent. Rises in PKT widen the range of economic opportunities by allowing access to more distant resources, intensifying competition, and thereby causing prices to decline and consumption—and GDP—to rise. This new, higher level of GDP per capita then translates into another increase in PKT.[25]

Table 2.1 summarizes aggregate meta-region data, selected regional data displayed in figure 2.5, along with absolute levels of population and PKT. Between 1950 and 2005, world average travel demand (using automobiles, buses, railways, and aircraft) has quadrupled from a yearly average of 1,420 kilometers to 6,020 kilometers per person (about 880 to 3,740 miles per person). Since the world population grew two and a half times during this period, world PKT multiplied by one order of magnitude, from 3.6 trillion to 38 trillion (2.2 to 23.6 trillion pmi). The highest growth in PKT, by a factor of about twenty-eight, occurred in the developing world, slightly reducing the "mobility divide" in relation to industrialized regions.

Table 2.1
Socioeconomic and travel characteristics in three metaregions, selected world regions, and the world in 1950, 2005, and the projection for 2050

	1950				2005				2050			
	Pop mill	GDP/ cap[a] $(00)	PKT/ cap km	PKT bill	Pop mill	GDP/ cap[a] $(00)	PKT/ cap km	PKT bill	Pop mill	GDP/ cap[a] $(00)	PKT/ cap km	PKT bill
Industrialized regions	588	6,578	4,530	2,660	957	27,667	18,400	17,600	1,093	57,336 (41,450)	42,200 (29,500)	46,100 (32,200)
North America only	168	11,075	11,100	1,860	333	36,263	27,400	9,120	444	64,743 (52,367)	48,000 (39,000)	21,300 (17,300)
Western Europe only	326	5,308	1,920	625	472	22,302	14,100	6,630	504	50,171 (32,940)	39,100 (23,700)	19,700 (11,900)
Reforming economies	268	2,402	947	254	401	10,725	5,620	2,260	343	23,642 (25,215)	15,000 (16,300)	5,160 (5,590)
Eastern Europe only	88	1,892	1,240	109	117	9,709	7,200	843	98	23,295 (26,649)	18,500 (21,400)	1,820 (2,110)
Developing economies	1,657	941	388	643	4,952	4,972	3,660	18,100	7,674	8,009 (15,238)	6,800 (14,600)	52,200 (112,000)
World	2,513	2,415	1,420	3,560	6,311	8,779	6,020	38,000	9,109	14,514 (18,757)	11,400 (16,400)	104,000 (150,000)

Note: Projections for 2050 are based on the economic growth rates of the MIT EPPA model reference run (EPPA-Ref) and those of the IPCC SRES-B1 scenario (in parentheses).
[a] GDP is measured in purchasing power parity (PPP).

In 1950 three out of four passenger kilometers traveled were in the industrialized countries, while in 2005, PKT in industrialized countries accounted for only about half of the world total. However, given that industrialized countries accommodate only 15 percent of the world population, differences in per person PKT remain substantial. In 2005, residents in the industrialized world traveled 18,400 kilometers (11,400 miles) per capita on average—about 50 kilometers per day—five times as much as traveled by those living in the developing world. The largest gap is between the highest and lowest income regions. While North Americans traveled some 27,400 kilometers (17,000 miles) per year on average (75 kilometers per day), residents in Centrally Planned Asia traveled only 2,300 kilometers (some 6 kilometers per day), a ratio of twelve. The mobility levels of the remaining nine world regions lie between these two extremes, largely following a sequence of their respective average income levels.

GDP is the most important but not the only determinant of PKT. As can be seen in figure 2.5, the average amount of travel per person can differ significantly across regions, even for people at the same income level. North Americans traveled nearly 20,000 kilometers (some 12,400 miles) at a GDP per capita level of $20,000, while Western Europeans traveled only 12,500 kilometers (7,770 miles) at that GDP level. Residents of the Pacific OECD having the same per capita GDP traveled fewer than 10,000 kilometers (6,200 miles) on average. These differences in levels of mobility can be attributed partly to differences in transportation and real estate prices. For example, lower transport costs and government support for (suburban) housing have led to comparatively high levels of PKT per capita in North America. The differences in mobility levels can also be explained by the fact that the comparison made in figure 2.5, in which expenditures are calculated at market exchange rates, may not accurately reflect the relative purchasing power of measured GDP in the different regions.[26] Also note that some of the regional trajectories in figure 2.5 are not smooth; yet, the variations can be explained.[27]

Historical Trends in Mode Choice

The combination of roughly fixed travel-expenditure shares and saturating travel costs translates increasing per capita GDP into roughly proportional growth in per capita motorized mobility. Since the travel-time budget is also fixed, this increasing mobility has to be satisfied within roughly the same amount of time. Since each transport mode operates within a limited range of speeds, the increasing per person PKT can be satisfied only by a shift toward ever-faster transport systems. Thereby, the rising relative importance of air travel has been reinforced through its lower transport costs (see figure 2.1).

Figure 2.6 shows this change in mode shares toward faster transport systems, again for a fifty-six-year historical time horizon. Three distinct phases of motorized mobility development can be distinguished. Below an annual mobility level of 1,000 pkm (620 miles) per person (fewer than 3 kilometers per day) low-speed public transportation accounts for nearly the entire traffic volume. These modes include urban public transport systems, commuter railways, regular intercity trains, and intercity buses. At the subsequent stage of development, between a yearly travel demand of 1,000 and 10,000 pkm per capita (about 3 to 30 kilometers per day), the relative importance of low-speed public transport modes declines to between 10 and 30 percent of total PKT—because of the more strongly rising automobile traffic volume. In that stage of mobility evolution, the share of LDV travel reaches its saturation point somewhere between 60 percent and 90 percent of total PKT.

The large dominance of automobile travel, however, may be only temporary. Mobility levels of 10,000 pkm (6,200 miles) per capita and above require a substantial share of air traffic or high-speed rail-based ground transportation systems (which we've aggregated into one generic "high-speed transport" mode). Thus, enabled by continuously rising income and declining air travel costs, the relative importance of these modes rises while that of automobile travel drops. This development constitutes the third stage of mobility evolution.

An example of a region in the third stage of mobility evolution is North America, where nearly all passenger traffic is split between automobiles and aircraft. The latter, which offer still greater mobility through significantly higher speeds, have strongly increased in market share at the expense of passenger cars over the last five decades. At the same time, ordinary rail and bus services continue to hold a low share in special market niches, mainly high population-density areas.

While factors such as different travel costs and different patterns of land use cause regional variations in overall mobility, they also cause regional differences in the relative importance of transport modes. In North America, where both the costs for owning and operating an automobile and urban population density (about eight hundred people per square kilometer in U.S. urban areas) are lowest, the share of automobile

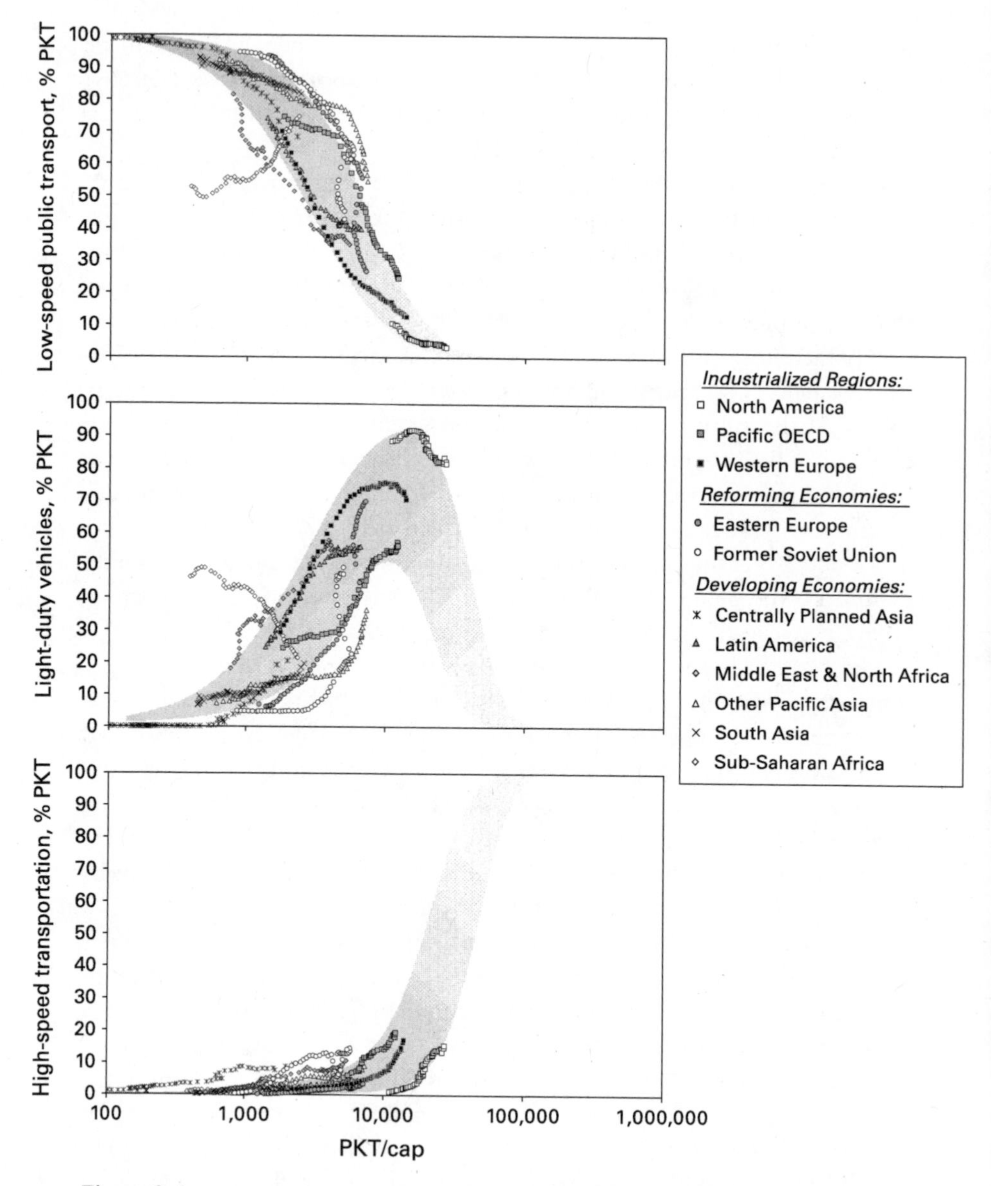

Figure 2.6
Three stages of the evolution of motorized mobility: declining share of public transport modes (top), the growth and relative decline of the automobile (middle), and the rise of high-speed transportation (bottom). Source: updated dataset of Schäfer, A., 1998. The Global Demand for Motorized Mobility, *Transportation Research A*, 32(6): 455–477.

traffic volume saturated at above 90 percent in 1960. In Western Europe, where automobile travel costs are higher and population densities are about four times that of U.S. cities, automobiles achieved only a 75 percent share of total traffic volume by that time. Correspondingly, in the Pacific OECD region, where historically the data refer predominantly to Japan and which generally has comparatively high costs for automobile travel and high population density (five to six times the U.S. level), the relative importance of automobile traffic volume is likely to peak below the Western European level.

Similarly, better access to low-speed public transportation and limited ownership of "private" automobiles in the formerly socialist countries of Eastern Europe and the former Soviet Union, allowed a higher share of buses and railways to accommodate the same per capita traffic volume as has existed in automobile-dominated Western Europe. However, in light of a fixed travel-time budget, per capita PKT can rise sharply only through enhanced use of faster transport modes. Already, a similar trend toward faster modes of transport can be observed there. The change of the political system in the early 1990s brought the removal of constraints on private automobile access. As shown in the middle and upper panels of figure 2.6, this produced a dramatic increase in the share of automobile travel and a decline in the low-speed public transportation share. Simultaneously, the previously highly subsidized air travel market in the former Soviet Union dropped overnight to levels more in keeping with market economy development (figure 2.6, bottom).

The low-speed public transport and automobile shares of Sub-Saharan Africa seem to counteract the natural trends discussed above: with rising travel demand, the relative importance of low-speed public transport modes increases at the cost of light-duty vehicles. This historical deviation is a result mainly of aggregating distinct socioeconomic classes from countries with significant differences in income. While the LDV travel of a small high-income class has dominated the transportation sector in the past, it has been outpaced by growing bus travel, which can increasingly be afforded by the much larger lower-income group. (Thus, if it were possible to separate out the travel patterns of the two socioeconomic groups, they would likely match the trends observed in figure 2.6). In addition, the shown trends may result from data uncertainties, which are especially large for this region.

This description of travel demand and changes in transport mode shares raises the question of what "ultimate force" pulls people out of their homes and pushes them into ever-faster transport systems. That force can be partly explained by the reasons given for traveling, which change with increasing income. Our comparison of travel surveys from around the world suggests that people at low income levels undertake a weekly average of one to two trips per day; their average trip distance is about 5 kilometers (3 miles).[28] (A "trip" is usually defined as a one-way movement from an origin to a final destination that takes place in a public setting, such that a person who goes to a grocery store and returns home has made two trips.) At such comparatively low mobility levels, three reasons for traveling can be identified: averaged over an entire population, which includes people who stay at home on a given day, one trip in the day is dedicated to a combination of work (short-term survival) and education (longer-term well-being), and about half of a trip per day is dedicated largely to personal business (such as shopping at local markets).

With rising income, the demand for additional trips and the average trip distance increase simultaneously. These trips are dedicated to satisfying needs regarding personal business (for more shopping, health care, and religious services) and leisure (including holidays). At the high-income levels enjoyed in industrialized countries, people make more than three trips per day. They still devote one trip to the combination of work and education, one to two trips to personal business, but now devote one trip to leisure. The highest rate—with an average of about four trips per day—and the longest average travel distance (16 kilometers or 10 miles) can be observed in the U.S.; there, personal business accounts for nearly half of all trips.

While the change in trip purposes explains the increase in trip rate, the increase in the second component of mobility, the trip distance, can be explained by various factors, including human curiosity to experience new environments, increased economic opportunities via the integration of more remote labor and product markets, and a greater spatial separation of home and work. Such increases in trip distances have been enabled by mainly technological improvements, most notably the increase in travel speed.

When the various daily trips are multiplied by their respective distances, the resulting traffic volume associated with personal business travel, leisure travel, and the combination of work, education, and business-work travel each account for one-third of the total PKT in the United States. In Western Europe, where travel distances by purpose differ from those in the United States, leisure traffic already accounts for 40–50 percent of total PKT.

Future Levels in Travel Demand and Changes in Use of Transport Modes

After having discussed the historical trends in global passenger mobility, we now describe plausible futures in travel demand and mode choice. As outlined at the beginning of this chapter, our aim is not to conduct sophisticated projections but to capture the main trends.

Future Trends in Total Mobility

That the expenditure shares of money and time are basically fixed not only helps explain the historical development of travel demand and mode choice; it also constrains the range of possible futures. For example, if travel expenditures continue to be a roughly stable fraction of income and the average costs of travel remain about level, then any future increase in per capita GDP would come with an approximately proportional increase in PKT. At the same time, as travel demand increases, the fixed travel-time budget would continue to "push" travelers toward faster modes of transport. Hypothetically, the maximum level of travel demand would then be achieved if travelers used the fastest transportation mode for their entire daily travel-time budget every day of the year.

Assuming that improvements in air travel could increase the current "door-to-door" speed of nearly 270 kilometers (168 miles) per hour (which includes getting to and from the airport) to ultimately 660 kilometers (410 miles) per hour (the current U.S. average commercial aircraft speed from one airport to another), a travel-time budget of 1.2 hours per day, during which—hypothetically—only this fastest transport mode is used every day (365 days per year), the annual per person traffic volume would be approximately 289,000 kilometers (180,000 miles).[29] Since at this high mobility level, most travel would be international and interregional, prices and income levels would adjust. Thus, regional differences in per capita traffic volume (at a given GDP per capita level) would decline, and the eleven trajectories shown in figure 2.5 would ultimately intersect at one single point in the distant future. Given the data we have from past development, we can project the GDP per capita value of that "target point" to correspond to $289,000 (in 2000 dollars). This *hypothetical* world of high-speed transportation helps us project future levels of PKT. We can approximate each of the eleven world-regional trajectories by one and the same type of regression equation. Simply by constraining one parameter of that equation, the estimated trajectory will necessarily pass through the target point.[30] Future levels of PKT can then be determined by the predicted levels of population and per capita GDP. (Later in this chapter, we discuss several limiting factors that may prevent such an unconstrained growth.)

While the size of the world's population is commonly projected to rise to between 8 and 11 billion people by 2050, the development of world-regional GDP per capita over the same period has a significantly larger uncertainty.[31] The *Special Report on Emission Scenarios* (SRES) prepared by the Intergovernmental Panel on Climate Change (IPCC) contains an analysis of the GDP projections of many studies of world energy use and emissions.[32] The underlying projections of the GWP range from 2 to as much as 8.5 times the 1990 level by 2050, with a mean of about 4.6 times the 1990 level.[33] Given the comparatively small differences among projected population levels in these studies, most of the (large) differences in GWP projections must result from different assumptions concerning per person GDP. This spread of per person GDP assumptions increases at the world-regional level and as the time horizon extends.

To take account of this large uncertainty with respect to future levels of economic development, we have used two contrasting projections. One projection is derived from the Emissions Prediction and Policy Analysis (EPPA) model, developed by the MIT Joint Program on the Science and Policy of Global Change, which was run without any policy constraints.[34] The EPPA model uses market exchange rates (MER) to run its reference projections of GDP per capita. Converting the MER-based projections to projections based on purchasing power parity (PPP), the GWP per capita would increase from $8,780 in 2005 to $14,510 in 2050, an increase by 1.1 percent per year.[35] (In MER terms, the corresponding growth in GWP per capita would be 2.2 percent per year.) In this projection of the EPPA reference run (EPPA-Ref), regional disparities in per person incomes remain large. Per person GDP would increase by more than 100 percent in the industrialized world (from $27,670 in 2005 to $57,340 in 2050) but by less than 60 percent in the developing world (from $4,970 in 2005 to $8,010 in 2050).

Other projections are based on the assumption that today's lower-income regions would reduce the income gap with the industrialized world over time and eventually catch up. The basis for our second projection is the SRES-B1 scenario. This scenario describes a future in which the PPP-adjusted per capita GDP rises by only 50 percent within the industrialized regions through 2050 but 200 percent in the developing world. Compared with the 60 percent increase of the 2005 GWP per capita in the EPPA-Ref projections, per capita GWP more than doubles in the SRES-B1 scenario, an increase by 1.7 percent per year. (The corresponding MER growth based GWP per person would be 2.6 percent per year.) Given the expected 44 percent growth in world population—as suggested by the medium variant of the United Nations population projections—GWP would increase by 140 percent (EPPA-Ref) to 210 percent (SRES-B1) by 2050.[36]

Considering these changes in socioeconomic conditions, a stable relationship between growth in GDP and traffic volume implies that world travel demand would grow approximately in proportion to the projected level of income, from 38 trillion pkm (24 trillion pmi) in 2005 to either 104 trillion (EPPA-Ref) or 150 trillion (SRES-B1) in 2050 (65 trillion pmi or 93 trillion pmi, respectively). Since the two scenarios of economic growth differ by metaregion, so does the increase in mobility. In 2005, industrialized and developing regions accounted for nearly half of world PKT each, with the reforming economies of Eastern Europe and the former Soviet Union region accounting for the remaining 6 percent. Using the EPPA-Ref rates of per capita GDP growth, the world-regional shares in PKT will remain almost unchanged through midcentury. In contrast, since developing regions' economic growth rates are higher in the SRES-B1 scenario, these regions would account for a rising share in world PKT, accounting for three out of four PKT in 2050.

Table 2.1 displays the historical data we have gathered and the scenarios for estimated future levels of PKT per capita and total PKT, along with the underlying socioeconomic projections. The scenarios were developed using both the EPPA-Ref and SRES-B1 economic growth rates for three metaregions, selected world regions, and the world as a whole. As noted earlier, these projections were derived under the assumption of an unconstrained transport system; later in this chapter, we discuss several limiting factors that may lead to lower levels of future travel demand. Also recall that these projections don't take into account the volatility in the world oil price, observed during the early twenty-first century. However, as we discuss in chapter 3, the impact of such oil price changes on travel demand are small and declining.

Future Trends in Mode Choice

What combination of transportation modes will be needed to satisfy the projected travel demand within a given travel time? In light of a fixed travel-time budget, the relative importance of faster means of transport would continue to increase. Thus, future traffic shares of low-speed public transportation modes, LDVs, and high-speed transportation systems would need to follow the development of the shaded portions (envelopes) of figure 2.6. In particular, the relative importance of passenger mobility provided by high-speed transportation systems would continue its historical increase largely within the envelope of this mode and ultimately account for the entire traffic volume in a *hypothetical world* where the target point can be reached. At the same time, the share of low-speed public transport modes supplying PKT would have to continue its historical decline and ultimately account for zero at the target point traffic volume of 289,000 pkm (180,000 pmi) per capita. In addition, after an initial increase, the share of LDV travel also would have to decline and ultimately converge to zero PKT at the target point traffic volume.

Note, however, that travel demand and mode shares are unlikely to get close to their respective target points' enabling conditions. Instead of representing a realistic situation, the hypothetical target point has to be considered a construct that ensures internal consistency for our model. Nevertheless, as long as the overall transportation system is sufficiently far away from that (hypothetical) condition, this model is a useful tool for understanding the fundamental dynamics underlying the evolution of travel demand.[37] A more sophisticated model would recognize two separate transportation markets, each having its own distinct modal substitutions. In the geographically expanding (sub-)urban transportation market, LDVs continue to displace low-speed public transportation modes, while in the other market, automobiles are displaced in intercity transport by high-speed transport modes.[38]

The internally consistent shift in the shares of transport modes necessary to satisfy the projected travel demand through 2050 can be derived in a number of ways. Perhaps the most rigorous method would be to estimate the parameters of statistical consumer-choice models. Unfortunately, such models would require time-series data (ideally, for 1950 to 2005) of speeds and travel costs for each transport mode. These data can be derived for the United States (see figure 2.1) and, to a limited extent, for a few European countries and Japan, but they are not available for most countries. Not having these historical data, we have performed simplified projections by determining plausible future shares of each transport mode for each modal envelope at the projected level of per capita PKT.

For the three industrialized regions of North America, Pacific OECD, and Western Europe, we extended the historical decline of the share of low-speed public transport modes through the zero share at the target point PKT. The future shares of the remaining two modes (automobiles and high-speed transport) were derived using balancing equations of travel time and distance, assuming current average transport speeds per mode to continue.[39] Because of the stringent travel-time budget constraint, the resulting projections of mode shares are very similar to those derived from more sophisticated consumer-choice models (as we verified for the North America region).[40] Should current speeds per mode remain unchanged, our projections suggest the share of high-speed transportation in North America will increase to 50 or 64 percent of total PKT by 2050, for the economic growth rates of the SRES-B1 scenario and the EPPA reference run, respectively. Similar shares for high-speed transportation in 2050 have been derived for the Pacific OECD and Western Europe regions.

For all other regions, we projected the different shares depending on whether a particular region is an early adopter or a latecomer regarding the diffusion of automobiles. Latecomers achieve lower shares of LDV travel (here assumed to develop along the lower boundary of the automobile envelope in figure 2.6), because they have already "leapfrogged" into high-speed travel and thus develop along the higher boundary of the high-speed transportation envelope.[41]

Figure 2.7 reports mode shares for 1950 and 2005 for the three meta-regions and the world, and two projections for 2050. The two 2050 projections are again based on the economic growth rates produced for the SRES-B1 scenario and by the EPPA-Ref. In the industrialized regions, LDVs and high-speed transportation are projected to account for nearly the entire traffic volume. Compared with the SRES-B1 scenario, the higher economic growth rates derived from EPPA-Ref lead to a stronger growth in travel demand and—because of our assumption of a fixed time budget—to higher shares of high-speed transportation modes. In this scenario, high-speed transportation would account for nearly 70 percent of all PKT. Although the relative importance of low-speed public transportation is projected to further decline, this mode would not

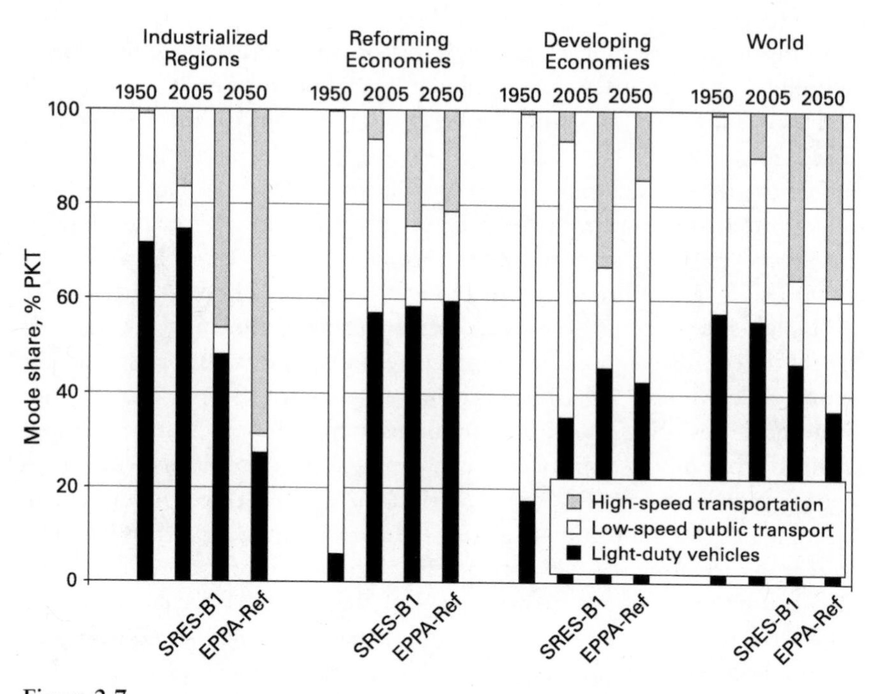

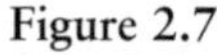

Figure 2.7
Relative importance of transport modes in the three metaregions and the world, history (1950 and 2005) and projections (2050).

become insignificant. For example, North Americans, who traveled an average of 860 kilometers in 2005 (2.4 kilometers per day) using buses and railways, would roughly maintain that level of mobility supplied by low-speed public transport modes through 2050.

Within the reforming economies and developing regions, automobiles would supply a rising share of PKT in 2050, accounting for about 60 percent of PKT in reforming economies and for 40–50 percent in developing regions. In both of these metaregions, high-speed transportation would also rise by 2050, at that time accounting for about 20 percent of PKT in the reforming regions and for 15–30 percent of PKT in the developing regions, depending on the rate of economic growth. Only within the developing world would low-speed public transportation continue to maintain a market share significantly larger than 20 percent. Globally, automobiles are projected to continue to satisfy the largest share of mobility demand in 2050, followed closely by high-speed transportation; low-speed public transport modes would supply only a small share of world mobility.

Based on these projections, we conducted several sensitivity tests for the industrialized regions. These analyses showed that the travel-time budget has the largest impact on future mode shares. Should, for example, the travel-time budget increase from 1.2 to 1.5 hours per person per day (a 25 percent increase), the projected 2050 shares of high-speed transportation for North America would decrease from 52 percent to 30 percent if using the economic growth rates from the SRES-B1 scenario, and from 66 percent to 49 percent using the EPPA-Ref economic growth rates. Although such decline would be significant, a 30 percent and 49 percent share would still be roughly twice and three times the share of that transport mode in 2005. On a global scale, an increase in the average travel time budget from 1.2 to 1.5 hours per person per day would maintain the clear dominance of automobile traffic through 2050 (46–57 percent), followed by high-speed transportation (24–30 percent), and low-speed public transportation (18–24 percent).

The second largest determinant of future shares of high-speed transportation is the average speed of LDVs. In contrast, an increase in mean aircraft speed would have a negligible impact on future air traffic shares, given the already significant speed difference from other modes.[42]

Figure 2.8 illustrates PKT by major mode of transport for 1950, 2005, and the projected 2050 levels for the three metaregions and both GDP scenarios. The significant historical and projected growth in total PKT (area) is shown as a consequence of an increase in population (horizontal axis) and PKT per capita (vertical axis). The strong growth in high-speed transportation (gray boxes) is already visible over the past five decades. The 2005 traffic volume of high-speed transportation modes already matched the total PKT by automobiles, buses, railways, and aircraft in 1950. Assuming a constant travel-time budget of 1.2 hours per person per day, the global traffic volume of high-speed transportation in 2050 would similarly correspond to the 2005 world PKT (when the economic growth rates of the EPPA-Ref are used). In the case of the SRES-B1 economic growth rates, the high-speed transportation-related PKT would be 40 percent higher.

Finally, having until now discussed the technical aspects of how we projected future levels of travel demand and mode shares, what are the implications of the hypothetical future we have projected? Perhaps surprisingly, the societal implications of a world giving greater emphasis to

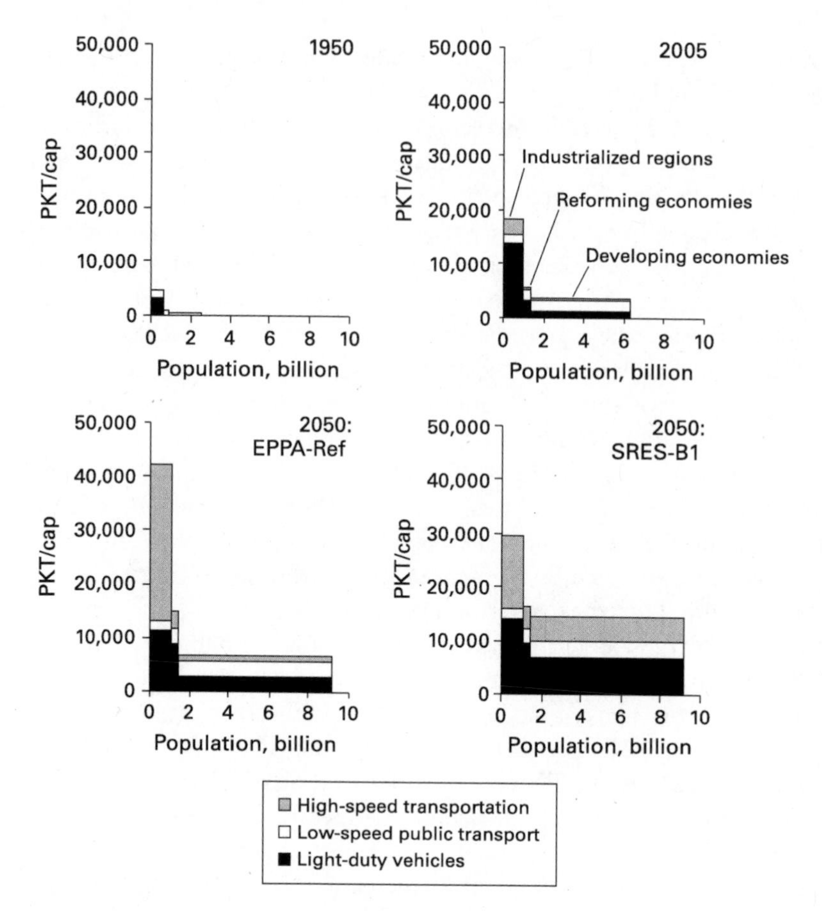

Figure 2.8
Passenger kilometers traveled (PKT) by transport mode and metaregion as a function of population and per person PKT, history (1950 and 2005) and projections (2050).

high-speed transportation would not be radically different. Assuming a travel time budget of 1.2 hours per person per day, in the industrialized world, high-speed transportation would account for between half and two-thirds of the projected 2050 total traffic volume, with LDVs satisfying nearly the entire remaining travel demand. Because of the comparatively low travel speed of LDVs, North Americans would still spend most of their travel time driving—depending on the rate of economic growth (EPPA-Ref or SRES-B1), an average of 44–51 minutes per day (down from slightly more than one hour in 2005). The remaining part of their time budget would be spent (on average) in low-speed public transportation modes (about 8 minutes, as in 2005) and high-speed transportation systems (12–19 minutes, up from 3 minutes), including feeder systems.

Many people in industrialized countries already live in a world dominated by high-speed transport systems. More than half of today's annual per person PKT of 18,400 kilometers could already be taken up by just one round-trip flight between Boston and London. Thus, what may be considered rare behavior today could become more routine by 2050. Whether it be a day trip to watch the World Cup final, or a night on the Côte d'Azur, long-distance trips with a short stay at the destination may become more normal. Simultaneously, the projected amount of travel (and its shift toward high-speed travel modes) could also cause unprecedented geographic relocations. Already, the home computer allows 13–19 million Americans to have more flexible work schedules and thus to work from home at least once a week.[43] Higher speed transportation systems would allow people to live in Paris and work in New York City, enabling them to spend most of their time in one of the French capital's brasseries with a laptop, while traveling to the office only once every couple of weeks.

Since high-speed transportation systems and telecommuting are complementary enablers of such a dispersed world, a future based on such high-speed physical and electronic links is very plausible from today's perspective. While it is likely that this lifestyle of a projected global citizen will become more and more the norm for an increasing share of the population within high-income countries, constraints on the further expansion of today's transportation system may allow only a rising fraction of the middle class to enjoy the projected increase in mobility. We discuss such potential limiting factors in the following section.

Can Growth in World Passenger Travel Endure?

Our projection of world travel demand and mode choice omits several potentially limiting factors. Overlooking these constraints may lead to exaggerated future levels of passenger mobility, and thus energy use and GHG emissions. Hence, in this section we will test the plausibility of our projections by examining several potential restrictions on travel growth.

Given a fixed travel-time budget, any increase in passenger travel requires an increase in mean travel speed. As discussed earlier, mean travel speed can increase significantly only through a rising relative importance of faster modes and through an increase in the mean door-to-door speed of each mode. Should these enablers be constrained, for example by significant levels of traffic congestion at infrastructures accommodating higher-speed modes (e.g., waiting times in airports) or the lack of technological opportunities and operational strategies to increase the mean vehicle speed, the rise in mean travel speed could stagnate and travel demand saturate.

Another potential limitation of the future growth of travel demand is an enhanced use of telecommunication means, as they could—in theory—substitute for most travel needs. However, even if mean travel speed can increase, congestion be bypassed, and transportation substitutes become ineffective, could the limited supply and asymmetric distribution of oil deposits constrain transportation's free movement? Unlike most other sectors of society, transportation is nearly completely dependent on oil products.

And finally, what of the rising political significance of the natural environment? Could global warming policies limit air travel or force car drivers to shift toward lower-speed public transportation? If any one or, more likely, a combination of these constraints becomes binding, travel demand may grow at a slower rate than per capita GDP, saturate, or even go into decline.

In the discussion that follows, we do not consider the impact of potential catastrophic events on the transportation system in general nor on air travel in particular. Although terrorist attacks, viral epidemics, and similar events may occur more frequently in the future, given their unpredictability in terms of time, space, and scale, a study like this, which uses projections based on past regularities, is not useful for understanding the impacts of these events.

Increasing Levels of Traffic Congestion

In 45 BC, as a means to fight traffic congestion, Julius Caesar declared the center of Rome off limits to all common private vehicles between 6 a.m. and 4 p.m. Traffic jams were also reported in nineteenth-century London, as described in chapter 1. And even before horses were fully replaced by automobiles at the beginning of the last century, Manhattan was already clogged with motor vehicles. Today, 40 percent of all urban areas in the United States regularly experience severe or worse congestion levels.[44] Similar levels of congestion are reported in European cities. Traffic congestion is especially severe in the developing world, where the vehicle fleet often grows by 20–30 percent per year, a rate impossible for infrastructure extension to match.

Despite the long history of traffic congestion, travel demand has continuously increased. Governments have learned to cope with congestion by adding infrastructure—a process that today can be observed in rapidly developing Asian cities.[45] Travelers have adjusted to congestion in several ways: by changing their travel patterns—departure times, routes, and modes of travel—by making more productive use of vehicle time through entertainment and communication devices (the "office on wheels"), and—as a last resort—by relocating workplaces and residences. In addition, employers have adjusted to traffic congestion by offering more flexible work schedules and by relocating from city centers into more easily accessible suburban office buildings. These adaptations can be seen from average U.S. commuting times, which have increased slightly from 21.0 to 23.3 minutes (a rise of 11 percent), while the average commuting distance has increased from 15.6 kilometers in 1969 to 19.5 kilometers in 2001 (a rise of 25 percent).[46] In addition, emerging technologies may allow travelers to overcome congestion in great measure. Intelligent transportation systems are being designed and implemented to increase mean vehicle speeds by directing the driver to less congested roads and by managing traffic more efficiently. In the long term, automated highways—where vehicle movements are computer controlled and traffic flow has few gaps—may greatly liberate additional roadway space.

Accommodating the future growth in high-speed transportation (mainly air travel) may turn out to be a particular challenge. According to our projections, which are based on a travel time budget of 1.2 hours per person per day, high-speed travel for North America would multiply by a factor of six to ten by 2050, which translates into an average annual growth rate of 4.2–5.3 percent. (Assuming a travel time budget of 1.5 hours per person per day, high-speed travel for North America in 2050 would still be four to seven times the 2005 level, an average annual growth of 3.0–4.5 percent.) This growth is the weighted average of intraregional and international travel with unknown proportions. Our range of growth rates is comparable to those from the passenger traffic forecasts for the aggregate of North American domestic and international flights from Airbus and Boeing, albeit these 2007 industry forecasts cover only twenty years.[47] According to the Airbus and Boeing projections, most of the growth in North American air travel would be associated with international and especially interregional travel. Industry forecasts over the next twenty years also agree regarding traffic projections for all industrialized regions together and for the world as a whole.[48]

How can such growth in PKT be accommodated? The key strategy would be to increase the capacity of the air traffic system, both in the en-route airspace and at the airport. Already, measures have been implemented to increase the capacity in the en-route airspace: in 2005, the U.S. Federal Aviation Administration reduced the minimum vertical separation for properly equipped aircraft cruising at altitudes between 29,000 and 41,000 feet from 2,000 to 1,000 feet, nearly doubling en-route airspace capacity.[49] Similar changes in altitude restrictions have been implemented around the world. An additional significant capacity increase might be provided by implementing "free flight" procedures: largely independent movements by aircraft that optimize their trajectories between airport areas. Advanced onboard aircraft-control systems would ensure self-separation from other aircraft, thereby reducing the need for tactical control by ground-based controllers.

With these measures in place or on the horizon, the binding constraint to a significant expansion of air travel is likely to remain capacity around and at airports, both the terminal area airspace (where the arriving and departing aircraft are accommodated) and, most importantly, the runway system itself.[50] However, several options also exist to increase airport capacity. The most obvious one is the extension of the airport infrastructure. Building new airports (or runways) in the industrialized world, however, is constrained by public acceptance unless new airports are built in truly remote areas. As a consequence, aircraft operations have begun to spill to secondary and tertiary airports. Further unlocking the potential of lower-order airports could have significant effects, as can be witnessed by the growing number of metropolitan areas with multi-airport systems around the world. In the future, very light jets (VLJs) may significantly increase operations among tertiary airfields (see chapter 1). However, given the comparatively small size of VLJs, even several thousand microjets as expected by the U.S. Federal Aviation Administration and the aircraft industry, will add only a few percent of extra capacity in terms of passengers.[51]

Changes in aircraft operations bear an additional significant potential for increasing the passenger flow to and from airports, such as encouraging a more uniform traffic mix and larger aircraft. Aircraft of a similar size allow the use of smaller separation requirements during landing and takeoff (as compared to a highly variable traffic mix), permitting a larger number of operations in any given time, while larger aircraft clearly allow more passengers to be accommodated in any one operation.[52] In 2005, the average U.S. aircraft had 114 seats in domestic traffic and 193 seats in international operations (see chapter 3). In contrast, significantly larger aircraft operate on the ultra-high-demand 450 kilometer-long route separating the capacity-constrained airport systems of Tokyo and Osaka, that is, a combination of Boeing 777 (with a seating capacity in excess of 300) and Boeing 747 (with the 747-400 version having a seating capacity greater than 400) vehicles. Airport passenger flows could be increased further through more directly routed flights, thus reducing the number of intermediate stops. Such a strategy not only liberates airport capacity at especially the busiest hub airports, but also satisfies an increasing consumer demand in the higher income, industrialized world.

A third family of strategies for unlocking unused airport capacity are changes in airport operations. Already, airports experiencing capacity constraints are smoothing the peaks and valleys of the incoming and outgoing flights and stretching operations over more early morning and late evening hours of a day. In theory, aircraft operations could be extended to twenty-four hours a day through the use of significantly quieter aircraft, which might be subject to fewer airport nighttime curfews (although more travelers would need to get used to departing or arriving at night). For this purpose, the Cambridge–MIT Institute's Silent Aircraft Initiative designed an advanced-concept aircraft that would be significantly quieter than today's air vehicles. By using a combination of advanced airframe and engine technologies coupled to advanced operating techniques, the peak noise level outside the airport perimeter was calculated to be comparable to the background noise in a typical urban environment—making the aircraft almost imperceptible to the human ear on takeoff and landing.[53] And as another technology measure, high-speed rail could substitute for air travel in suitable city-pair connections.[54]

Importantly, some growth in PKT is likely to occur with only little or without any additional stress on the airport system. Because either takeoff or landing will occur in a different region, the anticipated stronger growth in interregional travel requires only half the airport capacity than intraregional operations do. The stronger anticipated growth in interregional travel may also cause the average passenger trip distance to grow even faster than in the past. Between 1950 and 2005, this component of PKT has more then doubled and without the stronger growth in interregional travel a continuation of this trend would lead to an increase in average trip distance by 40 percent through midcentury.[55]

While it is not clear from today's vantage point that all the discussed options could be implemented (in time), the potential for additional increases in airspace capacity seems large. The U.S. Joint Planning and Development Office of the Next Generation Air Transport System is targeting an increase in airspace capacity up to three times the 2004 level by 2025.[56] Similarly, the Single European Sky Air Traffic Management Research (SESAR) initiative and the Advisory Council for Aeronautics Research in Europe have both set targets to achieve by 2020 a threefold capacity of the air transport system over the 2000 level.[57] At the same time, technology is likely to continue to develop beyond 2025, further pushing the capacity boundaries of the world's aviation system.

Increasing Mean Travel Speed

The second option for increasing mean travel speed is a further increase in the mean door-to-door speed of air travel. Over the past five decades, the average door-to-door speed of U.S. aircraft has increased from less than 130 kilometers per hour to nearly 270 kilometers per hour. This increase in speed was made possible initially mainly by a move from propeller to turbojet, turbofan engines and subsequently chiefly by an increase in the average distance passengers travel (average trip distance), allowing passengers to travel a larger fraction of their trip at high speed. These door-to-door speeds include the feeder travel to and from the airport, check-in time, time for transfers, and the time required for baggage collection—all together accounting for about four hours per trip.[58]

Aircraft door-to-door speed could be further increased through a combination of measures, including higher aircraft speeds through a critical number of supersonic aircraft, a further increase in passenger trip distances, a restructuring of the air transport system toward direct flight routings, a reduction in the average airplane access and egress time (as offered by VLJs) through more advanced security screening procedures to reduce the time required before a flight, and through the substitution of high-speed rail for aircraft in short-distance intercity travel. All these strategies are already being pursued by airlines and the rail industry, at least on a conceptual level. While the large-scale introduction of a supersonic commercial aircraft fleet is unlikely, at least over the next several decades, the combination of growth in mean trip distance, more direct routings, and the increased use of secondary and tertiary airports could increase mean aircraft speed to about 370 kilometers per hour by 2050.[59]

Transportation Substitutes

Instead of travelling farther and faster, will bytes of information increasingly substitute transportation needs? In theory, much of the travel demand shown in figure 2.5 could be satisfied through telecommuting. This is especially true in the industrialized world where the economies' shift toward the service sector makes an electronic alternative to people moving increasingly feasible. The classic example of replacing physical transport through information is the commute to and from work. Today, in the United States, an estimated 13–19 million employees telecommute to work at least once a week, about 10 percent of the employed labor force.[60] Assuming that the share of telecommuters could triple and that they might even work an average of 2.5 days per week at home, the total reduction in travel would amount to 15 percent of all work trips.

Unfortunately, even as the substitution potential of telecommuting grows, the share of work trips declines. As people increase their trip rates with rising income, additional travel is dedicated to personal business and leisure. In the United States, work trips accounted for only 14 percent of all trips in 2001. Thus, the combined effect of an increase in telecommuting on *total* trips would be 15 percent (the possible increase in telecommuting) of 14 percent (the percentage travel taken up by work trips), or only about 2 percent of all trips. (Since work trips are among the longest trips, the impact on PKT might be larger.)

Given this small potential for reducing travel to and from work, could telecommuting also substitute for other types of trips? Already, for example, electronic commerce has become a multibillion dollar business, and no saturation is yet visible. Similarly, virtual reality technology, by immersing "travelers" in a computer-simulated artificial three-dimensional environment of choice, eventually may have the potential of substituting for an increasing number of leisure trips once it is available on a sufficiently large scale. And, in theory, business trips by aircraft could be substituted by teleconferencing facilities, at least in part.

However, the substitution potential may only seem large in theory. Instead of being a substitute, telecommunication systems may—over the longer term—turn out to be a complement. After all, what would people do with their gained time when not traveling to work or when shopping online? Would extensive users of virtual reality technology at some point like to compare their virtual experience with the real world? If travel-time budgets continue to remain stable, people ultimately may substitute their reduced commuting time and time given to trips to shopping malls with other kinds of travel, such as longer-distance leisure trips. Also, personal contacts often prove to be the basis for further dealings in the business world. Thus, while telecommunication systems may reduce some air travel as a first-order effect, they may ultimately lead to an increase in aviation.

Historical analogies seem to confirm that expectation. In the past, revolutionary changes in communications technology (the telephone, fax machine, email, teleconferencing, mobile phone) have occurred, but travel has continued to grow at the same rate. Rather, it appears that while new means of communication have replaced some travel, they have also generated new linkages and opportunities for additional travel, at least over the longer term. Analogies to the paperless office are also apparent. Forecasts in the 1970s predicted the end of paper use, when most offices and households would own a personal computer. Thirty years later, the penetration of personal computers is even larger than was anticipated, but paper consumption per person has continued to grow. What has changed, however, is the way we use paper; instead of typing paper copies, we now electronically mail documents—which the recipient prints out, on paper.

Just as the increased use of the personal computer has changed the way we use paper, the large-scale use of advanced telecommunication systems could reinforce the trend toward ever-faster modes of travel. In the past, the automobile allowed commuters to move farther away from the workplace; the choice of workplace location was always limited, however, by the driving distance to that workplace. By contrast, advanced telecommunication systems may increasingly enable a nearly complete decoupling between home and workplace if the *frequency* of physical commuting can be reduced. Living in Paris and working in Boston may thus become easier.

According to extensive research by Patricia Mokhtarian of the University of California at Davis, a leading expert in travel behavior research, incorporating all determinants of telecommuting, including the willingness and frequencies of those commuters who could telecommute, the commuting patterns and residential locations of those commuters, and the possible substitution of nonwork trips for work trips, "telecommuting is likely to reduce only a fraction of 1 percent of household travel in the U.S., even well into the future."[61]

Oil Deposits

In March 1919, an article appeared in *Scientific American* titled "Declining Supply of Motor Fuel." The author expressed strong concern that the progression of gasoline consumption would ultimately result in a declining supply and rising price of motor fuels.[62] According to that article, only 6.5 billion barrels of oil (of an original total of 11.5 billion barrels) were left in the United States, and given the extraction rates at that time, less than twenty years of oil supply could be guaranteed.[63] Nevertheless, after ninety years of continued growth in oil consumption and numerous other warnings that the world would soon run out of (cheap) oil, the threat of imminent shortages and unavoidable drastic rises in oil prices still governs news headlines. Increasingly, analysts have warned that the world is about halfway to exploiting available oil, and once exploitation of the second half is begun—probably before 2010—they expect prices to increase sharply.[64] (Oil prices have increased sharply since the early 2000s, but this hike was related more to a range of factors—including instability in some exporting regions, rapid demand growth [particularly in China] in relation to the level of investment in extraction capacity, and insufficient U.S. refining capacity—than resource exhaustion.)

Many of the alarming predictions of oil supply shortages and rising prices are based on the work of M. King Hubbert, an American geophysicist who predicted in a 1956 paper, which has become renowned in the field, that U.S. oil production would peak some time between 1965 and 1970.[65] Hubbert's approach was simple. He made two educated guesses about the magnitude of U.S. oil reserves that were ultimately recoverable: the amount of oil that had already been exploited plus those deposits still in the ground that could be exploited profitably. Using projected maximum feasible oil production rates, he then fitted a bell-shaped curve to the historical production data such that the total cumulative volume of oil extracted would correspond to his two guesses, which were either 150 or 200 billion barrels of oil.

Following this approach, Hubbert's fitted curves peaked at about 1965 and 1970, depending on the assumed size of the reserves. Fourteen years after his projection, U.S. oil production did peak—and Hubbert became immortal. Ever since, the eventual peak in *global* oil production has been called "peak oil" or "Hubbert's peak."[66]

The accuracy of Hubbert's prediction was surprising but raises a number of questions. Perhaps most critical was his assumption concerning the size of the ultimately recoverable U.S. reserves, putting them at 150 or 200 billion barrels. Since Hubbert's 1956 paper, these estimates have been adjusted toward ever-larger amounts. In its most recent assessment, the U.S. Geological Survey estimates the size of ultimately recoverable U.S. reserves to be 362 billion barrels, roughly twice the amounts used by Hubbert.[67] Hence, if Hubbert had available more complete information about the original size of oil reserves, his methodology would have led to a later but still wrong prediction for when "peak oil" would be reached. (Conversely, if Hubbert had based his projection on the 1919 estimate of initial reserves, 11.5 billion barrels, hardly any oil would have been left at the time of his prediction.)

Thus, the decline in U.S. oil production after 1970 may have been caused, as suggested by MIT's Morris Adelman and Michael Lynch, by cheaper oil imports rather than by resource constraints.[68] In fact, in his 1956 paper, Hubbert also projected that the point of maximum oil production for the world would occur in 2000. However, instead of reaching the predicted peak, oil production has continued to rise and was already 8 percent higher in 2005 than in 2000.[69] Since Hubbert, several petroleum geologists have used similar methods to try to determine the time of peak oil, but they have assumed the reserves are larger, if still fixed.[70] Perhaps not surprisingly, up until the present, all projections have proved to be incorrect. Common to all these projections is the understanding—or assumption—that there are fixed reserves of oil.

If not, why have initial reserves "increased" over time? The basic answer lies in the definition of "reserves," which generally is meant as those quantities of oil that can be *profitably* extracted with existing technology at prevailing market conditions. Oil reserves are thus only a part of *total* deposits. In addition to reserves, a much larger amount of oil exists. Using the existing technology of a particular period, these larger deposits (resources) may be too expensive to extract for various reasons: because they are difficult to access, because their size at a given location is too small to justify large investments in extraction and transportation technology, or because they consist of lower-grade, carbon-rich substances such as heavy oil, bitumen, and oil shale—commonly known as unconventional oil—that need to be "upgraded" into synthetic light oil using sophisticated and costly refinery technology. (We discuss these types of resources and conversion processes in more detail in chapter 6).

However, intra-industry competition and rising oil prices induce the oil industry to research and develop oil detection and extraction technology and to drill for more oil, in more remote areas, and for less accessible and smaller deposits. In other words, resources are continuously being converted into reserves. Thus, extraction costs decline and make the exploitation of deposits economical that previously were considered too expensive to extract. Initial reserves have also increased because of the investments made to build up extraction equipment and pipeline transportation systems at larger wells. Once these are in place, oil extraction from smaller, neighboring wells becomes profitable.

The increase in reserves, however, is also limited at any particular time. Oil companies have no interest in expanding their reserves above a critical size, because too large an inventory could reduce prices. The resulting "reserve-to-production ratio"—the number of years of oil supply at the current production rate—has been about forty years since the mid-1980s and thus mainly reflects the planning horizon of the oil industry rather than an ultimate limit on the oil supply. MIT's Michael Lynch has examined in detail the various analyses of Hubbert scholars in their attempts to predict the time of maximum oil production and concludes that their forecasts typically ignore political and economic factors and commit a number of methodological errors.[71]

While Hubbert's followers have been overly pessimistic in projecting when maximum oil production would take place, they must be correct about the long-term trend. Since it takes several hundred million years to convert plant matter into fossil fuels, but the extraction of those fuels takes only several hundred years, the amount of oil reserves is ultimately finite from a less-than-geological-lifetime perspective. Thus, the amount of initial reserves eventually will decline, and light oil consumption increasingly will need to be complemented by the use of unconventional oil. Based on the current annual consumption level of 30 billion barrels, today's estimated reserves of 1,200 billion barrels of conventional oil are expected to last for about forty years.[72] However, ongoing technological progress in oil recovery in existing fields, together with the discovery of new (albeit smaller) fields, are likely to maintain that ratio at the current level for some (unknown) time, even taking into account the growing demand for oil.

Drawing on the available literature, table 2.2 summarizes estimates of proved reserves and resources of light oil. In addition, resource estimates of bitumen, heavy oil, and shale oil are shown. Dividing the global reserves of light oil by the 2005 consumption levels leads to a reserve-to-production ratio of about 41 years, that is, the period over which current consumption levels can be maintained. In combination with the estimated amount of conventional oil resources (the amount of oil that is uneconomic to exploit at current market conditions), the resource base (reserves and resources) of conventional oil would last for more than eighty years at current production rates. This time span can be increased fivefold if the heavy oil, oil sands, and shale oil resources are added. But even if the size of the resource base were more limited, the fundamental principle of demand and supply mandates that the world will never run out of oil. Rising demand causes the price of oil to rise, which in turn accelerates the search for alternatives, including unconventional oils. These alternatives will become ever more profitable and will thus increasingly add to oil reserves over the next five decades and beyond.

If the size of the resource base won't be a concern over the next five decades and more, what then is the problem? As can be seen from table 2.2, the geographic distribution of conventional oil reserves is inhomogeneous; the reserves are located mostly in politically less stable countries. Reserves in the Middle East alone account for more than 60 percent of the world's total reserves. In contrast, less than 3 percent of conventional reserves lie within the United States, which currently accounts for nearly one quarter of global consumption. Therefore, these regions' dependence on imported oil is growing continuously. At present, the United States satisfies two-thirds of its oil demand by using imported oil, about half of which comes from the Persian Gulf.

Over the short- to medium-term future, this dependency on oil, which comes so much from a single, unstable region, constitutes a threat to a low-cost, uninterrupted, and large-scale fuel supply. And this threat is augmented by two additional concerns. One is that of an "oil weapon" that could be used against the industrialized world. Following the rise in the world oil price in the early 2000s, several governments of resource-rich countries increased their control over their oil and gas reserves. Even while these governments demand a fair share of the exploitation

Table 2.2
Reserve and resource estimates of light oil and unconventional oil by world region (EJ), global consumption in 2005 (EJ), and associated reserve-to-production ratios (years)

	Reserves	Resources	Original resources in place		
	Light oil	Light oil	Bitumen	Heavy oil	Shale oil
North America	269	900	13,700	3,700	12,000
United States only	171	900	100	1,200	11,900
Latin America	665	800	12,900	6,400	500
Europe	96	300	100	400	500
Middle East	4,242	3,200	0	5,500	200
Former Soviet Union	728	1,300	4,400	1,300	1,600
Other regions	901	1,700	300	1,900	1,300
World total	6,901	8,100	31,400	19,400	16,100
2005 production levels	168	—	3	—	0.02
Reserve-to-production (RP) ratio	41	—	—	—	—
RP ratio based on 2005 oil consumption	—	47	183	113	94

Sources: Reserve estimates for light crude oil are derived from British Petroleum, 2007. *BP Statistical Review of World Energy*; www.bp.com/statisticalreview. Resource estimates for light oil are based on Paltsev, S., J.M. Reilly, H.D. Jacoby, R.S. Eckaus, J. McFarland, M. Sarofim, M. Asadoorian, M. Babiker, 2005. *The MIT Emissions Prediction and Policy Analysis (EPPA) Model: Version 4*, MIT Joint Program on the Science and Policy of Global Change, Report 125, Massachusetts Institute of Technology, Cambridge; http://web.mit.edu/globalchange/. Resource estimates for heavy oil and bitumen are based on Meyer, R.F., E.D. Attanasi, P.A. Freeman, 2007. *Heavy Oil and Natural Bitumen Resources in Geological Basins of the World*, U.S. Geological Survey, VA. Shale oil data are derived from Dyni, J.R., 2005. *Geology and Resources of Some World Oil-Shale Deposits*, U.S. Geological Survey, VA.
Notes: North America consists of Canada and the United States; Mexico is part of Latin America. Divide EJ (10^{18} joules) figures by 5.7 to obtain billion barrels of oil equivalent.

of their countries' natural resources, they simultaneously exert some level of control over to whom they sell their energy and at what price, at least until supply constraints cease to exist. A second concern is associated with their use of some oil revenues against Western interests. Since oil revenues can be a significant fraction of GDP in oil-rich countries, even a small percentage of GDP leaking to terrorist groups can be a threat to Western countries.

Governments in the industrialized world have taken various actions to mitigate their oil dependency, including mandating minimum fuel economy standards, subsidizing the production of biofuels, mandating the use of biofuels as gasoline-alcohol blends, investing in the research and development of synthetic fuels and more fuel-efficient transport systems, and setting up strategic oil reserves.[73] Although these policies contribute to reducing the risks of oil dependency, they cannot eliminate the fundamental problem, at least in the near term. The exact state of the world oil market at any particular moment is necessarily uncertain—and must be expected to remain so. Hence, oil price hikes resulting from political instabilities in oil-exporting countries cannot be excluded from happening over the next fifty years.

Environmental Policies

Because of its large scale and nearly complete reliance on carbon-containing oil products, the transportation sector is a major emitter of greenhouse gases and in 2005 accounted for approximately 23 percent of the world's CO_2 emissions (see table 1.1). As discussed in chapter 1, the absolute and relative importance of this sector's GHG emissions is rising.

Thus, even if the transport infrastructure were supplied at a sufficiently fast rate and oil supply secured, transportation may be influenced by GHG-emission-mitigation policies. Already, transportation has been the target of policy makers, either through focused research and development (R&D) programs, instituting mandatory fuel-efficiency standards, or through passing tailpipe emission regulations. A policy's content and how strongly it is enforced could have a significant impact on consumer behavior and vehicle technology. The enactment of truly visionary and sound policies could initiate one of the most significant transformations in the history of transportation. Assessing the range of technological opportunities, possible policies that accelerate the adoption of these technologies, and their potential impact on passenger travel GHG emissions over the next thirty to fifty years is the focus of the following chapters.

Summary

Over the history of transportation, technological progress has not only enabled travel at higher speeds but also contributed to lower cost. These cost reductions in combination with rising income have greatly increased the affordability of and thus the demand for travel.

Over the past five decades, the combined increase in population and per capita GWP has gone hand-in-hand with global PKT increases by nearly one order of magnitude. With technology continuing to advance and both per capita income and population continuing to rise, global passenger mobility will further grow. By midcentury, travel by Earth's inhabitants could nearly triple or quadruple based upon the 2005 levels, depending on the assumed growth in GWP.

The growth in passenger travel may turn out to be highest in the developing world, where population and per person income are projected to grow fastest. Following the two chosen scenarios of future GDP growth that underlie our projections, passenger mobility in the developing world —through the use of automobiles, buses, railways, and aircraft—would increase to levels in 2050 that are three to six times what they were in 2005. Over the same period, travel demand in the industrialized world is projected to double or almost triple, while in the reforming economies, it is expected to be somewhere between the rates of growth of the industrialized and developing regions. Since the strongest growth in travel demand is projected to occur in the developing world, the developing countries could account for up to three-quarters of world PKT, depending on what rate of economic growth is assumed. However, because of the industrialized world's high income levels, the highest per person mobility levels would remain there. North America would remain the region with the highest per person mobility levels, rising from 27,400 kilometers (17,000 miles) in 2005 to 39,000–48,000 kilometers in 2050 (24,200–29,800 miles).

As in the past, the projected increase in passenger mobility is likely to be made possible by changes in the relative importance of faster modes of transport. The projected mode shares strongly depend upon the assumed travel time budget. If per person travel time remains close to its current level of about 1.2 hours per day, in the industrialized world high-speed transportation modes could rise to about two-thirds of total PKT, with LDVs accounting for nearly the entire remaining traffic volume. In contrast, in the developing regions, the growth of LDV travel would be strongest, accounting for 40–50 percent of total PKT in 2050. Globally, automobiles could account for more than 40 percent, high-speed transport for almost 40 percent, and low-speed public transport for about 20 percent of world PKT. In contrast, should the average per person travel time increase to 1.5 hours per day, the projected shares of air travel would rise less strongly in favor of higher shares of automobile travel.

The growth in passenger mobility as projected is enormous, possibly because it does not take into account the possibility that existing constraints may become more stringent in the future. Perhaps the most stringent constraint on the future growth of travel is the limited air transportation system capacity, but also here a combination of infrastructure extensions, changes in operational strategies, and reductions in aircraft noise could unlock significant extra levels of capacity. Should such measures fall short of providing extra levels of air transportation system capacity sufficient to satisfy the projected demand, travel demand in the industrialized world would begin to saturate. However, it is plausible that the strategies required to enhance airport capacity, or to expand to secondary and tertiary airports, will generate enough new system capacity so that past trends toward ever-higher levels of air travel will continue, at least through midcentury. Should that be the case, the only remaining limitations on travel growth would be price spikes associated with the unequal distribution of oil reserves and, potentially, the impact of global warming policies on the world transportation system. Before we return to this question, however, we first estimate the GHG-emission implications of the projected travel demand.

3

Greenhouse Gas Emission Implications of Travel Demand

In the previous chapter, we survey the past fifty-six years of world passenger travel and projected PKT (passenger kilometers traveled) through 2050. After examining factors that could limit demand for more travel, we conclude that—provided the air traffic system capacity can be sufficiently increased over the next five decades—only radical policies to reduce greenhouse gas (GHG) emissions or a strenuous response to the risk of oil disruptions might cause future travel demand to grow less than we project.

In this chapter, we focus on the effects of travel demand on GHG emissions through analyzing passenger travel energy intensity (E/PKT) from a human behavior and airline operations perspective. This factor translates the projected passenger travel demand from chapter 2 into GHG emissions, all other factors being equal. Our approach at this time is to take the characteristics of today's technologies and fuels as given; opportunities for technological change will be analyzed in subsequent chapters.

Since history can be a useful guide to the future, we begin with a look at past trends in energy intensity. We then examine the determinants of passenger travel energy intensity, for both automobile and air travel. Finally, we bring together the changes in energy intensity induced by human behavior and airline operations and our projection of travel demand to estimate the resulting energy use and GHG emissions through midcentury.

Historical Trends in Energy Intensity

History books are filled with examples of the twentieth century's impressive technological achievements. They have affected all sectors of the economy. Consider transportation and the industries that design, construct, supply, and support it. Technological progress has reduced the amount of energy used to produce one ton of steel, decreased the amount of steel required to produce an automobile of a given size, and increased the efficiency of automobile engines. And overall, technological progress has increased the quality of each intermediate and final product. In light of these and many other successes, it may come as a surprise that passenger travel energy intensity—the total amount of energy used for travel by all modes of transportation divided by total PKT—has remained roughly constant or even increased over at least the past three decades in many countries. Figure 3.1 shows this development for France, Germany, Great Britain, Japan, Switzerland, and the United States. Only in the United States has the combination of high oil prices and fuel economy standards reduced passenger travel energy intensity. During the 1980s it fell from around 3 to roughly 2.5 mega-joules (MJ) per passenger-kilometer (pkm). Yet, the 2005 level is again close to what it was five decades ago.

However, since the energy efficiency of the various components of contemporary transportation technology has *increased*, the question arises: why has passenger travel energy intensity remained roughly *constant* or *grown* throughout the past half-century (except, briefly, in the United States)? The effect of technological improvements must have been offset by other factors. Among these factors is the increasing relative importance of ever-faster and more energy-intense modes of transport: in particular, the shift from low-speed public transport modes to automobiles and aircraft, as discussed in chapter 2. Table 3.1 shows the average levels of energy intensity by mode of transport in urban and intercity travel in the United States. As can be seen, energy use generally rises with speed within each of the two market segments. For example, a complete shift from low-speed public transport modes to automobiles increases the average energy intensity in intercity travel by some 60 percent; a complete shift toward aircraft would roughly double the energy intensity of the low-speed public transport level.

The same table also shows that energy intensities are higher in urban travel than in intercity operation. For both automobiles and public transport, urban transportation is roughly twice as energy intense as intercity travel (for reasons we discuss below). Thus, any increase in urban travel would lead to higher average energy intensities, all other factors being equal.

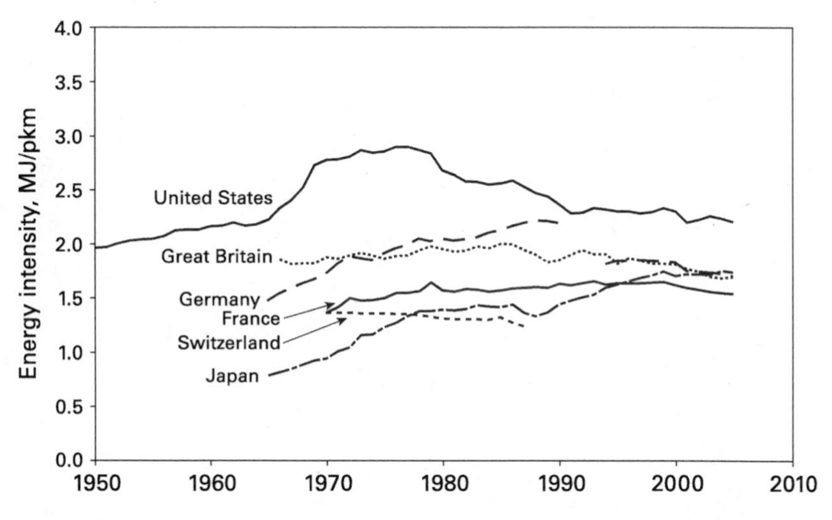

Figure 3.1
Passenger travel energy intensity in six industrialized countries. Sources: American Public Transportation Association, 2007. *Public Transportation Fact Book*, Washington, DC. U.S. Environmental Protection Agency, 2007. *Light-Duty Automotive Technology and Fuel Economy Trends: 1975 through 2007*. Compliance and Innovative Strategies Division and Transportation and Climate Division, Office of Transportation and Air Quality, Washington, DC. Davis, S.C., S.W. Diegel, 2007. *Transportation Energy Data Book*, Oak Ridge National Laboratory, Oak Ridge, TN. U.S. Department of Transportation, *Air Carrier Summary Data (Form 41 and 298C Summary Data)—Air Carrier Summary: T2: U.S. Air Carrier Traffic and Capacity Statistics by Aircraft Type*, Bureau of Transportation Statistics, Washington, DC. Oak Ridge National Laboratory, U.S. National Household Travel Survey; http://nhts.ornl.gov/tools.shtml. U.S. Department of Transportation, various years. *Highway Statistics*, Federal Highway Administration, Washington, DC. U.S. Census Bureau, various years. *Statistical Abstract of the United States*, U.S. Department of Commerce, Washington, DC. Lee, J.J., S.P. Lukachko, I.A. Waitz, A. Schäfer, 2001. Historical and Future Trends in Aircraft Performance, Cost, and Emissions, *Annual Review of Energy and the Environment 2001*, 26: 167–200. Bundesverkehrsministerium, various years. *Verkehr in Zahlen*, Deutscher Verkehrsverlag Hamburg. Institut National de la Statistique et des Études Économiques, various years. *Les comptes des transports*, Commission des Comptes des Transports de la Nation, Paris. Institute of Energy Economics, 2007. *Handbook of Energy & Economic Statistics in Japan*, The Energy Data and Modeling Center, Tokyo. Department for Transport, various years. *Transport Statistics Great Britain*, London. Bundesamt für Statistik, 1992. *Schweizerische Verkehrsstatistik 1990*, Bern.

Table 3.1
Average Characteristics of Transport Modes in the United States at about 2005

	Urban traffic				Intercity traffic			
	Door-to-door speed	Usage characteristics		Energy intensity	Door-to-door speed	Usage characteristics		Energy intensity
		Occupancy rate	Passenger load factor			Occupancy rate	Passenger load factor	
	km/h	pkm/vkm[a]	pkm/skm[b]	MJ/pkm	km/h	pkm/vkm	pkm/skm	MJ[c]/pkm
Low-speed public transport	19	15	—	2.4	35	24	—	0.92
Bus	13	9.1	—	2.7	35	23	—	0.56
Rail	28	22	—	1.9	34	26	—	1.9
LDV[d]	31	1.4	0.26	3.2	78	2.2	0.41	1.5
Car	31	1.4	0.28	2.9	78	2.2	0.45	1.3
Light truck	31	1.5	0.25	3.7	78	2.4	0.39	1.8
Aircraft	—	—	—	—	270	99	0.78	1.8
Domestic	—	—	—	—	260	88	0.77	2.4
International	—	—	—	—	410	150	0.79	1.6

Sources: American Public Transportation Association, 2007. *Public Transportation Fact Book*, Washington, DC. U.S. Environmental Protection Agency, 2007. *Light-Duty Automotive Technology and Fuel Economy Trends: 1975 through 2007*, Compliance and Innovative Strategies Division and Transportation and Climate Division, Office of Transportation and Air Quality, Washington, DC. Air Transport Association of America, 2007. *Economic Report*, Washington, DC. Davis, S.C., S.W. Diegel, 2007. *Transportation Energy Data Book*, Oak Ridge National Laboratory, Oak Ridge, TN. U.S. Department of Transportation, 2008. *Air Carrier Summary Data (Form 41 and 298C Summary Data), T2: U.S. Air Carrier Traffic and Capacity Statistics by Aircraft Type*, Bureau of Transportation Statistics, Washington, DC. Oak Ridge National Laboratory, Online Analysis Tool of the 2001 U.S. National Household Travel Survey; http://nhts.ornl.gov/2001.

[a] vkm: vehicle-kilometers.

[b] skm: seat-kilometers.

[c] MJ: million joules.

[d] Travel speeds and energy intensities of light-duty vehicles are averages of the Federal Test Procedure driving cycle components.

In addition to factors influencing the structure of passenger travel, fuel-efficiency improvements have also been offset by changes within transport modes. This is especially true for automobile travel, since consumers have increasingly valued larger, more powerful vehicles that also provide more passenger amenities. In the United States, the shift toward larger but less fuel-efficient vehicles has been especially pronounced in the increasing popularity of personal light trucks (vans, minivans, sport-utility vehicles, pickup trucks), and this is one important reason for the comparatively high levels of passenger travel energy intensity seen in figure 3.1.

While these and other "human factors" have partially offset, even undermined, improvements in fuel efficiency for automobile travel, economic factors have driven airline operations toward higher levels of fuel efficiency. However, that development may be partly compromised in the future by similar consumer trends toward more comfort and higher door-to-door speed.

Human Factors in Automobile Travel

In the most general sense, human factors in automobile travel include all areas of interaction between people and the road transportation system. That interaction starts with the purchase of a motor vehicle, continues with the changing environment in which it operates, includes the vehicle's operation and maintenance, and ends with finally giving it up through sale or otherwise.[1] While operation and maintenance can influence passenger travel energy use by about 5 percent, we will focus our analysis on the dominant human factors: consumer preferences for larger vehicles with higher engine power, the phenomenon of declining occupancy rates, increasing urban driving, and the implications of little consumer reaction to fuel prices. As we will show, in the past these behaviors have worked against efforts to reduce passenger travel energy intensity.

Consumer Shift toward Larger and More Powerful Vehicles

Ever since the introduction of the first automobiles over a century ago, motor vehicles have become larger and more powerful. At the beginning of the automobile era, Henry Ford's Model T, first introduced in 1908, had an empty weight of about 550 kilograms (kg) or about 1,210 lbs and an engine power of 15 kilowatts (kW), translating into a ratio of engine power to vehicle weight of 27 kW per ton. The Model T's fuel consumption was about 15 liters (L) per 100 vehicle-kilometers (vkm) or 15.7 miles per gallon (mpg) when measured according to the U.S. combined (urban and highway) driving cycle.[2] (Since fuel consumption in this driving cycle is only an approximation of real-world driving and is measured under laboratory conditions, which do not take into account other determinants of fuel consumption such as power for lights, windshield wipers, air conditioning, and other auxiliary uses, it underestimates fuel use by about 20 percent).[3] By the mid 1950s vehicle weight had tripled to 1,710 kilograms (3,770 lbs); engine power had grown more than fourfold to nearly 90 kW; while fuel consumption (measured according to the same driving cycle) remained largely unchanged. Over the subsequent two decades, vehicle weight and engine power continued to increase. By 1973 the average new automobile sold in the United States weighed 1,860 kilograms (4,100 lbs) and had an engine power of 110 kW, doubling the Model T's power-to-weight ratio and raising fuel consumption to above 18 L/100 vkm (13.1 mpg).

This growth trend was temporarily reversed by two major events. The OPEC oil embargo of 1973 and the Iran-Iraq war in 1979 led to sharp increases in the price of oil, and thus of gasoline. In the United States, government limits on the price of motor fuel caused local supply shortages and long queues at gas stations. The pressure on industry and consumers to save fuel was reinforced by legislation in the form of the Corporate Average Fuel Economy (CAFE) standards for automobiles and light trucks. First introduced in 1978 with the intention of doubling the fuel economy (that is, halving the fuel *consumption*) of new motor vehicles within a decade, the regulations required manufacturers to satisfy increasingly tight fuel-efficiency levels under the threat of economic penalties.

The manufacturers complied with the CAFE standards. While higher oil prices initially pushed Americans toward smaller and less powerful vehicles, maintaining the higher level of fuel efficiency was ensured by the legislated fuel economy standards. By the mid 1980s, fuel consumption by the average new American automobile had declined to 8.7 L/100 vkm (27.0 mpg)—half the level of the early 1970s.

However, as gasoline prices declined in the latter half of the 1980s and fuel economy standards remained unchanged, consumers started to rebalance their choices in favor of larger and more powerful vehicles. Since extreme increases in automobile size and performance were constrained

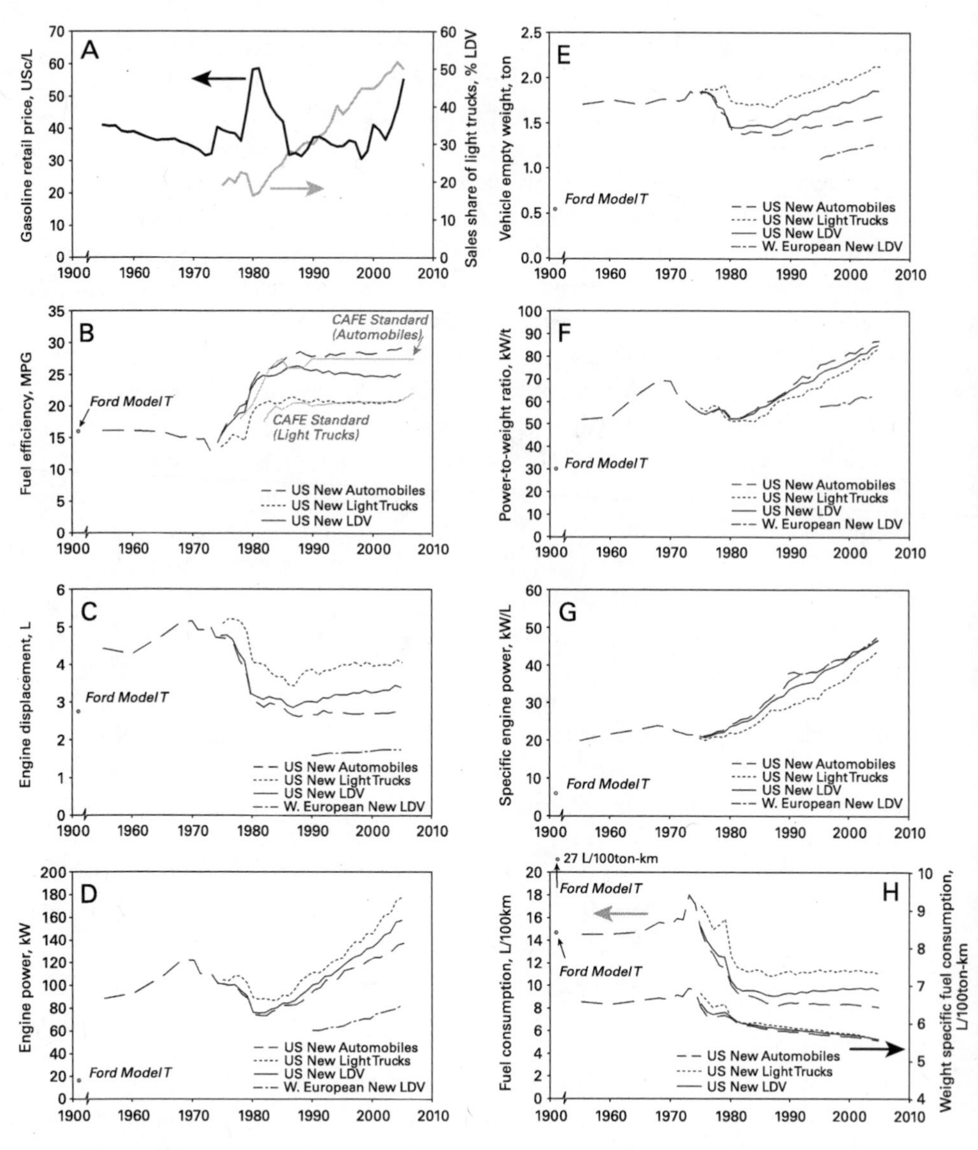

Figure 3.2
Historical development of light-duty vehicle characteristics in the United States. Fuel efficiency and fuel consumption levels are measured on the combined U.S. driving cycle and are not adjusted for real-world driving conditions. Sources: U.S. Department of Transportation, *Historical Passenger Car Fleet Average Characteristics*, U.S. National Highway Traffic Safety Administration, Washington, DC. U.S. Environmental Protection Agency, 2007. *Light-Duty Automotive Technology and Fuel Economy Trends: 1975 through 2007*. Compliance and In-

by the (continuing) fuel economy standards, consumers increasingly moved toward light trucks. Between 1985 and 2005, these larger vehicles, which enjoyed more relaxed fuel economy standards and were thus not as severely space-constrained as automobiles, doubled their share to 50 percent of all new light-duty vehicles (LDVs) sold in the United States. The relative growth in passenger travel by light trucks has been even larger, since the use of these vehicles for personal travel as opposed to commercial purposes has also increased, from about half of all personal travel in the mid 1970s to 77 percent in 2002.[4] Because of this greater share of light trucks, the weight of today's (2005) average new LDV has again reached the level of the average new automobile sold in 1975, while engine power has risen by more than 50 percent to 158 kW, yielding an average power-to-weight ratio of 85 kW/ton. This shift toward larger vehicles, which has again increased fuel consumption of the average new LDV to nearly 10 L/100 vkm (23.5 mpg), has offset most of the gains in fuel efficiency that have been achieved since 1980 within each vehicle size segment.

Figure 3.2 shows these and other long-term developments. For the sake of comparison, the corresponding characteristics of the Ford Model T, the dominant automobile during the 1910s, are also shown. In addition, figure 3.2 reports the corresponding characteristics from the sales-weighted Western European new LDV fleet.

Panel A indicates the rising share of light trucks in periods of low gasoline prices or declining fuel prices from high levels. Panel B shows that the fuel economy standards (shown by the gray continuous lines) made sure that the higher levels of fuel efficiency, mainly achieved by higher fuel prices, were maintained. The binding constraint of these standards is especially prominent for light trucks, because their fuel efficiency has developed in compliance with the standard since the mid-1990s. The rising consumer preference for light trucks has caused average LDV fuel efficiency to decline since the mid-1980s, although the fuel efficiency of

novative Strategies Division and Transportation and Climate Division, Office of Transportation and Air Quality, Washington, DC. British Petroleum, 2007. *BP Statistical Review of World Energy*; www.bp.com/statisticalreview. European Automobile Manufacturers' Association, *Trends in New Car Characteristics*; www.acea.be. Commission of the European Communities, various years. *Monitoring of ACEA's Commitment on CO_2 Emission Reductions from Passenger Cars*, Joint Report of the European Automobile Manufacturers' Association and the Commission Services, Brussels.

automobiles has increased slightly and that of light trucks has remained constant since then.

Panels C, D, E, and F show trends in engine size (displacement), engine power, vehicle weight, and the resulting power-to-weight ratio (a proxy for vehicle acceleration capability) for automobiles, light trucks, and the sales-weighted LDV fleet. The consumer shift toward smaller vehicles, induced by the two oil crises of the 1970s and the fuel economy standards in 1978, is clearly visible in terms of the decline of all four indicators. However, with the beginning era of lower fuel prices in the mid 1980s, engine size, power, vehicle weight, and acceleration capability increased. Compared with the Model T, today's new LDVs are three to four times as heavy, and engine power is nearly ten times as high. Thus, the acceleration capability of today's vehicles is nearly three times that of the Model T. The development of the dot-dashed lines indicates similar trends in Western Europe. Yet, differences in absolute levels remain significant. For example, in 2005, engine displacement of the average new LDV was about 1.8 liter in Europe, which is only half the size of its U.S. counterpart.

Despite these differences in vehicle and engine size, there is little difference in technology between U.S. and Western European vehicles. One example demonstrating such similarity is the uniform level of, and increase in, specific engine power, that is, engine power divided by engine size (panel G). Compared with a Ford Model T, specific engine power of today's new vehicles is nearly ten times as high.

The left vertical axis of panel H shows the developments of vehicle fuel consumption, expressed in liters of gasoline consumed per 100 kilometers. It thus mirrors the developments in fuel efficiency from panel B. Since this measure, liters of gasoline per 100 kilometers, scales with CO_2 emissions per unit of travel, we work primarily with this indicator in this book. Panel H also shows fuel consumption over time divided by vehicle weight over the same period. This relationship shows that despite the stagnating *fuel*-efficiency levels shown in panel B, vehicles have continuously become more *energy* efficient. However, much of this gain was lost in the shift toward larger vehicles, including vehicles within each of the two market segments looked at here, automobiles and light trucks. While the fuel consumption of a Ford Model T at the beginning of the twentieth century was almost twice that of today's average *automobile*, adjusting this comparison to account for vehicle weight results in a ratio of five to one: the Model T consumed five times as much fuel per unit vehicle weight. Significant fuel savings that could have been realized have been sacrificed in order to drive heavier (i.e., larger) vehicles.

How can this apparent consumer preference for larger vehicles be explained? For one thing, all other factors being equal, larger vehicles typically provide greater safety to their occupants. A recent analysis of U.S. accident statistics suggests that the risk of traffic death to drivers of compact vehicles is 40 percent higher, and to drivers of subcompact cars as much as 60 percent higher, compared with the risk drivers of midsize and large cars face when different driver behavior and the travel distances for each specific type of vehicle are taken into account.[5]

In addition, greater engine power provides higher acceleration capability and more power for running vehicle accessories (climate control, entertainment systems, navigation systems, etc.). Thus, larger and more powerful vehicles typically offer higher levels of driving comfort and convenience. As discussed in chapter 2, most people enjoy some travel in the course of a day, but longer trips often involve some discomfort. That "disutility" can be reduced by traveling in a larger and thus potentially more luxurious vehicle. As pointed out in a statistical analysis by Sangho Choo and Patricia Mokhtarian at the University of California at Davis, Northern California drivers who travel longer distances and those who dislike travel in general are both more likely to choose larger vehicles.[6] Finally, people also choose larger vehicles to demonstrate their position within society. The automobile has become more than simply a means of transport; it is also a means of self-expression and an emblem of social status. The analysis by Choo and Mokhtarian confirms that "status seekers" typically decide on a luxury vehicle (or sports car).

While most consumers would prefer larger, more comfortable, safer, and higher-performing vehicles, they need to balance these attributes against higher initial costs and higher lifetime-operating costs. However, these financial influences on consumer choice have become less burdensome over much of the history of the automobile. One reason is that income has grown faster than the average vehicle price. In 1967, a typical U.S. household had to work for about five months (twenty-one weeks) to afford the average new automobile of that time. That amount of work declined to about four months (seventeen weeks) by 2005.[7] At the same time, vehicle technology and comfort have advanced. Largely driven by the vehicle manufacturers' competition for market share, loyal buyers of a particular car model enjoy more space, engine power, and comfort with each model update.[8] The Toyota Camry, one of the most popular (midsize) automobiles in the United States, is an example. Introduced into the U.S. market in 1983, the first-generation Camry had a curb weight of 1,110 kilograms (nearly 2,450 lbs). Its two-liter engine provided a maximum power of 69 kW, and vehicle fuel consumption was 7.5 L/100 vkm (31.4 mpg) over the U.S. combined driving cycle. Over time, slightly larger, increasingly safe, and more comfortable updates have followed, increasing both vehicle weight and engine power. The weight of the completely redesigned 2007 model has reached 1,500 kilograms (about 3,300 lbs), while its most popular (2.4 L) engine provides 118 kW (70 percent higher than that of the 1983 model). This significant growth in vehicle weight and engine power caused fuel consumption to increase by 20 percent to 9 L/100 vkm (26.1 mpg). Similar trends have occurred with respect to other vehicles in all segments of the market.

Additionally, the auto industry has facilitated the consumer shift toward larger vehicles through the use of financial terms in the competition for market share. This struggle has intensified with the ever-growing number of entrants into the replacement markets of the industrialized world. The practice of providing incentives for purchasing vehicles with the aim of increasing market share, however, dates back almost to the beginning of mass production. Already in 1919, William Durant, the founder of General Motors, set up the first vehicle-financing firm, the General Motors Acceptance Corporation, to provide car loans to ordinary people. This innovation contributed to the second phase of motor vehicle adoption in American society (see chapter 1). By 1948, one-third of all new vehicle purchases in the United States were financed, a share that nearly doubled by 1956.[9]

Over the past few decades, the conditions of vehicle financing have continuously improved. Between 1980 and 2003, the average interest rate for car loans declined from around 15 percent a year to 3.4 percent; the average maturity term lengthened from forty-five to sixty-one months; while the average amount financed has remained roughly unchanged at about $22,000 per vehicle in 2003.[10] Because of an increase in interest rates initiated by the U.S. Federal Reserve, the average financing rate increased slightly during the post-2003 period, while the other characteristics remained essentially unchanged. These financing plans brought benefits to both consumers and manufacturers. While consumers were able, first, to "motorize" themselves, and later to purchase larger vehicles, the automobile industry made significant profits on the financial transactions alone, sometimes exceeding their profits from producing the vehicles.[11]

Another vehicle manufacturer strategy that "moves more metal into the market" and leads to an increase in average vehicle size is motor vehicle leasing. At the beginning of the 1990s, the auto industry responded to sluggish demand for new vehicles by creating attractive leasing plans to reduce their inventory. In 1985 leasing accounted for only 3.5 percent of all new vehicles introduced into the market. Yet, by 1990 the share of leased vehicles had already doubled and, by the late 1990s, had reached 30 percent of all new vehicles put into the market. Although leasing has declined slightly in recent years, it still accounted for 28 percent of all vehicles transacted in 2002.[12]

While financing plans typically allow people at lower levels of income to purchase and ultimately own a vehicle, leasing also enables lower-income drivers to operate a larger vehicle than they otherwise could afford.[13] Of course, leasing is not limited to customers with a constrained vehicle budget; leasing rates are highest (above 50 percent) for near-luxury and luxury vehicles—probably because of the ease of changing to a new vehicle at the end of the lease.

In addition to rising income and the various incentives provided by the automobile industry, a final enabler of the shift toward larger and more powerful vehicles has been stable fuel prices through this period of generally rising income. As can be seen in panel A of figure 3.2, the gasoline retail price has remained between 30 and 40 cents per liter (roughly $1.10 to $1.50 per gallon) for much of the past five decades. This narrow price range translated into fuel expenditures that accounted for only 10–20 percent of the total costs of owning and operating a new automobile (the exact value of which depends on the specific average gasoline price and the distance driven over a year—here between 16,000 kilometers and 24,000 kilometers, or 10,000–15,000 miles).[14] During the two oil crises, however, when the average gasoline retail price was nearly 60 cents per liter (about $2.25 per gallon), fuel expenditures accounted for roughly 30 percent of total vehicle costs, significantly influencing the tradeoffs among vehicle attributes as described above. The same figures also show that consumers in the 1970s reverted away from light trucks to automobiles when the price of gasoline rose above 40 cents per liter (around $1.50 per gallon).

Over time, however, rising income and declining vehicle ownership costs seem to have weakened consumer response to fuel economy. In

2004, when the average U.S. price for gasoline was around 50 cents per liter (approaching $2 per gallon), one market survey found that interest in fuel efficiency had just tied with interest in the number of cup holders in a vehicle.[15] Not surprisingly, in the 2006 to 2008 period, when fuel prices rose substantially, there again has been an apparent shift in consumer choice away from the larger vehicles.

Influences leading to larger vehicles also exist in Europe as well as in the United States, but higher gasoline prices, in the range of $1–2 per liter (around $3.80–7.60 per gallon), long existed there. As a result, fuel costs account for about 30 percent of the total costs of owning and operating a vehicle.[16] These higher shares have a profound effect on vehicle size—and thus on fuel consumption. Table 3.2 compares the major characteristics of the average new automobile, light truck, and resulting sales-weighted average LDV in the United States to the average new automobile sold in Western Europe. Compared with its Western European counterpart, the American automobile has a more than 60 percent larger engine size and power. Taken together with its nearly 25 percent greater vehicle weight, the average new U.S. automobile power-to-weight ratio and fuel consumption are about 40 percent higher than those of the average new vehicle operating in Western Europe. Overall, the average new U.S. automobile is about one size segment larger than its Western European counterpart, a difference between a midsize and a compact vehicle. The difference in vehicle size to the average European LDV is even greater when the U.S. average light trucks are taken into account.

Although higher fuel costs cause consumers to rebalance their priorities in favor of reduced vehicle size, power, and fuel consumption, to date they have not restrained an ongoing European trend toward larger vehicles. In Germany, the average automobile of the 1965 fleet had an engine power of 31 kW. By 2005 the average engine power of the vehicle fleet had more than doubled to 74 kW, while the engine power of the average new vehicle sold that year had a power output of 90 kW.[17] Panel D in figure 3.2 shows that the engine power of the average new Western European vehicle has also increased, from 61 kW in 1990 to 83 kW in 2005.[18] Yet, these levels are still only half the U.S. values.

This trend toward larger and more powerful vehicles also in Western Europe underlines the fact that low fuel costs are only one of the factors that have encouraged a move toward larger vehicles. Rising income and declining vehicle ownership costs with manufacturers' financial incentives (that make possible choosing larger vehicles that otherwise would

Table 3.2
Characteristics of the average new LDV in 2005

	United States			Western Europe
	Automobile	Light truck	Average LDV	Automobile
Engine displacement, L	2.8	4.1	3.4	1.7
Engine power, kW	134	179	160	83
Vehicle weight, kg	1,580	2,140	1,860	[a]1,270
Power-to-weight, kW/ton	87	84	85	[a]63
Interior space, m^3	3.8	—	—	—
Fuel consumption, L/100 km (mpg)	10.0 (23.5)	13.7 (17.2)	11.8 (19.9)	7.1 (33.1)

Sources: U.S. Environmental Protection Agency, 2007. *Light-Duty Automotive Technology and Fuel Economy Trends: 1975 through 2007*, Compliance and Innovative Strategies Division and Transportation and Climate Division, Office of Transportation and Air Quality, Washington, DC. An, F., A. Sauer, 2004. *Comparison of Passenger Vehicle Fuel Economy and GHG Emission Standards Around the World*, Pew Center on Global Climate Change, Washington, DC. European Automobile Manufacturers' Association, *New Passenger Car Registrations in Western Europe, Breakdown by Specifications: Average Power, Historical Series: 1990–2005*; www.acea.be/. Commission of the European Communities, 2005. *Monitoring of ACEA's Commitment on CO_2 Emission Reductions from Passenger Cars*, Joint Report of the European Automobile Manufacturers' Association and the Commission Services, Brussels.
Notes: Fuel consumption is measured on the U.S. combined driving cycle and inflated by 25 percent for real-world driving. We adjusted the original value of 6.8 liters of gasoline equivalent per 100 km for the average new Western European automobile for differences between the European and U.S. driving cycles plus the 25 percent difference in real-world driving.
[a] Figure relates to 2004.

be more expensive) are additional important factors that enable consumers to acquire larger, more comfortable, and safer vehicles that often represent higher status.

The rising affordability of the automobile has transformed American attitudes toward the vehicle. As summarized by an automobile industry consultant recently, "In 1980, 80 percent of new car buyers were forced to buy new cars when their old ones died. By 1990, that figure had dropped to 60 percent; today, only 18 percent of buyers purchase new cars out of necessity. The rest buy new cars to get the coolest, most exciting, or innovative models on the market—or because of financial deals they just can't pass up."[19]

Declining Occupancy Rates

In the United States in 1969, one automobile had an average of 2.2 occupants. This occupancy rate declined to 1.9 in 1977, to 1.75 in 1983, 1.64 in 1990, and 1.59 in 1995. The most recent information available, the 2001 U.S. National Household Travel Survey, found an occupancy rate of 1.63, suggesting that the at least three-decade-long decline in vehicle occupancy may have leveled off. Although this plateau can still angle upwards or downwards slightly, partly because of the different methods of reporting household travel used in the underlying surveys, data from many other countries are in keeping with this development.[20]

The decline in occupancy rates has been mainly the result of a decline in average household size and an increase in the size of the vehicle fleets, which reduced the need for sharing a vehicle. Both trends in turn have been partly stimulated by the rising share of women in the labor force.[21] In the industrialized world, these forces were strongest during the second half of the twentieth century. Between 1969 and 2001, the share of women in the labor force increased from 38 percent to 47 percent, the average household size declined from 3.16 to 2.63, and the number of vehicles per household rose from 1.16 to 1.85.

Occupancy rates also depend, however, on several other factors, including the purpose for which a vehicle trip is taken. In the United States and in many other industrialized countries, automobiles carry on average only 1.15 occupants in work-related travel; in contrast, the average vehicle occupancy rate is significantly higher for trips taken for "personal matters" (1.8) and highest for leisure travel (2.0).[22] Since much longer-distance travel is leisure related, occupancy rates are higher for intercity travel. According to the 2001 U.S. National Household Travel

Survey, the average vehicle occupancy rate is only slightly above 1.5 persons per vehicle for trip distances up to 50 kilometers (31 miles), but nearly 1.8 for trips over 50 kilometers.[23] Occupancy rates are highest for trips longer than one day, averaging two passenger kilometers (pkm) per vehicle kilometer (vkm) for distances greater than 160 kilometers (100 miles), and around 2.7 passengers for distances greater than 1,000 kilometers (around 620 miles). As discussed in chapter 2, the relative importance of vehicle trips changes with rising affluence. While work-related trips account for slightly less than one trip per person per day across all societies, increasing income adds to a traveler's daily trip schedule more and more trips for personal matters and leisure. Thus, the relative importance of work trips declines, from about one-third of all trips in very low-income countries to only 14 percent of total trips in the United States today. These changing shares of types of travel also may have contributed to the recent leveling off of vehicle occupancy rates.

Since energy intensity (E/PKT) is the product of vehicle energy use (E/VKT) and the inverse occupancy rate (VKT/PKT), a declining occupancy rate (growing VKT/PKT) directly leads to higher levels of energy intensity. The past decline of U.S. occupancy rates from 2.20 in 1969 to 1.63 in 2001 has thus led to an increase in energy intensity of about 35 percent (2.20/1.63). (The partly compensating effect of a slightly reduced vehicle weight, resulting from an average of 0.6 fewer passengers—roughly 50 kilograms [110 lbs]—being transported, is nearly negligible).[24]

While the decline in occupancy rates has leveled off in most industrialized countries, a similar process is about to begin in parts of the developing world. In these countries, economic growth will also result in a decline in household size, rising motorization, and thus decreasing occupancy rates. Typical occupancy rates in the developing world are currently about 2.5 pkm/vkm. Based on our two projections of economic growth outlined in chapter 2 and factoring in decline in household size and increase in vehicle ownership, the vehicle occupancy rate in the developing world would remain roughly constant (in the case of the EPPA reference run [EPPA-Ref], which assumes less aggressive income growth in the developing world), or would decline to about 2.0 pkm/vkm by 2050 (in the case of the SRES-B1 scenario, which assumes a more aggressive income growth in the developing world). A decline to 2.0 pkm/vkm would result in an increase in energy intensity by 25 percent over the current level. When all world regions are combined, the average global occupancy rate of about 1.8 pkm/vkm in 2005 would remain either roughly constant (in the SRES-B1 scenario of income growth rates), or would increase to about 2.0 pkm/vkm through 2050 (in case of the EPPA-Ref) because of the less strongly growing automobile traffic but significantly higher occupancy rates in the developing world.

Increasing Share of Urban Driving

In the United States, between 1940 and 1960, light-duty VKT were distributed roughly equally between urban and rural roads.[25] Thereafter, the share of urban travel has continuously increased, reaching 60 percent of all VKT in 1980 and 67 percent in 2005. This increase in urban travel may have partly resulted from the expansion of urban areas over this period. Long-term historical data from the U.S. Department of Agriculture suggest that urban sprawl accelerated moderately after the Second World War—the period of U.S. mass motorization. This expansion process likely urbanized roads that formerly were classified as intercity highways.[26] In addition to this potential relabeling of existing roads, the declining share of intercity VKT after 1960 was caused by air transport taking the place of automobile travel, a phenomenon we describe in chapter 2.

The rising share of urban travel influences LDV energy intensity in two ways. First, based on the U.S. combined driving cycle, LDVs consume almost 40 percent more fuel per VKT under urban driving conditions compared with relatively smooth intercity driving. In addition, and as discussed in the previous section, the driving cycle–based difference in energy use is amplified by urban travel's lower average occupancy rates. Thus, combining an almost 40 percent higher energy use per VKT and an almost 40 percent lower occupancy rate in urban travel leads to an energy intensity in urban travel that is more than twice that of intercity travel (see also table 3.1).

The impact of the ongoing urbanization of car traffic on automobile energy intensity, however, will be modest. In the industrialized world, most LDV-related VKT already occur over relatively short distances. According to the 1995 U.S. Nationwide Personal Transportation Survey, nearly 70 percent of all VKT occur on a trip distance of fewer than 100 kilometers (62 miles), and nearly 90 percent of all VKT occur on a trip distance below 400 kilometers (249 miles).[27] Thus, the potential for aircraft to take the place of intercity automobile travel is relatively small. Based on the results of a range of sensitivity tests concerning the size of the specific market segment within which projected growth in light-duty VKT is most likely to occur, our analysis suggests that average automobile energy intensity in the industrialized world may increase over current levels by about 10 percent by 2050. Within the developing world, the increase in LDV energy intensity will likely be smaller since automobile traffic there is more concentrated within the urban market.

Limited Consumer Reaction to Changing Fuel Prices

In chapter 2 we omit the consumer responsiveness to changes in fuel prices, the price elasticity of fuel demand, when we project world-regional travel demand. This choice is motivated by the lack of long-term historical data for gasoline retail prices; such information is available for only a few countries, among them the United States. However, the error associated with the omission is small.

Consumer responsiveness to price changes varies greatly depending on the consumer goods experiencing the changes. While the response to changes in movie and restaurant meal prices is strong, the response to price changes for transportation fuel is as small as for bread and potatoes. Studies using datasets from the 1970s through the 1990s have concluded that the short-run price elasticity of gasoline is about −0.2 to −0.3—meaning that a 10 percent increase in fuel price would lead to a 2–3 percent decline in fuel use.[28] In the long run, after consumers have had time to adjust to a higher gasoline price (such as by buying a more fuel-efficient automobile or substituting public transportation for car trips), the elasticity increases to about −0.6 to −0.7.[29]

Studies using more recent datasets have found the short-run elasticity to be even smaller, significantly below −0.1. This even smaller consumer response can be explained by rising income, increasing dependence on the automobile for urban travel, and the rising vehicle fuel economy from the mid-1970s to the mid-1980s.[30] In both the short and long run, the percentage reduction in consumer response is smaller than the percentage increase in price; economists denote such demand as "inelastic."

For a given increase in gasoline price, empirical studies conclude the reduction in vehicle travel to be still smaller than the decline in gasoline demand.[31] The need and desire for travel may induce consumers to mitigate the rising fuel costs through driving in a more energy-efficient style, by using a more energy-efficient household vehicle more frequently, or through other mechanisms. We will discuss the reasons for these low price elasticities along with their policy implications in chapter 7.

Relative Importance of Human Factors in Light-Duty Vehicle Travel

To what extent may human factors have contributed to changes in LDV energy intensities? For the sake of simplicity, we will examine only three contributing factors: the change in energy use (E/VKT) for automobiles and light trucks, the change in occupancy rates (PKT/VKT) for the two vehicle categories together, and the relative change in travel demand (PKT) between the two vehicle categories.

U.S. history provides an illustrative example. Over the entire 1970–2005 period, there were three significant developments: vehicle fuel consumption declined; car occupancy rates decreased; and light trucks increased their share in the makeup of PKT. The net result of these interconnected factors was a decrease in LDV energy intensity over this time. Using a formula similar to that presented in chapter 1, elementary mathematical manipulations can disentangle the various human factors that went into producing this result during this period.[32] The outcome of this decomposition, shown in figure 3.3, suggests that the initial *increase* in energy intensity between 1970 and 1979 resulted mainly from a decline in vehicle occupancy rates (an increase in VKT/PKT), which more than offset the modest reduction in fuel consumption over that period by the U.S. LDV fleet.

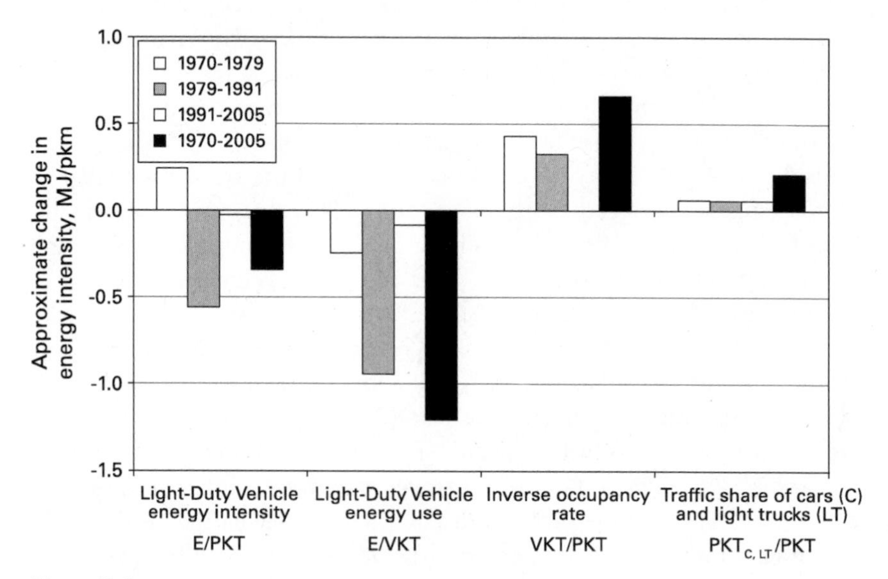

Figure 3.3
Determinants of light-duty vehicle energy intensity in the United States between 1970 and 2005.

In contrast, the subsequent (1979–1991) decline in the LDV fleet's energy intensity was caused by a reduction in the fleet's fuel consumption (a decline in E/VKT). This reduction was significantly greater than the increase in energy use caused by the simultaneous decline in car occupancy rates (an increase in VKT/PKT). Thereafter, with vehicle occupancy rates practically saturated from the beginning of the early 1990s, an almost constant level of fuel consumption resulted in constant levels of energy intensity. Throughout this period, a shift toward light trucks continuously contributed to rising energy intensity.

Overall, a decline in LDV energy intensity between 1970 and 2005 by some 0.3 MJ/pkm consisted of a loss of nearly 1.2 MJ/pkm from reductions in LDV fuel consumption. However, most of this reduction was offset by an increase in energy intensity of 0.7 MJ/pkm caused by a declining occupancy rate and by another increase of nearly 0.2 MJ/pkm because of a shift from automobiles to light trucks. A similar analysis for Germany shows that between 1970 and 1990, a decline in the occupancy rate and an increase in vehicle fuel consumption (resulting from a shift toward larger and more powerful automobiles) similarly contributed to an increase in LDV energy intensity. After 1990, both occupancy rates and vehicle fuel consumption remained approximately stable—as, consequently, did LDV energy intensity.

Economic Factors in Air Transportation

Whereas purchasers of personal automobiles seek to maximize their utility by balancing a vehicle's attributes—including size, occupant safety, and aesthetic appearance—against purchase costs, fuel consumption, and various other determinants of costs, the choices made by the owners and operators of commercial airplanes are driven almost exclusively by the desire to maximize profits. Since there is little difference in passenger service among the major airlines, they compete largely by satisfying consumer demand at minimum costs. Also, "consumer demand" is not limited to passengers, since a significant fraction of airfreight (50 percent in 1990, declining to 30 percent in 2005, for the United States) is flown in the cargo holds of commercial passenger aircraft rather than in dedicated cargo aircraft. Because of the intense competition among airlines and the resulting low profit margins, small differences in operating costs—whether caused by their own behavior or external factors (such as labor agreements, fuel prices, and airport fees)—can mean the difference between profit and loss for carriers. In fact, the U.S. airline industry as a whole has operated at a net loss over the past six decades.[33] Thus, it may not come as a surprise that the entire network structure of the airline industry has become oriented toward minimizing costs.

In much of today's air traffic network, longer-distance trips along less-frequented routes are broken down into individual segments that have stronger demand. In this kind of hub-and-spoke network, where large airports (hubs) serve as transfer points to connecting flights (spokes), airlines take advantage of economies of scale by using a smaller number of larger aircraft. Travel can thus be provided more cheaply. However, with rising income and the increasing value of travelers' time, airlines are beginning to identify a small but growing fraction of travelers who would be willing to pay higher fares for a faster, direct connection.[34]

Operators of commercial aircraft have additional ways to reduce total costs. One way is to buy cheaper aircraft of a given size. An airline could, for example, purchase a used, ten-year-old, 250-seat aircraft instead of a new next-generation aircraft of similar capacity. However, exclusively pursuing a reduction in investments typically leads to higher direct operating costs and vice versa.[35] In the case of the used 250-seat aircraft example, its advanced age and outdated technology inherently would require higher operating expenses.

Similar trade-offs between upfront investments and annual operating costs, however, also apply to new aircraft since higher fuel efficiency or lower operating costs require more advanced (and more costly) materials and technologies. Consequently, prices for newer aircraft tend to be higher than those of their predecessors (see chapter 5). In the past, the profit maximization behavior among airlines has translated into demand for more fuel-efficient aircraft technologies as well as improvements in the way aircraft are operated. As technology has improved, the costs of operating these aircraft—usually termed the direct operating costs (DOC)—have declined, while purchase prices, generally called investments (I), have increased. (Together, DOC+I account for about half of total airline expenditures; the other half of an airline's entire operating budget consists of indirect operating costs, such as ticket commissions, ground operations, various fees, and administrative costs.)

Once an airplane has been purchased, airlines employ various strategies for maximizing its utilization (increasing the average flight hours per day) and minimizing the number of its empty seats (maximizing the "passenger load factor," that is, the percentage of seats filled). Maximizing utilization and load factor leads directly to increased profits (as long as ticket fares are such that the airplane *can* make a profit—some airlines operate routes that regularly lose money, occasionally even with high load factors, in order to enhance their competitiveness over a network of flight services).

To minimize operating expenditures, airlines today employ a wide variety of cost-saving procedures: they give discounts for web-based ticket purchases versus those made through an airline agent; remove food service on all but the longest flights; taxi using one engine instead of two to save fuel; carry additional fuel if there is a significant difference in fuel prices at different destinations or as a way to reduce turnaround time and thus increase aircraft utilization; wash airplanes and engines more frequently to reduce drag and improve performance; and regularly change flight speeds by 1 or 2 percent to best balance drag-related fuel costs with time-related crew costs. While the immediate purpose of all of these measures is to reduce operating costs, most of them also reduce an airplane's energy intensity.

Figure 3.4 summarizes the direct operating costs and investments, or DOC+I, of three distinct classes of commercial aircraft in 2005: large jets, regional jets, and turboprops. Historically, fuel costs have ranged from 25 percent of DOC+I (or 12–13 percent of total airline costs) in

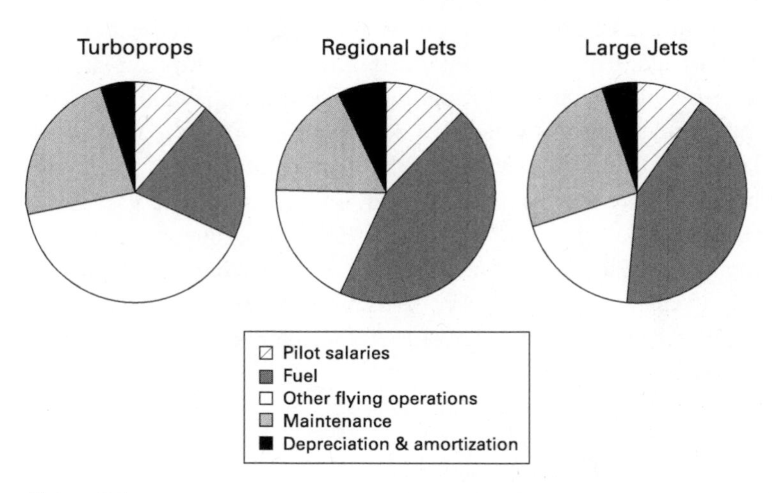

Figure 3.4
Direct operating costs and investments (DOC+I) for three classes of U.S. aircraft in 2005. Source: U.S. Department of Transportation, *Air Carrier Financial Reports (Form 41 Financial Data), Schedule P-52*, U.S. Bureau of Transportation Statistics, Washington, DC; www.bts.gov.

periods of low oil prices to a high of 65 percent of DOC+I (or 32–33 percent of total airline costs), which they rose to during the second oil crisis. With the crude oil price in excess of $100 per barrel, fuel accounted for about half of DOC+I of large jets in 2007.[36] Maintenance accounted for 25 percent, and pilot salaries for about 10 percent. The relative importance of these cost items differs among the various classes of aircraft, mainly because of the way the aircraft are operated. Maintenance costs are closely related to the length of time the engines are operating at maximum power during takeoff; they thus tend to reflect the number of flight operations rather than the flight time. By contrast, crew and fuel costs more closely mirror flight time. Therefore, planes that are flown on shorter routes tend to have a higher fraction of maintenance costs relative to fuel costs since they have more operations per distance flown.

In the following sections, we examine several features of airline operations that determine aircraft energy intensity: trends in aircraft size, changes in stage length (distance of a flight), passenger load factors, and infrastructure characteristics. We also discuss consumer response to changes in ticket prices.

The Rise and Fall in Aircraft Size

During much of the century-long history of aircraft technology development, the main goal of aircraft engineers has been to improve aircraft performance by increasing flight range, altitude, speed, payload capacity, and fuel efficiency. Panel A of figure 3.5 shows the change in energy use per available seat kilometer (ASK) for the U.S. aircraft fleet. The transition from propeller to more powerful jet engines led to an increase in energy use per seat kilometer during the 1960s. Subsequent technology improvements then led to a continuous decline in that indicator—by approximately 50 percent since 1970. As we discuss in more detail in chapter 5, that reduction in energy use was achieved mainly through improvements in engine design.

Driven by the desire to lower unit costs, airlines have employed larger aircraft within their hub-and-spoke network to exploit economies of scale. Between 1950 and the early 1980s, the average number of seats per U.S. aircraft more than quadrupled, going from 36 per plane to 165. More recently, in competing for more passengers, airlines have employed smaller regional jets that are able to service short- and medium-haul routes operating at higher frequency. (Improved service

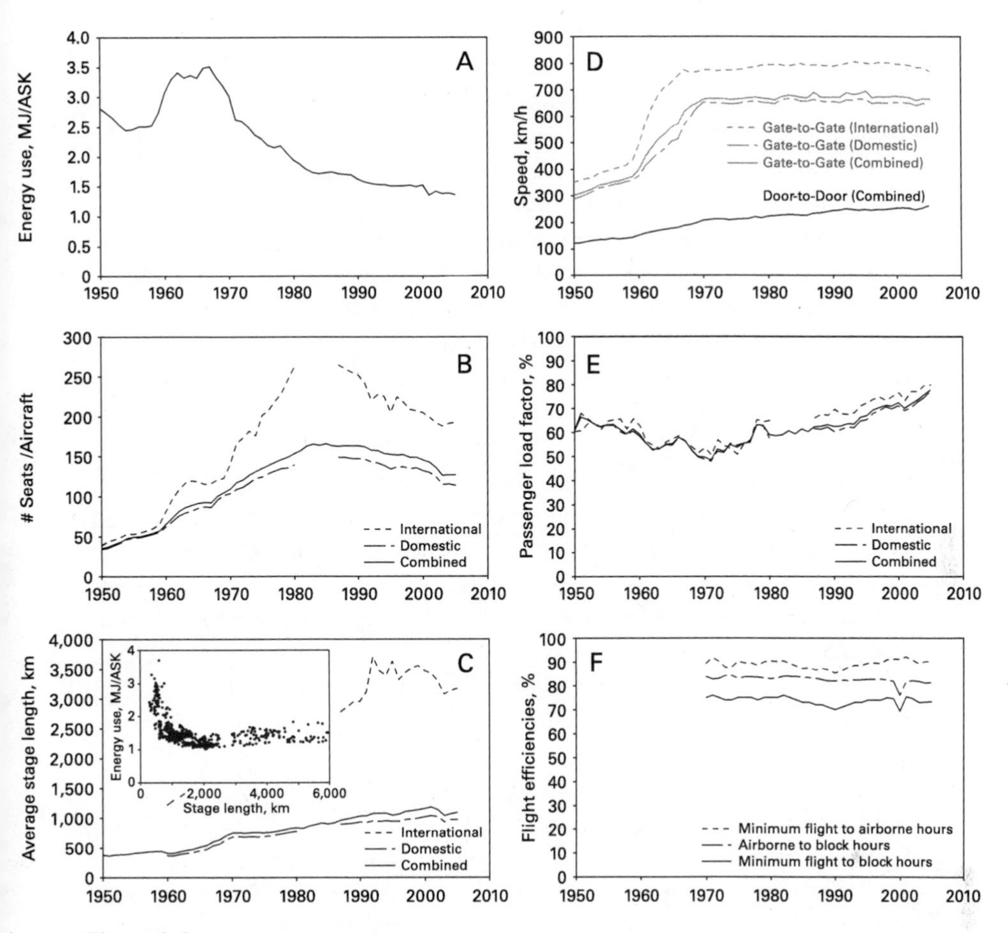

Figure 3.5
Characteristics of the U.S. air transportation system between 1950 and 2005. Sources: Air Transport Association of America, 1950–2007. *Economic Report* (formerly *Air Transport Facts and Figures*), Washington, DC. Babikian, R., S.P. Lukachko, I.A. Waitz, 2002. Historical Fuel Efficiency Characteristics of Regional Aircraft from Technological, Operational, and Cost Perspectives. *Journal of Air Transport Management*, 8(6): 389–400. U.S. Department of Transportation, *Air Carriers: T-100 Domestic Segment (U.S. Carriers)*, Bureau of Transportation Statistics, Washington, DC. Lee, J.J., S.P. Lukachko, I.A. Waitz, A. Schäfer, 2001. Historical and Future Trends in Aircraft Performance, Cost, and Emissions, *Annual Review of Energy and the Environment 2001*, 26: 167–200. U.S. Census Bureau, various years. *Statistical Abstract of the United States*, U.S. Department of Commerce, Washington, DC.

frequency, characterized by more flights per day, historically has been an important driver of market share in the airline industry—even more than pricing.[37]) The rising share of smaller aircraft has caused a decline in the average size of carriers to 127 seats per vehicle at present. Note that improvements in aircraft technology have also enabled such trends, because they allow smaller twin-engine aircraft to operate on long oceanic routes. While such decline in aircraft size itself has no significant impact on aircraft energy use per ASK, the associated increase in flight frequency contributes to air traffic congestion and thus, as a second-order effect, to an increase in energy use and GHG emissions.[38] Panel B of figure 3.5 includes this development in aircraft size in its display of decreasing seats per aircraft for international traffic.

A further, significant decline in average aircraft size may be caused by a large number of very light jets (VLJs), which offer complete independence of airline time tables and allow operation between several thousand underutilized airfields. The U.S. Federal Aviation Administration and the aircraft industry anticipate a fleet of several thousand microjets over the next decade.[39] Given a current U.S. fleet of nearly eight thousand commercial aircraft having an average seat capacity of 127, an equally large VLJ fleet would essentially cut the current average aircraft seat capacity in half (given that the typical VLJ capacity is three to six passengers).

All other factors being equal, such large differences in aircraft size would influence energy intensity. As we discuss in more detail in chapter 5, several design parameters determine an aircraft's fuel consumption. One is the ratio of an aircraft's "operating empty weight" (the plane's weight without payload and fuel) to its maximum takeoff weight. The denominator of this ratio differs from the numerator by including the weight of the passengers, cargo, and fuel. For today's large commercial aircraft, this ratio is around 0.5, while for VLJs, it is on the order of 0.6. This structurally determined higher weight share, together with other scale-related impacts on engine and airframe performance, result in two to three times the energy use per ASK for VLJs compared with large commercial aircraft under similar operating conditions (this is not shown in figure 3.5).[40]

Although energy use per ASK is considerably higher for VLJs, their lower seat capacity and lower utilization results in significantly lower passenger-traffic volumes and shares of energy use. For example, in a fleet of nearly eight thousand large commercial carriers and the same number of VLJs, the latter would account for only some 2 percent of total PKT. The VLJ's higher energy use per PKT, however, would translate into a 4–8 percent share of total air traffic energy use.[41]

Changes in Aircraft Stage Length

For a given aircraft and payload, its energy intensity is determined mainly by the way it is operated. Since the energy intensity of each flight stage (taxi out, takeoff, climb, cruise, descend, landing, and taxi in) differs, differences in stage length (flight distance) change the relative importance of each flight stage and thus the overall energy use per ASK. Aircraft operated on longer routes spend a shorter fraction of total flight time taxiing, climbing, and descending. The longer, less energy-intensive cruise phase thus contributes to a lower average energy use per ASK compared with aircraft operated on shorter routes. These operational differences are the main reason for the lower seat-kilometer specific energy use of large aircraft, which tend to be operated on long routes, compared with smaller aircraft (such as regional jets and turboprops), which are typically employed on shorter ones.

Since the beginning of commercial aviation, the average aircraft stage length has grown, thus increasing the relative importance of the low-energy-use cruise phase and reducing overall energy use per seat kilometer. The historical increase in aircraft stage length is shown in panel C. In the early 1950s, the mean distance traveled by an airplane was below 400 kilometers (about 250 miles). With the continuing expansion of formerly regional economies, the average stage length grew to more than 1,100 kilometers (about 680 miles) by 2005. This distance is the weighted average of slightly shorter aircraft travel distances common in domestic travel (about 1,000 kilometers or 620 miles in 2005) and significantly longer distances that are more the norm in international travel (above 3,000 kilometers or 1,860 miles in 2005).

The associated aircraft energy use per ASK as a function of stage length is shown in the inlay of panel C.[42] At stage lengths below 2,000 kilometers (1,240 miles), energy use is determined mainly by the more energy-intensive takeoff and climb stages. At stage lengths above 2,000 kilometers, energy use is influenced mainly by the weight of the extra fuel that needs to be carried. Together, both trends produce a U-shaped curve, with energy use being 1–1.5 MJ per seat-kilometer (skm) at a stage length of about 2,000 kilometers, and can be more than twice that at lower stage lengths. (Note that the energy use of 1–1.5 MJ/skm relates to the aircraft fleet operating during the 1990s. The penetration of newer, more fuel-efficient aircraft will result in a lower energy use at each stage length.)

In what direction is the average stage length going to evolve? One factor contributing to a reduced length is the increasing use of air transport for shorter-distance intercity travel instead of automobiles. Other developments, such as rising globalization and more direct flight routings, push toward an increase in the average stage length, at least in international travel. Considering the U-shaped relationship between aircraft energy use and stage length, each of these trends would cause aircraft energy intensity to increase. However, absent any more detailed analysis and a forecasting model of air transport activities, which takes the different stage lengths into account, the potential increase in average energy intensity cannot be quantified.

Given that aircraft energy use strongly depends on stage length, one proposal for reducing both fuel use and GHG emissions is to split long-distance flights into refueling segments of about 2,000 kilometers.[43] However, although this strategy would reduce GHG emissions, the increased number of takeoffs and landings would increase local air pollutant emissions and community noise and lead to additional air traffic congestion around airports. In addition, such a policy would work against the economic interest of airlines; it would cause a substantial increase in travel time, thereby threatening the loss of many consumers.

An increase in average stage length does not only contribute to lower levels of energy use. Panel D of figure 3.5 shows the continuous increase in mean door-to-door speed, that is, the mean passenger trip distance divided by the gate-to-gate travel time plus four hours for the transfer to and from the airport. Mean door-to-door speed has increased from about 130 kilometers per hour in 1950 to 210 kilometers per hour in 1970 (80 to 130 miles per hour), mainly because of the transition from propeller to faster turbojet aircraft. Therafter, travel speed has continued to increase to nearly 270 kilometers per hour (nearly 170 miles per hour) in 2005. This increase in travel speed can be attributed to the rising stage length, which causes the share of the high-speed cruise phase to increase, which in turn increases overall speed.

Increasing Passenger Load Factors

An airline decision to purchase a particular aircraft is based on profitability and the inherent trade-off between capital investment and lifetime operating expense. In order for an airline to recoup these costs (and be profitable), a seat must be utilized beyond a minimum threshold, sometimes stated as a "minimum load factor requirement." Panel E in figure 3.5 shows historical trends in passenger load factor. After an initial decline, the average load factor on domestic and international flights operated by U.S. carriers climbed by 56 percent between 1970 and 2005. Average passenger load factors were nearly 80 percent in 2005 (which is to say, eighty passengers on average when one hundred seats are available). This percentage, which is the average of higher load factors on lower-frequency long-haul routes and lower load factors on higher-frequency short-haul routes, contrasts sharply with the average load factors observed in personal automobile travel. In intercity travel, for example, dividing the typical vehicle occupancy rate (in the industrialized world) by a typical vehicle's occupancy capacity—an average of 2.3 passengers by 5.5 seats per vehicle—yields a passenger load factor of 41 percent: about half the load factor of a commercial passenger aircraft (as shown in table 3.1). Also, the aircraft passenger load factor does not take account of the fact that the same aircraft is frequently used to carry freight at the same time it is carrying passengers—so an additional "unrecorded" service is provided for the energy used.

Gains in passenger load factors have been attributed to deregulation in the United States and to global air travel liberalization, both of which contributed to the advent of hub-and-spoke routing systems.[44] Other major drivers of the increase in the passenger load factor over the years include a variety of influences: the availability of more aircraft of different sizes; improvement in the airlines' scheduling and fleet assignment capabilities—allowing them to match aircraft size to actual demand on each flight leg—and developments in pricing, revenue management, and distribution (largely enabled by the Internet) that have allowed airlines to better fill, using almost any low fare, what otherwise would have been empty seats. This latter capability in particular is apparent to almost all travelers, who can find daily variations in prices and seat availability as airlines attempt to optimize revenues and gain market share. These enabling factors are expected to continue and to lead to a further gradual increase in passenger load factors (which we have assumed to be 0.2 percent per year), and thus to a decline in energy intensity.

Improving Infrastructure Characteristics

The air transport-system infrastructure is another important determinant of aircraft energy use. Under congested conditions, for example, fuel is burned on the ground during various nonflying operations. On average, hours spent in the air (airborne hours) do not account for more than 75–90 percent of the total operational hours of an aircraft. This can be seen in panel F in figure 3.5 ("block hours" measure time from the moment an aircraft leaves its departure gate to the time it arrives at its destination gate). This ratio of airborne to block hours can be considered an indicator of ground-time efficiency.

Similarly, noncruise portions of a flight, poor routing, and various delays in the air constitute inefficiencies that cause an expenditure of fuel during a flight beyond that which would be required by the shortest flight path at constant altitude and cruise speed. This inefficiency can be measured as the ratio of minimum flight hours to airborne hours. Minimum flight hours represent a hypothetical shortest time required to fly a certain stage length (assuming that the aircraft flies steady and level at altitude at its nominal cruising speed with no time spent taxiing, climbing, or descending) and reveals any extra flight time because of nonideal flight conditions.[45]

The product of these two efficiencies is another measure of efficiency, the flight-time efficiency. All three of these measures, plotted over time, indicate that air traffic efficiency has remained constant since 1970. One can speculate that this has resulted from a consistently achieved balance between an ever-increasing demand for more flight capacity (through rising travel demand and higher-frequency use of smaller aircraft) and increasing air traffic management capabilities.

Passenger Response to Changes in Airline Fares

In chapter 1 we identify the declining cost of air travel as an important contributor to rapidly rising demand for that mode of transportation. Figure 1.1 shows that average air travel costs in 2005 were about one-third of those in 1950 when corrected for inflation. But what if the price of travel increased? How would consumers react?

The basic answer is that consumer response depends on the type of travel. Long-distance and business trips are more difficult to substitute or postpone than short-distance and leisure trips. Thus, demand for the former market segments is less elastic. David Gillen and his colleagues at Wilfrid Laurier University in Waterloo, Canada, have analyzed the results from more than twenty studies of demand and long-distance air travel and found that demand for international business trips is least elastic. Given a median price elasticity of −0.3—that is, a roughly 3 percent reduction in travel demand for a 10 percent increase in travel costs—a doubling of airfare would result in a 19 percent decline in travel demand.[46]

At the other extreme, the most elastic segment of the air travel market is short-haul leisure travel. Transportation alternatives exist (automobile, public surface transport), and such travel can be postponed relatively easily. In this market, based on a median price elasticity of −1.5, a doubling in airfare would result in a 65 percent decline in travel demand. The elasticities of most other flight market segments are close to −1.[47] The most recent (1995) U.S. long-distance travel survey suggests a three-way division of the long-distance air travel market: 43 percent of all trips are for business travel; 47 percent are for leisure travel; and the remaining 10 percent are for personal matters, which includes weddings, funerals, school-related activities, and obtaining medical treatment. Using these shares and the elasticities noted above, a doubling in airfare on all airline tickets would result in a drop in air travel demand by nearly 50 percent.[48]

Since fuel costs account for only a fraction of total airline costs, the price elasticity with regard to changes in fuel price is smaller. Historically, fuel costs accounted for between 12 percent and 33 percent of total airline costs. (Remember that DOC+I represent only about half of total airline costs.) Using the 2005 share of fuel costs to total airline costs of about 22 percent, and assuming that fuel cost increases were fully passed on to consumers, the price elasticities of air travel range from below −0.1 for international business trips to −0.3 for short-haul leisure travel. This range compares well with the price elasticity of ground vehicle travel with respect to fuel price of below −0.1 in the short run and −0.3 in the long run. Thus, the overall effect of a fuel price increase has a similarly small effect on the demand for automobile and air travel: a doubling in fuel price would lead to a reduction in travel demand by less than 20 percent even over the long term. With rising income and aircraft fuel efficiency, this effect is likely to decline.

Relative Importance of Economic Factors in Air Travel

Between 1970 and 2005, the energy intensity of the U.S. aircraft fleet declined from 6.1 MJ/pkm to 1.8 MJ/pkm, an average reduction of 3.5 percent per year. All the factors of airline behavior discussed above have contributed to this decline, albeit to a lesser extent than technological improvements. As described in more detail in chapter 5, technology improvements in aircraft fuel efficiency have accounted for about 80 percent of that decline. The remaining decline in energy intensity was the result of operational changes, including higher passenger load factors and longer average stage lengths.

Implications for World Passenger Travel Energy Use and GHG Emissions

We will now project passenger travel energy use and carbon dioxide (CO_2) emissions through midcentury. To do this, we use the travel demand projections we presented in chapter 2 and the findings stated above from our analysis of the determinants of energy use. Absent any change in vehicle technology, our projections show that the combination of a gradual shift toward larger vehicles and an increase in the share of urban driving would increase energy intensity of the LDV fleet within the industrialized world by about 20 percent by 2050.[49] In the developing world, depending on the projected growth in the economy and the associated decline in vehicle occupancy rates, we project the increase in LDV energy intensity to be 10 to nearly 40 percent by midcentury.[50]

During this same period, we assume the energy intensity of the aircraft fleet to decline by about 15 percent. This reduction results from the retirement of older, less fuel-efficient aircraft and the adoption of more modern aircraft currently in production. We assume other effects to compensate each other: the increasing passenger load factors causing average aircraft energy intensity to decline by about 10 percent and the counteracting effect of the introduction of more energy-intensive VLJ aircraft and a potential change in the stage length distribution toward shorter stage lengths in domestic air travel and longer stage lengths in international travel. Globally, in this constant technology scenario, we project the 2050 level of passenger travel energy intensity (of all modes together) to be about 5–20 percent over current levels, depending on the rate of economic growth.[51] (A slightly larger increase in world passenger travel energy intensity by 10–25 percent would result from a demand scenario with a travel time budget of 1.5 hours per person per day, instead of that based upon 1.2 hours per person per day. Because of the small difference, the resulting impact on energy use and CO_2 emissions is small.)

Multiplying these increases in energy intensity by the projected growth in travel demand yields a projected increase in passenger travel energy use. The 5–20 percent increase in passenger travel energy intensity, combined with a world travel demand that may rise to a level between 2.7

and 3.9 times the current level, produces a total energy use by 2050 that is between roughly 3 and 5 times today's level (when technological improvements are not taken into account). While the industrialized world was responsible for roughly two-thirds of the 53 exajoules of world passenger transport energy consumed in 2005, its share would decline to below 60 percent of the projected 2050 level (using the economic growth rates of the EPPA-Ref discussed in chapter 2) or to below 30 percent of the projected 2050 level of world passenger travel energy use (when the economic growth rate of the SRES-B1 scenario is employed).[52]

Assuming that oil products will continue to fuel nearly the entire world transportation system, the growth in CO_2 emissions and the share of total emissions contributed by each region will be identical to the projected growth and regional shares of energy use. Table 3.3 shows the development of passenger travel CO_2 emissions by metaregion for 1950, 2005, and our projections for 2050. CO_2 emissions released by passenger travel modes increased from about 370 million tons in 1950 to nearly 4 billion tons in 2005—and are projected to account for about 11 to some 18 billion tons in 2050. Again, it is important to note that these 2050 levels do not include any potential effects from possible changes in technology or transportation fuels. These technology and fuel possibilities will be discussed in the following chapters.

Summary

In a future scenario without any technological advances, an increase in total passenger travel energy use will be the result of growth in travel demand and an increase in energy intensity. We identified a range of non-technology-related factors that affect automobile and aircraft energy intensity similarly; several of these factors lead toward higher levels of passenger travel energy intensity.

While the size and performance of automobiles has increased, mean aircraft size has begun to decline, and the reduction in aircraft size could be accelerated by widespread introduction of VLJs. Both trends, however, satisfy the same objective: the satisfaction of consumer desires for comfort and speed. As a byproduct, both trends, directly or indirectly, also lead toward higher levels of energy use. Another similarity between these modes is the small and possibly further declining responsiveness in travel as a result of changes in fuel prices. (While this trend itself does not increase energy use, it lessens the potential for reducing energy use

Table 3.3
Direct CO_2 emissions (in million tons) from passenger mobility, past (1950), present (2005), and projections for 2050

	Historical development		Economic growth rates	
	1950	2005	EPPA-Ref 2050	SRES-B1 2050
Industrialized world	326	2,530	5,940	4,640
North America	264	1,440	2,960	2,630
Western Europe	44	818	2,330	1,560
Reforming economies	16	196	591	646
Developing economies	29	1,160	4,510	12,800
World	371	3,890	11,000	18,100
Relative amount, 2005 = 100	9	100	284	465

Sources: For EPPA and SRES economic growth rates, Paltsev, S., J.M. Reilly, H.D. Jacoby, R.S. Eckaus, J. McFarland, M. Sarofim, M. Asadoorian, M. Babiker, 2005. *The MIT Emissions Prediction and Policy Analysis (EPPA) Model: Version 4*, MIT Joint Program on the Science and Policy of Global Change, Report 125, Massachusetts Institute of Technology, Cambridge; http://web.mit.edu/globalchange/. Nakicenovic, N., J. Alcamo, G. Davis, B. de Vries, J. Fenhann, S. Gaffin, K. Gregory, A. Grübler, T. Yong Jung, T. Kram, E. Lebre La Rovere, L. Michaelis, S. Mori, T. Morita, W. Pepper, H. Pitcher, L. Price, K. Riahi, A. Roehrl, H.-H. Rogner, A. Sankovski, M. Schlesinger, P. Shukla, S. Smith, R. Swart, S. van Rooijen, N. Victor, Z. Dadi, 2000. *Special Report on Emissions Scenarios*, Intergovernmental Panel on Climate Change, Cambridge University Press, Cambridge, UK.
Notes: These projections result from our travel demand projections in chapter 2, which are based on two different sets of economic growth rates, those from the reference run of the MIT EPPA model (EPPA-Ref) and those from the B1-scenario of the *Special Report on Emission Scenarios* by the Intergovernmental Panel on Climate Change (SRES-B1). The travel time budget underlying the travel demand projections is 1.2 hours per person per day. The constant technology projections for 2050 *assume no improvements in ground and air vehicle fuel efficiency technology over the early 2000 level.* Numbers may not add up to the world total because of rounding.

as a result of fuel price increases.) Other similarities result from the direct interaction between these modes. Given these two modes' substitutability, the increase in the relative importance of urban automobile driving is a consequence of the substitution of higher-speed air travel for automobile intercity travel. Both trends, the rising share of urban driving and the relatively short aircraft stage length associated with the substitution of intercity driving, lead toward higher levels of energy intensity. But in long-distance air travel, aircraft stage lengths are rising, also leading to an increase in aircraft energy intensity. Even the only difference between these two modes, the trend in occupancy rates, may be deceptive, at least over the long term. In the industrialized world, the automobile energy intensity has declined and saturated between 1.5 and 1.6 pkm/vkm. While average aircraft occupancy rates are increasing, a small but growing share of travelers may increasingly rely on personal VLJ aircraft with likely occupancy rates significantly below those of large commercial carriers.

Overall, our assessment suggests that world average passenger travel energy intensity may increase by up to about 20 percent over current levels by 2050 (in a scenario without any technological advances relative to today's vehicles). In combination with a world travel demand that may nearly triple or quadruple, total passenger travel energy use could triple or quintuple. If oil products continue to fuel nearly the entire transportation sector, directly released CO_2 emissions would grow at the same rate as total passenger travel energy use: from about 4 billion tons of CO_2 in 2005 to 11 to 18 billion tons of CO_2 in 2050. Again, these projections ignore possible changes in transport-system and fuel technology. In the following chapters, we examine the opportunities for such changes.

4

Road Vehicle Technology

In the previous chapter, we conclude that in the absence of technology change, growth in travel demand and the continuation of certain other long-standing consumer and airline behaviors could push the emissions of greenhouse gases (GHGes) from passenger travel to between 11 and 18 billion tons of carbon dioxide (CO_2) a year by 2050. This is an amount roughly three to five times the level emitted in the year 2005. In this chapter, we discuss the opportunities improved technology may provide for reducing these projected 2050 levels of energy use and GHG emissions by light-duty vehicles (LDVs).

After a brief introduction to the fundamental relationships of energy use in road transportation, we examine the opportunities that improvements in current technology can provide for LDVs and, to a more limited extent, for other surface modes of transportation. At this time we accept transportation fuel as currently available and used as given. In chapter 6 we discuss in some detail alternatives to gasoline and diesel fuel, namely synthetic fossil fuel–based oil products, biofuels, compressed natural gas, electricity, and hydrogen.

Technology Fundamentals

Motion requires energy inputs. Running to a bus stop, driving on an interstate highway, cruising through the air—each of these activities burns a fuel: fats and sugars for the human body; hydrocarbons for a ground vehicle or aircraft. The fundamental laws of physics require that the higher the speed of movement, the greater the energy necessary to accelerate a body and overcome its friction and drag. This requirement has linear, quadratic, and cubic dimensions, so that doubling vehicle speed requires, on average, quadrupling the energy to be applied.

Inside the human body, fats and sugars are converted chemically into movement at an average efficiency of about 20 percent. Automobile engines convert fuel energy into vehicle movement at surprisingly similar efficiencies. However, unlike the human body, which converts chemical energy directly into motion, an engine's combustion process transforms the fuel's chemical energy first into thermal energy and then into useful work, which produces the motion.

Two fundamental principles of physics direct this process of converting thermal energy into motion, the first and second laws of thermodynamics. The first law of thermodynamics, the principle of energy conservation, governs the relationship between the energy of a system and the transfer of energy into and out of the system. Applied to internal combustion engines, the first law indicates that about 70–80 percent of the chemical energy released during combustion (from the fuel and the oxygen taken from the air) leaves the engine in a form different from usable power—as engine waste heat in the coolant and as unused chemical and thermal energy in the exhaust.

The useful work output of an engine compared to the total chemical energy input (released by burning the fuel) is called an engine's "brake efficiency." The basic engine cycle and the laws of thermodynamics restrict how much of the total energy input can be converted into usable work. For an optimum engine-operating cycle based on ideal thermodynamic processes (where there is no friction and no undesired heat transfer), the efficiency would be about 50 percent. Taking into account real-world engine heat losses and friction, the average (real-world) engine brake efficiency is only about 20–30 percent. Trying to reduce the gap mainly between the ideal efficiency and actual brake efficiency is one of the ongoing challenges faced by automobile industry engineers.

The efficiency with which an automobile engine converts input energy to useful work is important, but it is not the only determinant of overall vehicle energy use. The full amount of energy used also depends on how much power is required to move the vehicle, and thus depends on the design of the vehicle itself. The design characteristics that particularly affect energy use are those that relate to Newton's second law of motion. This fundamental principle of physics states that a vehicle's momentum (mass times velocity) can change only through forces applied to the vehicle. In the case of an automobile, in addition to the tractive force generated by the engine, these forces include at least three others: acceleration resistance, which depends on the vehicle's mass and the change in speed;

aerodynamic resistance, which is determined by the vehicle's shape, cross-sectional area, and speed, and the density of the air; and rolling resistance, which is attributable to the weight of the vehicle, tire characteristics, and the smoothness of the road surface. When the vehicle is being accelerated or is on an incline, the gravitational force, attributable to the vehicle's mass, also plays a role.

Vehicle energy requirements and engine energy supply can be linked by combining the equation of motion with the first law of thermodynamics as applied to internal combustion engines. In the case of road vehicles, the required tractive force needed to overcome driving resistances is derived from the engine, which converts fuel energy (E) into rotational energy at the engine driveshaft. In combination with the transmission and other drivetrain components, that rotational energy is transmitted to the wheels with an overall efficiency expressed as $\eta_{Propulsion\ System}$ (or rotational energy at the active wheels divided by the chemical energy used by the engine). On a horizontal surface, vehicle energy used per unit of distance traveled (vehicle kilometer) is

$$\frac{E}{VKT} = \frac{1}{\eta_{Propulsion\ System}} \cdot (A + D + R) \tag{4.1}$$

where A is acceleration resistance (or vehicle inertia), D is aerodynamic drag, and R is the tire rolling resistance, measured in newtons. These are defined as follows

$$A = m\dot{V} \tag{4.2}$$

$$D = c_D S\left(\frac{\rho}{2}V^2\right) \tag{4.3}$$

$$R = c_R mg \tag{4.4}$$

with m the vehicle mass; V the vehicle speed; and $\dot{V}$ the change in vehicle speed over time, that is, the acceleration; c_D the aerodynamic drag coefficient; S the vehicle's cross-sectional area; and ρ the air density (typically 1.2 kg/m^3); and where c_R is the rolling-resistance coefficient and g the acceleration due to gravity (9.81 m/s^2).

When climbing a hill, vehicle energy use requires an additional term, the gravitational or climbing resistance, which depends on the road gradient. The specific contributions of the resistances depend on the vehicle characteristics and operation. For example, a driver accelerating an automobile on a horizontal road from standstill at a constant rate of 1 m/s^2

would achieve a vehicle speed of 108 kilometers per hour (67 miles per hour) after thirty seconds. Using typical automobile characteristics, the total driving resistance after this period results in about 2,000 newtons, of which acceleration resistance alone accounts for 73 percent, aerodynamic drag for about 20 percent, and rolling resistance for the remaining 7 percent.[1]

Because of its quadratic nature, aerodynamic drag increases strongly with speed. Thus, if the driver were to continue to maintain that degree of acceleration, after another thirty seconds vehicle speed would double and aerodynamic drag would roughly equal the acceleration resistance.

Although equations 4.1 through 4.4 clearly identify the levers by which reducing energy use could be achieved, engaging those levers does not necessarily prove to be straightforward. One reason is that a change in any one of the determinants of vehicle energy use may cause undesired consequences. For example, engine efficiency could be increased, but that could bring an increase in local air pollutant emissions; automobile weight could be reduced, but that might adversely affect passenger safety, comfort, and capacity; and reducing aerodynamic drag is constrained by consumer desire for appealing design aesthetics and ergonomic changes tailored to the growing share of older drivers and passengers. Such trade-offs don't necessarily prevent making progress in fuel efficiency, but they can significantly slow down its pace since often the most obvious solutions simply cannot be directly implemented.

Perhaps the single largest trade-off in engineering design is cost. As we will discuss in the following sections, fuel-saving technology comes at a price, especially when the objective is to achieve significant reductions in energy use. Since drivers of personal automobiles in the past valued passenger amenities over fuel efficiency (see chapter 3), implementing a large fuel-saving potential can be seen as a significant financial risk to a vehicle manufacturer.

In addition to such uncertainties concerning consumer acceptance, industry also faces invisible costs with respect to safety and liability. Since the development costs amount to several billion U.S. dollars, when engineering a new (and hopefully successful and profitable) vehicle, the automobile industry tries to reduce each component of risk. Consequently, new vehicle technology is typically developed from existing, already proven designs.

This process of evolutionary improvement suggests that many of the advanced fuel-saving technologies we discuss here have a long history.

Some of these technologies have maintained their original purpose. For example, in 1898, Austrian-born Ferdinand Porsche, later founder of Porsche automobiles but then employed by electric vehicle producer Jacob Lohner & Co. in Vienna, developed an early version of today's hybrid car. Porsche's serial hybrid worked like a modern diesel locomotive. A spark-ignition engine ran a generator that produced electricity that charged batteries, thereby overcoming their low energy density and thus extending the vehicle range. The batteries in turn fed two electric motors. Only in 1997, one hundred years after Porsche's original design, did the first hybrid-electric vehicle, the Toyota Prius, enter mass production.

Other technologies were intended originally to increase engine power and vehicle speed, not to reduce fuel consumption. In fact, some of these technologies' primary motor vehicle application was in racing cars. However, after the two oil crises in the 1970s, such technologies have increasingly been looked at for ways to reduce energy use. Take the application of forced induction: in 1885, Gottlieb Daimler received a German patent for supercharging an internal combustion engine in order to increase its power. In 1902, Louis Renault obtained a French patent for a more sophisticated, centrifugal design for doing the same thing. Three years later, the Swiss inventor Alfred Büchi patented an exhaust-driven supercharger, the first true turbocharger. Already in the 1920s and 30s, many racing cars were supercharged to increase the power of a given size engine and thus vehicle speed. In the 1930s turbochargers were introduced into aircraft piston engines on a large scale. Their use in aviation was crucial to maintaining acceptable engine power at high flight altitudes, where air pressure is significantly lower. Down to the present, turbochargers have prevailed in applications where high engine power per unit of engine displacement is required. The technology is also widely used in large marine and stationary diesel engines, and in heavy-duty vehicles. Turbochargers have only been increasingly used in automobile diesel engines since the late 1970s, more than seventy years after Büchi's patent.

Early automobile technologies that were built with the intention to increase vehicle performance but alternatively can contribute to fuel savings also include technologies besides the vehicle engine. Aluminum-intensive automobiles, for example, were already produced in the early twentieth century. In 1913, vehicle producer NSU built the 8/24 car with an all-aluminum body, well before aluminum became the chief aircraft material. NSU, a company that actively and successfully participated in car racing at that time, used the lightweight component to reduce driving resistances in order to increase the vehicle's top speed. Yet, it took nearly eighty years until the first all-aluminum body automobile, the Honda NSX, went into mass production.

While the basic concepts behind early engines and other vehicle components have changed little, more sophisticated designs, along with progress in materials, controls, fuels, and manufacturing techniques have made these and other fuel-saving technologies increasingly attractive. Whether and when improved technologies enter large-scale production also depends on changes in market and regulatory conditions. Since these conditions can be different in various parts of the world, so will be some specific engine characteristics. In the United States, long-term historical low crude oil prices of some $20 per barrel, combined with only modest fuel taxes, have contributed to consumers' valuing large vehicle size, comfort, and performance over high fuel economy.[2] In contrast, significantly higher fuel taxes in Western Europe and Japan have moved consumer preferences toward smaller vehicles and turbocharged diesel engines, although higher income households in those regions increasingly value the attributes traditionally prevailing in the United States.

Thus, as we conclude from our discussion of figure 3.2, it is mainly the difference in vehicle size and the different levels of diesel engine penetration that account for the differences in vehicle fuel consumption across these regions—vehicle technology itself is similar. Therefore, many of the conclusions we derive below primarily from U.S. data also apply to other industrialized regions. (The situation, however, is different in the developing world, where infrastructure constraints, including different market conditions and the lack of clean transportation fuel, can limit the use of some technologies.)

The above examples from the long history of fuel-saving technologies show that the lead time between the first concept invention and the first large-scale commercial introduction of new technologies can be several decades. Spreading the new technologies over vehicle production and fully introducing these improved vehicles into the vehicle fleet can add another twenty to fifty years.[3] Thus, overall, it can take fifty years or more from the first feasible design to a significant fleet impact. This lag also applies to the technologies currently under development and discussed below. Yet, our focus is on technologies that have already undergone most of the historical development and could be introduced within the next two to three decades. Natural fleet turnover could then spread these technologies across the 2050 vehicle fleet.

Reducing Light-Duty Vehicle Greenhouse Gas Emissions

Since the two oil price shocks in the 1970s, the potential for reducing motor vehicle energy use has been one of the more researched challenges in automobile engineering. Over the last thirty years, the automobile industry, government agencies, independent researchers, environmental groups, and technology enthusiasts alike have conducted numerous fuel economy studies. These assessments typically examine what degree of improvement would be technologically feasible and how much it would cost to develop that technology. In their analysis of more than twenty such studies in the United States, David Greene of the Oak Ridge National Laboratory and John DeCicco of Environmental Defense demonstrate that most projections show a small, zero-cost fuel-efficiency improvement potential even when U.S. fuel prices remain at their historically low levels.[4] Since an increasing number of more and more complex fuel-saving technologies would need to be added to reduce fuel consumption to ever lower levels, large improvements in fuel economy result in increasingly higher extra costs. Thus, real-world policies never minimize energy use and emissions (which would be very expensive), but rather meet environmentally and socially acceptable emission limits that have minimum economic disruptions.

In their twenty-plus-study analysis, Greene and DeCicco also found that while rising fuel economy improvements are accompanied by increases in the retail price of vehicles, the study differences in the projected levels of fuel economy and costs can be large. Since the technology assessment methods underlying these studies are well established, the cost and fuel economy differences result mainly from the underlying assumptions concerning the projected technology characteristics and market conditions used in the studies. The data collected by Greene and DeCicco imply that nonindustry experts tend to be more optimistic with regard to these technologies' market potential and costs, perhaps because they haven't experienced the industry's constraints.

In contrast, industry experts, or studies with industry involvement, appear to be inherently more conservative in assessing the opportunities and costs for improving vehicle fuel efficiency. This sobriety likely results from industry's firsthand experience with the many requirements successful vehicles must satisfy. History suggests that actual development has fallen between the two general estimates of opportunities and costs, and that industry estimates may well deserve more credit than they often receive.

Greene and DeCicco also find that many studies assume that service characteristics (various vehicle consumer attributes, including interior size, vehicle acceleration capability, and maximum range with a full fuel tank) remain unchanged with respect to the assumed reference car. This convention does allow consistent comparisons to be made to the base year reference vehicle, but it does not take into account the various shifts in size, weight, and performance that have significantly compromised much of the potential for saving fuel in the past (see discussion in chapter 3). Neither does this convention allow for a move toward smaller vehicles. Such downsizing alone would provide noticeable reductions in energy use and GHG emissions.

To identify the major losses in vehicle energy use, figure 4.1 illustrates the energy flow within a typical U.S. midsize sedan.[5] Of the total fuel energy entering the engine, only an average of 16 percent actually drives the wheels; the remaining 84 percent is dissipated in the engine, transmission, drivetrain, and vehicle accessories, as well as when the engine idles. Since the engine operates more smoothly and in a more efficient regime during highway driving, the drivetrain efficiency is 20 percent under those conditions (compared with merely 13 percent in urban driving). Not surprisingly, given that more than three-fourths of the energy losses occur in the engine, engine efficiency improvements are a very important strategy for reducing vehicle fuel consumption. According to equation 4.1, a 10 percent increase in propulsion system efficiency leads to a 9 percent decline in vehicle energy use.[6] Of the 16 percent of the input energy

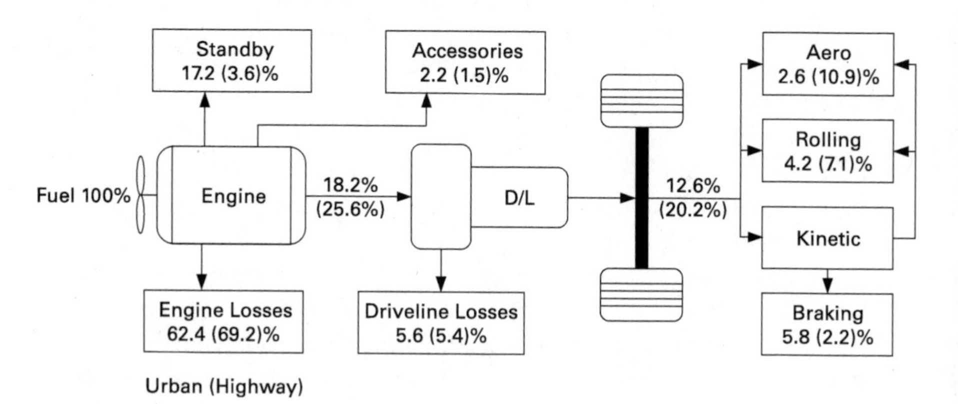

Figure 4.1
Typical energy flows in a light-duty vehicle. Reproduced from Partnership for a New Generation of Vehicles.

that drives the wheels, aerodynamic drag accounts for about one-third, rolling resistance for a second third, and acceleration resistance for the remaining third.

Using these shares of driving resistances together with equations 4.1 to 4.3, we can immediately estimate the impact changes in vehicle characteristics would have on energy use. Since mass affects both rolling resistance and acceleration resistance, a 10 percent decrease in vehicle mass causes energy use to decline by about 6 percent. Reducing the rolling-resistance coefficient or the aerodynamic drag coefficient by the same fraction leads, respectively, to a 3.4 and nearly 4 percent reduction in resistance and drag.

If we combine the fraction of input energy that drives the wheels with the low weight of the vehicle occupants, we see that energy efficiency is disappointingly low. Since occupants account on average for only 6 percent of the total vehicle weight, given the low propulsion-system efficiency, only about 1 percent (6 percent of 16 percent) of the total fuel-energy input is used to move the vehicle occupants.[7]

Because of the strong relative importance of engine efficiency improvements, we dedicate much attention to a propulsion system for ground vehicles. We examine improvements to existing engines and drivetrains, and to their operation together with an electric motor and electricity storage systems. We show that such combined, or hybrid, propulsion systems offer significantly higher levels of fuel economy than conventional mechanical-only systems. We then move to examining the determinants of vehicle energy requirements, the driving resistances. Finally, we project future vehicle characteristics by combining vehicles that have different levels of energy requirements with alternative propulsion systems.

Improving the Efficiency of Drivetrain Energy Supply

In this section we discuss the opportunities for increasing a vehicle's propulsion system efficiency. We begin with the two most common internal combustion engines—spark-ignition engines and compression-ignition, or diesel, engines—and continue with electric and hybrid-electric systems. Finally, we summarize the opportunities for improving transmission efficiencies.

Spark-Ignition Engines

When Nikolaus Otto demonstrated the first four-stroke internal combustion engine in 1876, transportation technology was about to change forever.[8] After decades of experimentation with various types of engines and fuels by inventors across the industrializing world, Otto's comparatively lightweight and robust design would finally realize the dream of self-propelled horseless carriages. In the course of the following 130 years, about 2 billion of these engines have been built. While the fundamental principles of internal combustion engines have not changed over their more than one century-long history, advances in technology, materials, improved design, ever more sophisticated control systems, and better and cleaner fuels have improved their performance in all dimensions. Today's engines are more powerful, durable, reliable, and energy efficient; are cheaper to produce and maintain; are significantly more compact; and operate more quietly with less vibration than their early predecessors.

Consider engine compression ratio as an example. While early spark-ignition engines had a compression ratio of about three or four to one, advances in fuel quality, engine design and controls, and materials have increased that ratio to ten to one and above.[9] That increase not only has deeply reduced fuel consumption, but also has led to more compact engines. Henry Ford's Model T engine had a power-to-volume ratio of 5.2 kW/L; that ratio is one order of magnitude higher for today's spark-ignition engines.[10] When today's engines are combined with an exhaust gas-cleaning system (the three-way catalyst), exhaust gases other than CO_2 and water vapor are reduced by more than two orders of magnitude compared with a century ago.[11] These and other engine characteristics are reported in the upper part of table 4.1. While much progress has been made during the past one hundred years, further significant improvements are possible and likely.

In general, the brake, or useful, efficiency of a spark-ignition engine is the product of two components, the indicated and the mechanical efficiency. The indicated efficiency, or the ratio that expresses the work transferred to the engine's pistons from the fuel's chemical energy, can be increased in several ways: through higher compression ratios, faster combustion through changes in cylinder flow, more effective combustion chamber design, and increased excess air in the air-fuel mixture. Today's indicated efficiency could be increased over the next two to three decades by as much as 8 percent. In contrast, a much larger potential exists for improving an engine's mechanical efficiency. This is the ratio of engine "brake work" (or useful work) and piston indicated work. It can be increased by reducing engine friction through a variety of design changes and material substitutions. A large part of this potential, however, lies in

Table 4.1
Evolution of midsize automobile characteristics

	1910–1920	2005	2020–2030
Spark-ignition engine			
Compression ratio	3–4:1	10:1	12:1
Power-to-volume ratio, kW/L	≈ 5	50	65
Power-to-weight ratio, kW/kg	<0.1	0.7	≈ 1.0
Average brake efficiency, %	10	22	27–33
Exhaust emissions of HC/NO_x, g/km[a]	10/2[b]	0.03/0.13	0.01/0.02
Oil change intervals, km	500	5,000	10,000
Vehicle			
Aerodynamic drag coefficient	≈ 0.8	≈ 0.33	0.25–0.30
Tire rolling-resistance coefficient	≈ 0.02	≈ 0.009	≈ 0.007
Vehicle mass pear seat, kg	136	320	200–300
Lightweight metals and plastics, % kg	≈ 0	15	20–50
Fuel consumption, L/100 km (mpg)	15 (16)	10 (24)	4–8 (59–29)

Notes: The representative vehicle for the period 1910–1920 is a typical Ford Model T, the most sold automobile at that time. Between 1915 and 1920, the Model T accounted for around 40 percent of all automobiles sold in the United States: U.S. Department of Commerce, 1941. *Statistical Abstract of the United States*, U.S. Department of Commerce, Washington, DC. Model T Ford Club of America, *Model T Encyclopedia*; www.mtfca.com/encyclo/index.htm. Fuel consumption of the Model T (13 mpg city, 21 mpg highway) released by Ford headquarters to *CBS News*, 2005. Could You Save Gas Money By Driving A Model T? 16 August; http://cbs5.com/seenon/local_story_228191600.html.
[a] CO levels are 10–100 times higher.
[b] Rough estimate.

reducing part-load pumping losses, which result from the fact that most of the time, spark-ignition engines operate with subatmospheric pressures in the intake (and which are controlled by the throttle).

In today's spark-ignition engines, the fuel-air ratio is stoichiometric, which means that the mix of hydrocarbon fuel and air is exactly matched, so that complete combustion will take place. That specific mixture composition is required by the three-way catalyst for effective emission control under all engine-operating conditions. While the stoichiometric fuel-air ratio is a crucial prerequisite for drastically reducing air pollutant emissions, it also constrains the spark-ignition engine's efficiency. During part-load operation, engine torque is reduced by throttling the continuously maintained stoichiometric air-fuel mixture, a control method that reduces engine efficiency through the associated pumping losses. Reducing these pumping losses offers substantial potential for improving the engine's efficiency.

Already, several technologies for reducing these losses have been developed and are entering the market. One approach varies, in effect, an engine's size to meet the required power output. It does this by selectively deactivating cylinders while operating at part-load, thereby reducing the engine's displacement volume. Another option reduces pumping losses by manipulating each valve's timing (and lift) to take in the required amount of air-fuel mixture at a higher intake pressure. Such flexible valve operation reduces the amount of throttling.

Also, in perhaps the most radical approach, instead of burning a stoichiometric gasoline-air mixture, the required amount of gasoline is injected directly into a cylinder filled with already compressed air—an approach that mimics the diesel engine cycle discussed below and lessens the pumping losses from throttling. A combination of the latter two options—gasoline direct injection and variable valve control—can improve the mechanical efficiency of an engine by up to one-third during part-load operation.[12]

Further gains in efficiency are possible through turbocharging, which would allow the engine to be downsized while maintaining the same engine power.[13] As shown in table 3.2, the average new automobile in the United States has an engine with a displacement of 2.8 liters that produces a maximum power of 134 kW. That same power can be produced by a significantly smaller turbocharged engine. Use of a smaller engine would offer several benefits. It would operate under less throttled conditions, so the pumping work that contributes to friction would be reduced (as would the relative importance of friction), and the engine weight would be lower. The comparatively rare periods of demand for high engine power, such as during rapid vehicle acceleration, would be satisfied by the turbocharger feeding in compressed air. By realizing such a combined increase in both the mechanical and indicated efficiency, an additional 15 percent reduction in fuel consumption could be obtained.

At typical operating conditions, the mechanical efficiency can be increased by 20–25 percent at the relevant engine loads. In combination with the 8 percent improvement in indicated efficiency noted earlier, average engine brake efficiency could be increased by close to one-third over the next two to three decades.

Table 4.2 summarizes the projected efficiency improvement potentials for current state-of-the-art spark-ignition engines from four recent studies: our two MIT studies on the comparative performance of various automobiles, the U.S. National Research Council study on fuel economy standards, and a study by Sierra Research on assessing possible improvements in automobile fuel efficiency.[14] These (and other) studies agree that the potential for improving spark-ignition engine efficiency is about one-third over current levels during the next twenty-five years or so, with the increase in mechanical efficiency being the largest component.

However, if technologies to improve engines are *not* used to reduce vehicle fuel consumption but go to increase vehicle performance, size, and weight (as was largely done over the past two decades), then these potential reductions in fuel consumption will not be realized. This trade-off between vehicle fuel consumption, size, weight, and performance, and thus the importance of how engine improvements are used, applies to all propulsion systems.

Compression-Ignition Engines

Rudolf Diesel demonstrated his first compression-ignition engine in 1897, having been motivated by the low efficiency of then available steam engines and the four-stroke spark-ignition engine developed by Otto. In contrast to Otto's engine, ignition in a diesel engine results exclusively from the spontaneous combustion of the fuel-air mixture within the diesel engine's cylinders. This happens when fuel oil is injected into the high pressure and high temperature compressed air.

The high compression and expansion ratio of the diesel cycle is one reason for the diesel engine's higher energy efficiency compared with the spark-ignition engine. Another reason is the engine's ability to modulate its output by controlling the amount of fuel injected into each cylinder (as opposed to the amount of fuel-air mixture). Brake efficiencies as high as 45 percent are achieved in turbocharged diesel engines in heavy-duty trucks. More recently, progress in turbocharging, direct fuel injection, and control technology has greatly increased the compactness, acceleration performance, and brake efficiency of smaller, higher speed automobile diesel engines. Today, state-of-the-art turbocharged, direct injection diesel engines are about 20–25 percent more energy efficient than their gasoline-fueled, naturally aspirated counterparts, and have become the latter's energy efficiency target.[15]

Not surprisingly, diesel engine vehicles enjoy high market shares in countries that have high fuel prices and where diesel fuels carry less tax.

Table 4.2
Efficiency improvement potential (in percent) of spark-ignition engines

	MIT	NRC	Sierra
First implementation	2030	2010–2015	2010–2020
Indicated efficiency	8	5–7	1–8
Mechanical efficiency	22	17–44	22–38
Reducing mechanical friction	—	2–6	3
Reducing pumping losses	—	15–36	17–32
Engine accessory improvements	—	1–2	2
Total (brake efficiency)	32	23–54	24–50

Sources: Weiss, M.A., J.B. Heywood, A. Schäfer, V.K. Natarajan, 2003. *A Comparative Assessment of Advanced Fuel Cell Vehicles*, MIT Laboratory for Energy and the Environment 2003-001 RP, January, Massachusetts Institute of Technology, Cambridge. Weiss, M.A., J.B. Heywood, E.M. Drake, A. Schäfer, F. AuYeung, 2000. *On the Road in 2020: A Well-to-Wheels Assessment on New Passenger Car Technologies*, MIT Energy Laboratory, October, Massachusetts Institute of Technology, Cambridge. National Research Council, 2002. *Effectiveness and Impact of Corporate Average Fuel Economy (CAFE) Standards*, Committee on the Effectiveness and Impact of Corporate Average Fuel Economy (CAFE) Standards, National Academy Press, Washington, DC. Austin, T.C., R.G. Dulla, T.R. Carlson, 1997. *Automotive Fuel Economy Improvement Potential Using Cost-Effective Design Changes*, Draft #3, prepared for the American Automobile Manufacturers Association, Sierra Research, Inc., Sacramento, CA. Austin, T.C., R.G. Dulla, T.R. Carlson, 1999. *Alternative and Future Technologies for Reducing Greenhouse Gas Emissions from Road Vehicles*, prepared for the Transportation Table Subgroup on Road Vehicle Technology and Fuels, Sierra Research, Inc., Sacramento, CA.
Notes: The baseline engine is a six-cylinder, naturally aspirating, four-stroke, four-valves-per-cylinder engine. The total (brake) efficiency improvement potential results from multiplying the improvements in indicated and mechanical efficiency. The MIT study indicates plausible averages of engine efficiency improvements. Ranges in the National Research Council (NRC) study indicate uncertainty. The lower numbers of efficiency improvement in the Sierra study represent no engine performance adjustment, while the higher numbers reflect downsizing the engine to maintain the same power output.

In Europe, with gasoline prices well above U.S. $1 per liter (about $4 per gallon), the share of diesel engine–equipped automobiles has reached half of all new vehicles registered.[16] By comparison, the share of diesel-fueled vehicles is close to 10 percent of the total automobile fleet in Japan and about half that percentage in the United States.[17]

Since diesel engines are already high compression-ratio lean-burn engines, the potential for increasing their efficiency is more limited compared with spark-ignition engines. Future improvements will come largely from faster combustion, additional intake air boosting (about 6 percent increase in indicated efficiency), and a reduction in friction losses (about 15 percent at part-load operation). These improvements translate into an 18 percent increase in brake efficiency.

However, about 5 percent of that 18 percent will be offset by more rigorous efforts to control emissions. Just as the stoichiometric air-fuel mixture contributes to the spark-ignition engine's poor part-load performance, the diesel engine's complex combustion process and high air-fuel ratio make emission control challenging. Carbon monoxide and unburned hydrocarbons can be easily oxidized using an oxidation catalyst, but increasingly tight regulations on particulate and nitrogen oxide (NO_x) emissions are a major challenge, especially for the U.S. LDV market. However, higher fuel injection pressures coupled with catalyzed filters have become one practical approach today to reducing particulate emissions significantly. And catalyst systems for reducing NO_x emissions ($DeNO_x$ systems) are making the transition from prototype to practical hardware.

An alternative to $DeNO_x$ catalysts is low temperature combustion. In this option a more fully mixed fuel-air mixture, diluted with recirculated exhaust and residual gas, burns with excess air at a lower temperature. It thereby avoids high soot and NO_x formation rates. Promising results are now being achieved with the partial use of low temperature combustion. If this development effort proves fully successful, it could significantly reduce diesel emissions.

However, even if such developments provide the required improvements, it remains unclear whether the diesel engine will be able to meet the most stringent future light-duty emissions regulations in California, since the cost of compliance is likely to be high. This raises the question of market competitiveness, since a state-of-the-art diesel engine with exhaust filters and catalysts might cost about twice as much as an equivalent spark-ignition engine. Thus, at present it appears unlikely that a diesel LDV market will develop in the United States to anything like the extent it has already developed in Europe.

A number of alternative engines have been proposed and studied for possible use in automobiles. To date, none of these engines has proved to be a viable alternative. And it is unlikely that this situation will change, at least during the next few decades. Two-stroke engines, for example, are used widely in small engine applications, motor scooters, and motorbikes. Although criteria emissions from two-stroke engines have been greatly reduced during the past decade, the engine's critical problem remains cooling—and thus durability. The Wankel rotary engine's longer combustion chamber and lower compression ratio inherently lead to higher emissions and lower brake efficiency, thus offsetting its advantage of smoother engine operation. And gas turbine engines, because of their higher costs and poor fuel consumption, have never experienced commercialization in automotive applications.

In addition to more energy-efficient engines, advanced *transmissions* can further increase the energy efficiency of a vehicle's powertrain. Transmissions are required to match an engine's speed and torque to the speed and torque demand at a vehicle's wheels. Independent of the type of transmissions—manual or automatic—essentially three complementary opportunities exist for reducing powertrain energy losses: increasing the efficiency of a transmission's components, modifying a transmission's shift logic toward more fuel-efficient engine operation, and increasing the number of gear ratios so that the engine operates closer to minimum fuel consumption conditions. Promising options for achieving these possibilities include automatic transmissions that have six or seven gears (instead of four); automated manual transmissions (with six or more gears), in which shifting is controlled electronically; and continuously variable transmissions, which could change the gear ratios continuously. Implementing these improvements could achieve transmission efficiencies of close to 90 percent, and each improvement alone could reduce vehicle fuel consumption by 8 percent.

Electric Systems

The history of the electric vehicle can be traced back almost as far as that of the internal combustion engine. At the end of the nineteenth century, when the future dominance of the internal combustion engine vehicle had not yet become clear, the few horseless carriages on the road were propelled by a steam engine, an internal combustion engine, or an electric motor. At the time, consumer interest in electric vehicles derived from the vehicles' comparatively superior comfort. Drivers of electric cars experienced less noise, smell, and vibrations than drivers of cars with an internal combustion engine. And while the internal combustion engines of the day had to be started with a hand crank, and steam engines required long start-up times before they could operate, electric vehicles could be started easily. In addition, the torque-speed characteristics of electric motors did not require shifting gears—whereas changing gears was an adventurous undertaking in the early days of the internal combustion engine automobile. Starting in about 1910, however, the disadvantages of electric cars became clear—and have remained foremost until today.

A battery's energy density determines, for a given size battery, the amount of electrical energy it can store and thus determines how far an electric vehicle can travel. The motors of the early electric vehicles were powered by lead-acid batteries that had an energy density of about 10 watt hours per kilogram (Wh/kg). Given the low travel speeds at that time, the resulting range of 40–60 kilometers (25–37 miles) was sufficient for common urban travel. However, with the improvement of intercity road conditions and a rising demand for longer trips, electric vehicles could not satisfy this growing segment of the market.

Over the course of the last century, much progress was made in storing more energy per unit of battery weight. The energy density of today's lead-acid batteries is more than three times as high as it was a century ago. Even larger gains have been achieved by using new materials, electrodes, and manufacturing technologies.

Today's nickel metal hydride (NiMH) battery offers an energy density of 60–70 Wh/kg (along with greatly improved discharge characteristics), and lithium ion (Li-ion) batteries already approach 100 Wh/kg. However, if one of these batteries were used in an all-electric vehicle to give it a driving range comparable to that of a gasoline-powered car, the battery system would be very large, and the vehicle's weight would approach twice that of a vehicle using an internal combustion engine. That greater weight would increase energy use—and thus a still larger battery would be required to achieve the original target distance. To overcome this multiplier problem, battery energy density needs to be about 300 Wh/kg. Nonetheless, with further progress in electrochemistry, material science, and manufacturing processes, batteries ultimately may be able to achieve such energy densities.[18]

In addition to vehicle weight and driving range, a battery's energy density also determines the battery's cost. Based on current costs for NiMH modules of U.S. $300/kWh, a battery having a storage capacity of 30 kWh would cost U.S. $9,000. At this price, it is highly questionable whether the lower electricity costs (for recharging the battery) compared to gasoline fuel costs would offset the much higher cost of the battery over any reasonable payback period. Also, until recently, the lifetime of advanced batteries was uncertain. If an electric vehicle were to need a new battery pack every 100,000 kilometers (62,000 miles), it would not be economically feasible. However, NiMH batteries recently have been successful in lasting for a driving distance of 160,000 kilometers (100,000 miles); they are now expected to achieve a driving capability of about 240,000 kilometers (150,000 miles) or so—the typical end-of-life odometer reading for an internal combustion engine vehicle.[19]

High battery weight and cost argue that electric vehicles should use as little energy storage as possible. One way to achieve this is to design a battery-run electric vehicle that has low driving resistances. This was done with the General Motors EV1 electric car, the most advanced electric vehicle ever brought to the market. By contemporary standards, the EV1 had very low driving resistances: an aerodynamic drag coefficient of 0.19, low rolling-resistance tires, and a curb weight of only 1,330 kilograms (2,930 lbs). However, even for such a sophisticated vehicle, the comparatively low energy density of the batteries it could use meant that the EV1's battery mass came to several hundred kilograms, or 20–40 percent of the vehicle's total weight.

Incorporating a more modest battery size at present requires designing electric vehicles exclusively for niche markets that have short-distance driving requirements. Since two-thirds of all automobile trips are below a trip distance of 10 kilometers (6 miles), electric vehicles could gain an increasing market share even with smaller-capacity batteries, at least as second or third household vehicles. In fact, all commercial electric vehicles to date have had short driving ranges; at about 160 kilometers (100 miles), depending on driving patterns and ambient conditions, the General Motors EV1 is at the upper limit.

While the disadvantages of using electric vehicles have remained essentially unchanged, so have the benefits. As with their early predecessors a century ago, electric vehicles produce neither emissions nor any smell at the point of operation, and they generate less noise than gasoline-powered vehicles. This is especially important in urban areas, where there are high concentrations of air pollutants. Batteries can be charged at home overnight, taking advantage of reduced power plant loads and thus lower electricity prices. Following this approach, if sufficiently high-energy-dense batteries can be developed, battery-only electric vehicles could supplant most of the U.S. LDV fleet without a major increase in power plant capacity.[20]

The emission characteristics of electric vehicles depend on the design of the vehicle and the fuel mixture used to generate the electricity in the first place. Measured on the U.S. combined driving cycle and using the 2005 U.S. electricity-generation characteristics, power plant NO_x emissions per kilometer driven would be about 0.17 g/kWh for a highly advanced electric vehicle having the electricity consumption characteristics of the EV1—a level about 20 percent above that of today's new gasoline-powered automobiles (see table 4.1).[21] In contrast, CO_2 emissions from the electricity-generating power plants over the same driving cycle would be about 110 g per vehicle-kilometer (vkm), which compares to about 119 g/vkm for the average new automobile on the road.

A largely unresolved issue remains the environmental impacts of battery production and recycling. Relatively sparse data exist for assessing this impact, especially for batteries more advanced than lead-acid. However, as challenging as it would be to reduce these emissions, no technical reason explains why they could not be limited to very low levels—probably, however, incurring high costs. Notwithstanding these benefits, the technical reality of the limited energy-storage capacity of batteries makes producing a market-competitive, standard electric vehicle a daunting task.

Hybrid-Electric Systems

The continuous but gradual progress in battery technology over the past century has not significantly changed the fundamental limitation electric vehicles have: the energy density (energy per unit weight or volume) of current batteries is only about 1 percent compared with that of gasoline. As a result, energy storage is heavy, and vehicle range is limited, typically to 100–200 kilometers (some 60–120 miles) for a vehicle with a fully charged battery. Hybrid systems combine the advantages of the internal combustion engine (high energy density) and the electric drivetrain (high motor efficiency and zero idling losses).

There are several different hybrid designs. One distinction among them is between mild and full hybrids. A mild hybrid has a more powerful starter motor, which allows the engine to be shut off during idling and to be restarted quickly. A large battery, typically 48 volts, supplies the power for the starter motor and the auxiliaries and is usually recharged through regenerative braking. However, an internal combustion engine propels the vehicle.

Common to all full hybrid systems is that an electric motor assists a downsized internal combustion engine. In the most fuel-efficient arrangement, the internal combustion engine either drives the wheels directly (through the transmission) when cruising at higher speed or acts as a generator to produce electricity. The electric motor typically propels the vehicle during most stop-and-go (urban) driving and acts as a generator during braking. Electricity is stored in an energy storage system, usually a high-power-density battery (or, in the future, perhaps an "ultra capacitor"), to rapidly transfer electricity from the storage device to the electric motor. Both propulsion systems, internal combustion engine and electric motor, work together when requiring high levels of acceleration and are turned off during standstill.

The efficiency advantage of hybrid systems results from four factors. Since the internal combustion engine is turned off during standstill, idling losses are eliminated; this feature alone yields energy savings of up to 17 percent in urban driving or 11 percent for "average" driving conditions (see figure 4.1). In addition, the electric system can recapture, by employing regenerative braking, some of the vehicle's kinetic energy that normally is dissipated in braking (which accounts for nearly 6 percent of the energy used in urban driving, or 4 percent under average conditions). Taking into account the conversion losses, regenerative braking offers additional energy savings of about 2 percent. Additionally, a full hybrid's downsized internal combustion engine operates mainly in more efficient regimes, at higher loads and lower engine speeds. This third advantage offers energy savings on the order of 15–20 percent. Finally, the continuously variable transmission provides energy efficiency gains of about 5–10 percent. All together, hybrid systems consume about 30–40 percent less energy than their mechanical counterparts do. The performance projections of future hybrid systems depend, of course, on the assumptions used. For hybrid technology, mechanical and electrical components are expected to improve, and thus it is plausible that the benefit hybrid vehicles enjoy in average driving will stay about the same.

An alternative hybrid configuration, the plug-in hybrid—a hybrid vehicle that charges its battery with grid electricity (mainly during off-peak hours)—is attracting substantial interest at present. One can think of it as a pragmatic version of the electric vehicle for today. Since about 95 percent of all vehicle trips are less than 50 kilometers (some 30 miles), this vehicle would operate much like an all-electric vehicle for most of its trips.[22] Its internal combustion engine would operate, then, mainly during longer-distance travel, thereby removing the range limitation of electric vehicles (and enabling electricity to substitute for a substantial fraction of the petroleum-based fuel that otherwise would be used during short-distance driving).

A recent study by the Pacific Northwest Laboratory concluded that nearly three-quarters of the U.S. fleet could be replaced by plug-in hybrid-electric vehicles, using the existing electricity-generation infrastructure to recharge their batteries. (In that study, the battery was assumed to allow average driving requirements of some 50 kilometers per day in an electricity-only mode.) Based on current levels of travel and oil demand, this partial electrification of the propulsion system could cut total oil imports by about half. Oil imports and GHG emissions could be further reduced if oil were replaced by biofuels—an option we discuss in chapter 6. However, for plug-in hybrid-electric vehicles to become a viable technology, the battery system's energy storage capacity has to be increased, while its durability and economic feasibility also have to be proved.

Several vehicle manufacturers, mainly Japanese, began producing commercial hybrid automobiles in the late 1990s, and more recently, other companies have followed suit. By the end of 2005, Toyota alone had already sold five hundred thousand of its Prius hybrid vehicle on the world market, despite the Prius's higher costs compared with similar-sized mechanical-drivetrain cars. The hybrid's cost differential is a function, first, of the fact that hybrid vehicles incorporate two engines instead of one. Control electronics, necessary for directing the interplay of the various components (engine, motor, and battery), add further costs. The result is a current price difference of U.S. $3,000–4,000. While this disincentive is likely to decrease as battery and power electronics costs are reduced, it will remain a significant negative factor for this technology.

The market share of hybrid technology, then, ultimately will depend on fuel price and government policies (see chapter 7). While at present all commercially available hybrids use a spark-ignition engine, diesel hybrids are also being studied as a possible alternative. However, diesel's emissions problems and still greater costs make it an unlikely near-term option, especially in the United States, despite its 15 percent higher energy efficiency than a gasoline-engine hybrid.

While internal combustion engine hybrid vehicles promise significant reductions in energy use in the near term, one long-term objective of vehicle propulsion technology is a fuel cell system coupled to an electric motor. Like a battery, a fuel cell converts chemical energy directly into electricity. However, unlike a battery, which contains a set amount of reactive chemicals within it, the only storage limitation for a fuel cell is the amount of fuel stored onboard the vehicle, in this way overcoming a battery's low energy density. The heart of a fuel cell system is the "stack," an assembly of electrochemical cells that converts hydrogen directly into electric power. Each individual cell of the stack has its own separator, anode, cathode, and electrolyte. When in continuous operation, the stack requires effective heat, air, hydrogen, and water management, which is taken care of by auxiliary equipment such as pumps, blowers, and control systems. As with the internal combustion engine hybrid, a high-power battery or ultra capacitor stores regenerated braking energy and releases electricity during acceleration or other driving stages.

For fuel cell vehicles to be at all practical, they will require a new fuel production infrastructure, which—in addition to efficient and low-cost production and storage of hydrogen—would also need to develop a low CO_2-emitting hydrogen production system in order to achieve major reductions in GHG emissions. When—or even whether—a large-scale hydrogen infrastructure could be built that could fuel a significant number of fuel cell vehicles remains unclear (see chapter 6).

One suggested alternative that would not require such comprehensive infrastructure change is the gasoline-fueled fuel cell system. It makes use of a fuel processor to convert gasoline chemically to hydrogen onboard the vehicle itself. Before feeding the stack, however, the system requires hydrogen cleanup to remove any carbon monoxide and hydrocarbons. Our analysis of the system and that of other research groups shows, however, that the energy conversion losses in the fuel processor outweigh the efficiency benefit offered by the fuel cell. The result is energy use and CO_2 emission levels very similar to the gasoline-fueled internal combustion engine hybrid vehicle.[23]

Several challenges to fuel cell technology remain to be resolved. One is to reduce the amount of platinum catalyst needed to enhance the cell reaction at the relatively low operating temperature of about 80°C. Given the high cost of platinum (and its relative scarcity as a resource), it remains to be seen whether costs of U.S. $20–30 per kW, which are required to compete with a mechanical drivetrain, can ever be achieved.

A similar challenge is how to achieve low-cost and energy-efficient hydrogen storage on the vehicle. (See chapter 6 for a more thorough discussion.)

Fuel cell vehicles have been designed and tested by research centers and various vehicle manufacturers for several decades. This effort, especially since the early 1990s, has shown encouraging progress, and each subsequent generation of vehicles has offered significant improvements in fuel efficiency, compactness, and robustness, as well as steady reductions in costs. Nonetheless, significant new improvements still need to be made. The ultimate potential of hydrogen fuel cells remains unclear.

Reducing Vehicle Energy Requirements

Equally important to efficient energy supply is reducing the tractive force a vehicle requires. As expressed in equations 4.1 to 4.4, three resistances are particularly significant: acceleration resistance, which represents the energy necessary to increase a vehicle's kinetic energy inertia while accelerating to a desired speed; rolling resistance, which is the work necessary to overcome the deformation of a vehicle tire when moving; and aerodynamic resistance or drag, which equals the work needed to move a vehicle through the air. Averaged over urban and highway driving, these three driving resistances are roughly equal. Three parameters are used to define them: vehicle mass, the coefficient of tire rolling resistance, and the coefficient of aerodynamic drag. One additional resistance is climbing resistance, which comes into play when a vehicle drives up hill.

Vehicle Weight

Since vehicle weight affects both acceleration and rolling resistance, reducing weight has a significant impact on total driving resistance and thus vehicle fuel use. Indeed, a 10 percent reduction in vehicle mass will yield a 6 percent drop in fuel consumption, all other factors being equal.

All the same, for vehicles of a given size, weight can be reduced through optimizing the effective use of standard materials, substituting lighter and stronger materials when structurally possible, and by creatively combining these approaches. Already, advances in body, engine, and drivetrain design have led to vehicles that use interior space more efficiently. More sophisticated computer simulations and material-forming techniques, by optimizing a vehicle's shape to better accommodate both operational and customer requirements, can identify more opportunities for incremental weight savings.

Realizing more drastic reductions in weight would require the use of lightweight materials and the exploitation of secondary weight savings, which would require reengineering the entire vehicle. A "body-in-white"—an empty car body without any closures—typically accounts for one-quarter of a car's mass. To perform the same as an "original-weight" vehicle, a lighter car body can use a lighter, less powerful engine. A lighter body and smaller engine can be integrated into a lighter chassis, and this overall more fuel-efficient structure could travel the same distance as a heavier vehicle with a smaller fuel tank. In such a holistic approach, these secondary weight gains are of the same order as the original reduction in body mass. However, this theoretical potential is constrained by economies of scale in actual vehicle production. Since different vehicle models often share platforms, engines, and other components to reduce costs, the secondary weight-saving potentials typically can be exploited only in part and over the long term. Thus, past progress has been only gradual.

The proportion of lightweight materials in automobiles has been rising continuously, although the pace has been slow, mainly for the cost reasons mentioned above. In 1975, the first year detailed statistics on material use became available, iron and steel accounted for 75 percent of the total weight of the average automobile produced in the United States, while aluminum and plastics accounted for 4 percent and 2 percent, respectively. Over the subsequent three decades, the share of ferrous metals declined to 64 percent of total vehicle weight, while that of aluminum and plastics rose to 8 percent each.[24] During the same period, there was a shift toward stronger and lighter steels within the various ferrous metals. Similar trends toward increased use of lighter materials have occurred in non-U.S. vehicles.[25]

Significant additional reductions in vehicle weight could be achieved, for instance, through the large-scale use of high-strength steel. Various studies have demonstrated that the weight of a conventional car body made of "mild" steel can be reduced by up to 20 percent if made of high-strength steel—and at comparable size and costs, and with a simultaneous increase in stiffness.[26]

Additional significant reductions are possible if aluminum is used. An aluminum vehicle body of similar size and stiffness as one of today's average steel auto bodies, and having comparable strength, would weigh about half as much.[27] Given, however, aluminum's four to five times higher price (a result mainly of the energy-intensive process by which it is produced), a sweeping shift toward its use would result in significant additional costs. At production levels of 300,000 vehicles per year, typical for large-scale automobile production, an aluminum body would cost about U.S. $1,000 more than a steel body. Several commercially available aluminum-intensive vehicles do exist, some of which have a strongly reduced mass compared with their (hypothetical) steel-intensive counterparts.[28]

Further significant weight reductions are possible through the large-scale use of plastic composites, offering a roughly 80 percent weight reduction compared with mild steel.[29] And since carbon fiber composites are shaped, more complex forms can be produced, thereby reducing the number of parts needed to construct an auto body. Expected cost savings, however, are more than offset by higher raw material costs and an inherently slow production process for making parts. To date, this has prevented composites from being used in the mass market. Absent technological breakthroughs in the foreseeable future, carbon fiber composites will be used only in luxury vehicles produced in low volume and for reinforcing vehicle components that require a combination of light weight, high strength, and high stiffness.

Also, glass-fiber reinforced plastics and, more recently, plant fiber composites, although they have less strength and stiffness than steel, are increasingly being used in components in the passenger compartment, including for acoustic and thermal insulation and for body panels. Overall, according to studies by the MIT Material Systems Laboratory, for mass-produced vehicle *bodies*, a 20 percent reduction in weight can be achieved at virtually zero cost, a 50 percent reduction at a cost of U.S. $1,000, and a 70–80 percent reduction at a cost of several thousand U.S. dollars.[30]

Throughout this discussion our assumption has been that the average size of a vehicle (as defined by its interior) remains constant. If so, significant weight reductions can be achieved only through major changes in the mix of materials. Another way to reduce weight, however, is to move toward smaller vehicles. Again, as shown in figure 4.2, weight strongly determines vehicle energy use. Reducing the size of the average new U.S. automobile—which has a weight of 1,580 kilograms (almost 3,500 lbs) as indicated in table 3.2—to that of the average Western European vehicle (1,270 kilograms or 2,800 lbs), together with engine

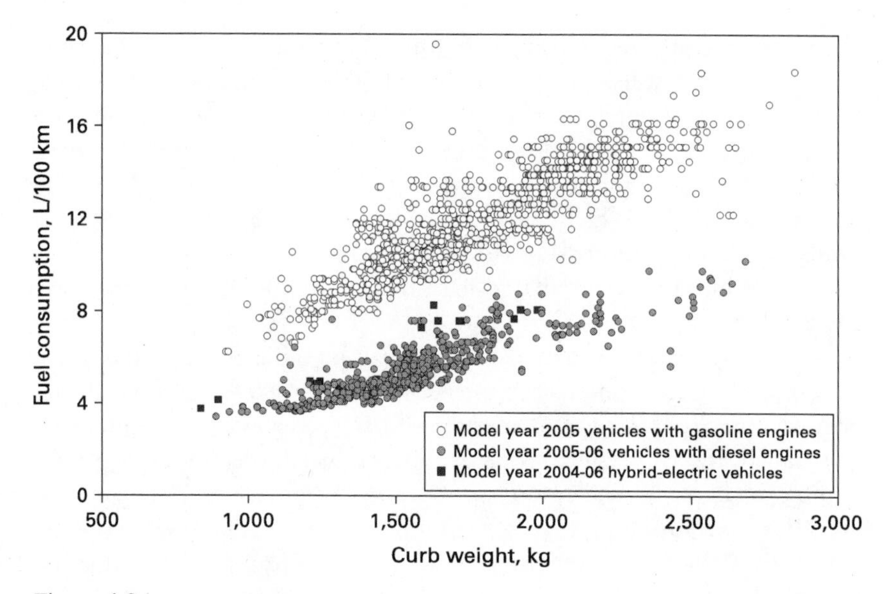

Figure 4.2
Vehicle fuel consumption in liter of gasoline equivalent as a function of vehicle weight. Source: Bandivadekar, A., K. Bodek, L. Cheah, C. Evans, T. Groode, J. Heywood, E. Kasseris, M. Kromer, M. Weiss, 2008. *On the Road in 2035: Reducing Transportation's Petroleum Consumption and GHG Emissions*, MIT Laboratory for Energy and the Environment Report, Massachusetts Institute of Technology, Cambridge.

downsizing, while roughly maintaining the engine power–to–vehicle weight ratio, would alone lead to a fuel consumption reduction of about 16 percent.[31] That decrease in energy use could also come with lower vehicle purchase costs.[32] The value of any loss in comfort, however, is not calculated.

Rolling Resistance

In addition to having low rolling resistance to reduce energy use, tires are expected to have good grip under varying weather and surface conditions to ensure vehicle safety, low wear to guarantee a long lifetime, smooth ride and low noise characteristics for better driving comfort, and all these requirements at affordable costs. Many of these qualities are interrelated. A tire's rolling resistance depends on the characteristics of the tire (tread design and composition, cross-section geometry, tire pressure, and diameter), but also on the nature of the road surface.

For a given tire design and material, minimizing rolling resistance may result in compromising other attributes, especially tread wear and grip. However, advances in tire design and tread composition have reduced rolling resistance to about 0.009 (which is half of what it was before 1980) and simultaneously improved lifetime and grip, without any increase in costs.[33] Further significant reductions in the rolling-resistance coefficient appear feasible. Many studies project that a tire rolling-resistance coefficient of 0.006–0.007 will be achieved by 2020. Preliminary data suggest that other innovations, such as the introduction of run-flat tires, which allow driving with a flat tire for up to 200 kilometers (125 miles) before repair, or a possible move toward airless tires, would not increase the tire rolling-resistance coefficient.

Aerodynamic Drag

Aerodynamic drag accounts for 40 percent of all driving resistances when a representative mix of urban and highway travel is taken into account. Thus, a 10 percent reduction in the drag coefficient would lead to a 4 percent decline in total driving resistance, all other factors being equal. Since aerodynamic drag increases with the square of vehicle speed, it is relatively small in urban stop-and-go traffic but becomes dominant at highway speeds.

Today's typical midsize vehicle has an aerodynamic drag coefficient of between 0.30 and 0.35, down from above 0.5 in the mid-1970s. There are significant opportunities for further reductions. The theoretical lowest drag coefficient of a three-dimensional body is that of an axisymmetric pear-shaped droplet, which would have a coefficient of 0.04. However, in making that shape practical for automobile use, the coefficient increases significantly. As demonstrated by concept vehicles, removing the droplet shape's axial symmetry with a flat floor, adding wheels, using the surrounding air for engine cooling and climate control, and braking off the tapered rear end causes the drag coefficient to rise to about 0.15. Further, when requirements for ground clearance, good visibility, cargo space, and a range of consumer preferences are taken into account, the drag coefficient rises to about 0.22 or 0.25. The latter value is already achieved by a few vehicles currently in production and thus represents a feasible target for the fleet average within the next two to three decades.

A second target of opportunity for reducing aerodynamic drag, a vehicle's cross-sectional area, is more difficult to realize because of people's side-by-side driver and passenger seating preference. Similarly, given the inherently less streamlined shape of light trucks (abrupt windshield slope, open rectangular bed for cargo, etc.) and their higher ground clearance, these vehicles' aerodynamic drag coefficients do not come close to those of automobiles. Add to this their larger cross-sectional area and light truck aerodynamic drag coefficients will remain substantially larger than those of automobiles. Several studies suggest a possible reduction to 0.4 from current values of about 0.5.[34]

Characteristics of Low Greenhouse Gas Emission Vehicles

Based on a projected reduction in vehicle energy demand because of lowered driving resistances and improvements in propulsion system efficiency, significant reductions in vehicle energy use can be expected over the next two to three decades. In keeping with equation 4.1, a reduction in vehicle energy demand by 20 percent, together with an increase in drivetrain energy efficiency by one-third, would be expected to cut vehicle energy use per distance traveled by 40 percent of the original value ($0.80/1.33 = 0.60$). To push fuel efficiency to the limit would require using a hybrid propulsion system instead of a mechanical system. In that case, energy use might be reduced by another 30+ percent, for a total reduction approaching 60 percent compared with the average new car on the road today.

In table 4.1, the right-most column lists a range of possible performance characteristics and qualities a mass-produced automobile could have within the next two to three decades. The more conservative numbers reflect assumed continuous incremental improvements encouraged by existing market forces. Expected to accompany these incremental improvements would be continuing increases in engine power that would improve driving performance. The larger changes shown would require supportive government policies. Although still larger reductions in fuel use could be achieved in theory—by using a carbon-fiber reinforced plastic body—at present it appears unlikely that such radically different materials and the processes needed to manufacture them will become commercially viable on a sufficiently large scale during the next two to three decades. While unlikely over the short term, drastic reductions in vehicle weight are an important longer-term possibility.

Table 4.3 offers greater detail for a range of potential future automobile technologies that we examined in several studies conducted at MIT.

A typical midsize automobile of the early 2000s was our starting point. The table shows that by reducing that vehicle's demand for energy—through incremental reductions in driving resistances, improvements in drivetrain components, and shrinking its engine from six to four cylinders (while maintaining the same engine power–to–vehicle weight ratio)—would reduce the vehicle's energy use by about 30 percent (in this case, from 2.5 MJ/vkm to 1.8 MJ/vkm). To have a consistent basis of comparison, all energy use numbers in the table are based on the U.S. Federal Test Procedure-75 driving cycle. However, this driving cycle underestimates fuel consumption in real-world driving by about 20 percent.

Additional significant reductions in energy use are possible. Integrating a hybrid-electric powertrain into a vehicle having greatly reduced weight and driving resistances would yield a further decline in energy use (and CO_2 emissions) by nearly 40 percent for spark-ignition engines. And still larger reductions in energy use of nearly 50 percent can be achieved with diesel engines. In this case, however, the accompanying potential for lowering CO_2 emissions would be less, about 44 percent, because of diesel fuel's higher carbon content.

The lowest level of energy use we can anticipate at this time—0.6 MJ/vkm—could be achieved by incorporating a fuel cell into a hybrid-electric powertrain. (The fuel cell would draw on onboard hydrogen storage and would be integrated into the low driving-resistance vehicle.) While GHG emissions of petroleum-driven vehicles roughly match the vehicles' energy use (more energy use, more emissions; less energy, fewer emissions), the fuel cell hybrid vehicle would release zero GHG emissions during use. However, as will be discussed in chapter 6, the amount of life-cycle emissions released then depends on how the hydrogen is produced. Hydrogen production today, which is primarily from natural gas, also produces significant GHG emissions.

Table 4.3 also shows that significant reductions in energy use result in strong increases in vehicle retail price (as more and more sophisticated and expensive fuel-saving technology is added). Compared with a mechanical propulsion system (the advanced gasoline-fueled automobile in the third data column of table 4.3), fuel cell vehicles would be about 30 percent more expensive, if fuel cell costs of currently about U.S. $3,000/kW can be reduced to U.S. $60/kW.

Figure 4.3a presents the projected costs of reducing the fuel consumption of automobiles from five studies, based on the characteristics of the average new automobile sold in the United States in the early 2000s.[35]

Table 4.3
Achievable characteristics of future automobiles

	Early 2000s	2020–2030			
Propulsion system	Mechanical	Mechanical	Mechanical	ICE[a] hybrid	FC[b] hybrid
Stored fuel	Gasoline	Gasoline	Gasoline/diesel	Gasoline/diesel	Compressed H_2
No. cylinders	6	4	3	3	—
Engine displacement, L	2.5	1.8	1.7/1.8	1.1/1.2	—
Propulsion power, kW					
Engine	110	93	85/89	58/59	—
Electric motor	—	—	—	29/30	95
Fuel cell system, kW	—	—	—	—	63
Battery, kW	—	—	—	29/30	32
Chassis and body					
Rolling-resistance coefficient	0.009	0.008	0.006	0.006	0.006
Aerodynamic drag coefficient	0.33	0.27	0.22	0.22	0.22
Cross-sectional area, m^2	2.0	1.8	1.8	1.8	1.8
Total vehicle					
Mass, kg	1,320	1,110	1,000/1,050	1,020/1,060	1,130
Energy use					
U.S. urban, MJ/vkm	2.8	2.0	1.8/1.5	1.2/1.0	0.7
U.S. highway, MJ/vkm	2.1	1.5	1.3/1.0	0.9/0.8	0.5
U.S. combined, MJ/vkm	2.5	1.8	1.5/1.3	1.1/0.9	0.6

L(gasoline equivalent)/100 km	7.7	5.5	4.6/4.0	3.4/2.8	1.8
Relative	139	100	83/72	61/50	33
Mpg (gasoline equivalent)	31	42	51/59	69/85	127
CO_2 emissions, g/vkm	183	131	110/97	80/67	0
Relative	139	100	83/74	61/51	0
Retail Price, U.S. $(2000)	20,000	21,200	22,450/23,500	24,400/25,400	29,050

Notes: All vehicles have an engine power–to–vehicle weight ratio in excess of 80 kW/ton and a range of 650 km with a full fuel tank. The main vehicle material is mild steel in the case of the early 2000s vehicle, high-strength steel in the case of the gasoline-fueled mechanical powertrain vehicle, and aluminum for all other future vehicles. The fuel cell vehicle costs strongly depend on those of fuel cells, here assumed to be $60/kW. All energy, fuel use, and CO_2 emission figures relate to the U.S. combined driving cycle. Due to the different energy content of gasoline, diesel, and hydrogen, fuel use values are expressed in liters of gasoline equivalent. CO_2 emissions only relate to vehicle use; the zero value of the hydrogen fuel cell vehicle remains zero only on a life-cycle basis if hydrogen production, transport, distribution, and storage is fueled by zero-carbon energy. See chapter 6 for details.

[a] ICE: internal combustion engine.

[b] FC: fuel cell.

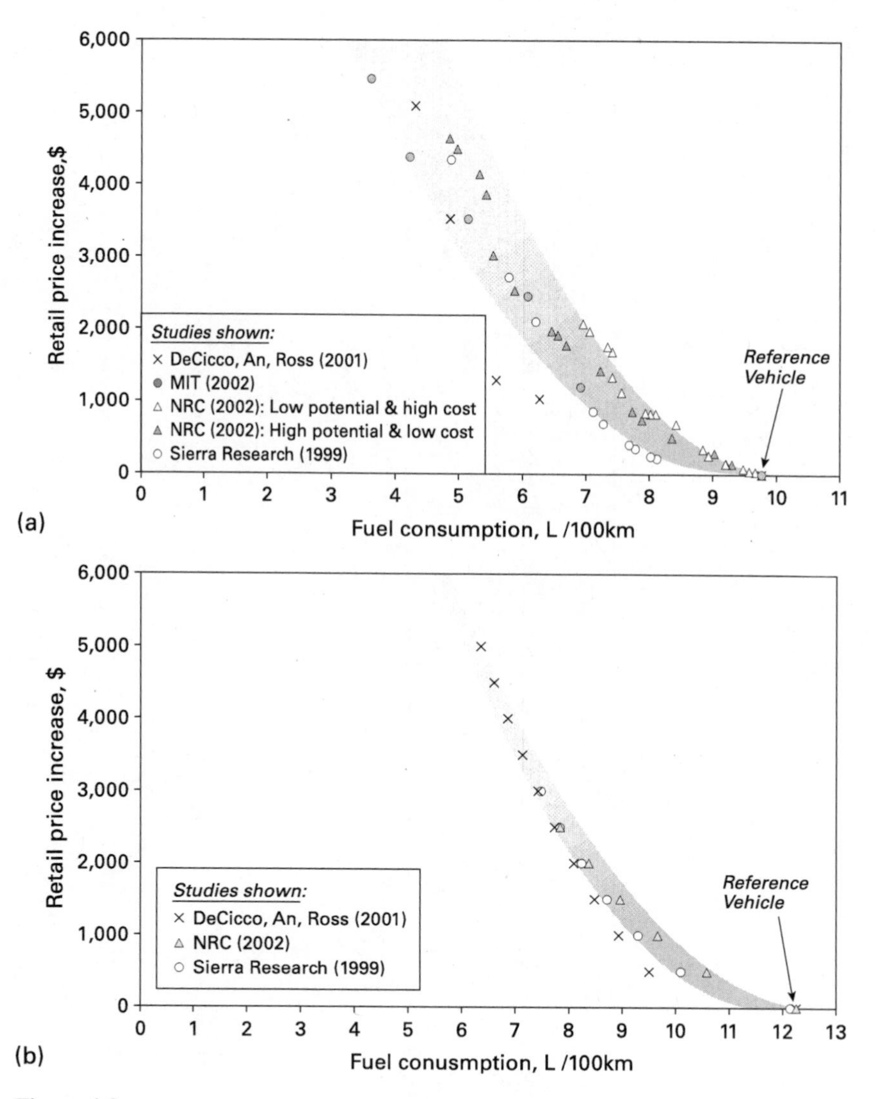

Figure 4.3
The costs for reducing light-duty vehicle fuel consumption: (a) midsize automobiles; (b) light trucks. Since fuel consumption levels of the respective reference vehicles differed slightly across studies, we scaled the fuel consumption of all reference vehicles and those with improved fuel efficiency to that of the National Research Council. Data sources: Weiss, M.A., J.B. Heywood, A. Schäfer, V.K. Natarajan, 2003. *A Comparative Assessment of Advanced Fuel Cell Vehicles*, MIT Laboratory for Energy and the Environment 2003-001 RP, January, Massachusetts Institute of Technology, Cambridge. Weiss, M.A., J.B. Heywood, E.M. Drake, A. Schäfer, F. AuYeung, 2000. *On the Road in 2020: A Well-to-Wheels*

All the studies show retail price increasing with decreasing fuel consumption. Ranking the fuel-saving technologies according to cost effectiveness, the figure shows that a first-step 25 percent reduction in fuel use can be accomplished by a relatively large number of low-cost, evolutionary measures. These include incremental reductions of all driving resistances (such as lower drag coefficients and high-strength steel auto bodies) and moderate engine and transmission improvements (direct fuel injection, slightly higher compression ratios, and the incorporation of variable valve lift and timing technology). These measures can cut fuel consumption by one quarter and would cost somewhere between U.S. $700 and U.S. $1,500.

Achieving a further reduction of the same size, however, requires more substantial measures and thus still higher costs. These measures would include using aluminum-intensive components to achieve substantial reductions in driving resistances and employing a hybrid propulsion system, which would provide significantly more efficient energy conversion. These measures would increase add-on costs to between U.S. $3,500 and U.S. $5,000. Intermediate options would be to switch to a more efficient turbocharged direct injection gasoline engine or to diesel engines.

The differences in cost estimates are mostly minor and can be traced to the different technologies included in the different analyses and to the assumptions the different researchers made concerning the potential a particular technology might have for reducing energy use and the price increase that would accompany that technology.[36] Such slight differences in the technology sets and the assumptions about them help explain the

Assessment on New Passenger Car Technologies, MIT Energy Laboratory, October, Massachusetts Institute of Technology, Cambridge. National Research Council, 2002. *Effectiveness and Impact of Corporate Average Fuel Economy (CAFE) Standards*, Committee on the Effectiveness and Impact of Corporate Average Fuel Economy (CAFE) Standards, National Academy Press, Washington, DC. Austin, T.C., R.G. Dulla, T.R. Carlson, 1997. *Automotive Fuel Economy Improvement Potential Using Cost-Effective Design Changes*, Draft #3, prepared for the American Automobile Manufacturers Association, Sierra Research, Inc., Sacramento, CA. Austin, T.C., R.G. Dulla, T.R. Carlson, 1999. *Alternative and Future Technologies for Reducing Greenhouse Gas Emissions from Road Vehicles*, prepared for the Transportation Table Subgroup on Road Vehicle Technology and Fuels, Sierra Research, Inc., Sacramento, CA. DeCicco, J., F. An, M. Ross, 2001. *Technical Options for Improving the Fuel Economy of U.S. Cars and Light Trucks by 2010–2015*, American Council for an Energy-Efficient Economy, Washington, DC.

two data points (the two x's) that fall outside the shaded envelope in figure 4.3a, where the points seem to suggest a retail price increase of less than U.S. $2,000 despite significantly less fuel consumption. In contrast to the other studies, the DeCicco, An, and Ross analysis includes the mild hybrid technology discussed above, known as an integrated starter-generator. This technology offers an effective additional reduction in fuel consumption at comparatively little extra cost.

Figure 4.3b shows a similar projected fuel consumption–cost relationship for light trucks. These data points are actually aggregates of data about sport-utility vehicles, pickup trucks, and minivans taken together. Since the aggregation process has, in effect, factored out information on individual technology measures, only the results from three studies are shown. Once again, we see that initial reductions can be accomplished at relatively low cost. Reducing light truck fuel consumption by one quarter requires extra investments of U.S. $1,000 to U.S. $1,500. But further reductions would increase the retail price strongly. For example, reducing fuel consumption by *another* quarter would increase light truck retail prices by U.S. $4,000 to U.S. $5,000. The Sierra study suggests that the first 25 percent reduction in fuel use can be accomplished with a number of low-cost measures, similar to those applied to automobiles. Any additional reductions would similarly require more substantial technology changes, such as aluminum-intensive vehicle components, turbocharging, and, in the case of minivans, powertrain hybridization.

To exploit the full potential of the fuel-efficiency improvements we discussed earlier would require additional investments of about U.S. $5,000. Given that about U.S. $20,000 is the typical retail price for a midsize automobile, the increase in price to maximize fuel efficiency amounts to one-quarter of the cost of the original vehicle. Since a typical light truck costs about U.S. $25,000, the U.S. $5,000 increase is about one-fifth of the original retail price. To amortize this cost of fuel-saving technologies over a modest period of four years, fuel prices would have to be nearly U.S. $1.50 per liter (U.S. $5.70 per gallon) to make these high-efficiency technologies marketable. In chapter 7 we will examine various policies that can help fuel-saving technology become competitive in the market.

In this discussion of vehicle costs, we have again assumed vehicle size to remain constant. However, a move toward smaller vehicles could provide a reduction in fuel use without any change in technology. Switching from an average new U.S. automobile having a mass of 1,580 kilograms

(3,480 lbs) to that of Western Europe, having a mass of 1,270 kilograms (2,800 lbs), in combination with engine downsizing would alone lead to a fuel consumption reduction of around 16 percent. However, given the trend toward larger vehicles and its underlying causes (described in chapter 3), many consumers would not readily embrace smaller vehicles. In the context of severe climate change, however, more drastic policy initiatives might well be accepted.

Other Surface Transport Modes

Much of the discussion above on reducing LDV fuel consumption also applies to the two remaining major modes of passenger surface transportation, buses and railways: the fundamental principles governing vehicle energy use, the technological opportunities for lowering fuel consumption (increased propulsion-system efficiency and lower drag coefficients and vehicle mass), and the various constraints that somewhat offset technology improvements (including rising expectations for passenger comfort). Nonetheless, there are some differences.

Unlike LDVs, buses and railways are typically designed to operate in either urban or intercity markets. This segmentation lends itself to tailored engineering solutions. For example, given that urban buses' driving cycle is determined by frequent stops and accelerations, they make ideal candidates for hybrid-electric propulsion systems. And because of their low maximum speeds, they don't require any advanced aerodynamic features. Just the opposite logic applies to intercity buses. Similarly, advanced urban railways are light in weight and regenerate their braking energy into the electricity grid—while the primary focus for high-speed intercity railways is on mitigating aerodynamic drag. All other factors being equal, this task segmentation reduces the number of technology parameters that need to be applied to each of these modes of transportation.

On the other hand, since fuel costs are a significant expenditure item on the balance sheet of public transportation agencies, various fuel-saving technologies have already been incorporated into many of these vehicles. For example, the turbocharged direct injection diesel engine had been operating in heavy-duty vehicles long before it was introduced into the LDV market; and hybrid-electric propulsion systems were introduced into urban buses beginning in the early 1990s. Thus, the overall potential for reducing energy use is likely to be lower than that of LDVs.

Technology Opportunities for the Developing World

Citizens of many developing countries currently enjoy mobility levels between 1,000 and 10,000 passenger kilometers (pkm) per person (620–6,200 passenger miles per person), a regime in which the automobile strongly penetrates the transportation sector. Compared with the replacement markets of the industrialized world, introducing fuel-saving technology into the growth markets of developing economies would have the fastest impact on reducing average fuel consumption by the global vehicle fleet. However, there are several barriers to implementing such a strategy.

Some of the fuel-saving technologies described above require clean fuels. While South Korea, Taiwan, and other higher income countries within the developing world are already making a transition toward clean, low-sulfur fuels, the fuel market in many countries still accommodates low-octane, high-sulfur fuels.[37] But poor fuel quality not only leads to increased levels of urban pollutants; it also limits opportunities for reducing vehicle fuel consumption.

The impact of poor-quality fuel can be both direct and indirect. For example, increasing an engine's indicated efficiency requires higher engine-compression ratios and, in turn, high-octane fuels to avoid uncontrolled combustion. Also, less clean fuels can hinder or prevent the use of fuel-saving technologies that rely on catalyst technology to reduce high engine emissions. Since sulfur can poison $DeNO_x$ catalysts, direct injection gasoline technologies would not be fully effective unless higher NO_x emissions were tolerated. Because improvements in fuel quality lead to improved urban air quality and allow engine-based reductions in fuel consumption, a strategy of increasing the availability of clean fuel in the developing world should have a high priority.

Similar to the lack of clean transportation fuels in developing countries, the generally lower quality of road surfaces increases vehicle rolling resistance—and thus the benefit low-rolling-resistance tires provide would not translate into reduced levels of vehicle fuel consumption in these countries. Additionally, even if both fuel quality and road surfaces were the same as in the industrialized world, the vehicle fleet is not. Our technology analysis was based on a reference vehicle that represents the average new automobile sold in the United States in the early 2000s. At the beginning of this chapter, we suggested that smaller size is the major difference between European and American vehicles and that otherwise, the standard technologies of both are similar. The similarity to U.S. technology, however, fades away in many parts of the developing world, where the characteristics of the LDV fleet differ widely. In the lowest income areas, the average vehicle age is likely to be well above twenty years. A recent survey in Ghana and Uganda conducted for the World Bank recorded that 90 percent of all imported vehicles are used and have an average age of fifteen years.[38]

Even in the higher-income developing countries of Asia and Latin America, the still lower average income (compared with the industrialized world) argues that the size of motor vehicles in those countries would be skewed toward the smaller end. These presumably predominant vehicles generally employ simpler, lower-end technology, while the larger vehicles are virtually identical to those that are common in the industrialized world. Thus, even in the higher income countries within the developing world, the technologies for potentially reducing energy use that we have identified can be applied only to a small part of the vehicle fleet, and there is little opportunity at present to reduce fuel consumption in the largest fraction of new vehicles.

Since it takes some time for income levels to increase and more sophisticated fuel-saving technology to be included in a larger share of new automobiles, do opportunities exist for developing countries to bypass the incremental improvement of oil-based fuels and the internal combustion engine and leapfrog to, perhaps, a nonfossil fuel infrastructure and dedicated vehicles? Just such a process occurred in telecommunications, where mobile phones made (capital-intensive) landline telephones obsolete. As we discuss in chapter 6, some developing countries have already leapfrogged to a biomass-derived fuel and vehicle infrastructure. In contrast, hydrogen fuel and fuel cells are still far from being cost competitive and would require billions of dollars of investment just to produce a working product and initial support system.[39] And as we conclude, the ultimate role of hydrogen and fuel cell systems remains to be seen.

Summary

In this chapter, we identified several significant opportunities for reducing fuel use in LDVs. Over the next two to three decades, technologies can be developed to reduce fuel consumption and CO_2 emissions of the average new midsize automobile in the United States of the early 2000s by up to 60 percent without significantly compromising performance, size, and safety—important criteria for the vehicles to be marketable. If all vehicles achieved that target on average, natural fleet turnover would translate these gains into the road vehicle fleet by midcentury. Comparable reduction levels are technically feasible for light trucks.

Reducing fuel consumption and CO_2 emissions to increasingly lower levels, however, causes vehicle costs to rise sharply because of the requirement for more sophisticated fuel-saving technologies. Reducing vehicle fuel consumption by about one-quarter below the original (early 2000s) level can be achieved through a combination of incremental measures that aim to reduce vehicle driving resistances and improve propulsion system efficiency. Such changes will result in an increase in vehicle costs by up to U.S. $1,500. However, reducing fuel consumption by another quarter would require more sophisticated technologies, including, at the least, a hybrid propulsion system and an aluminum-intensive vehicle, thereby adding additional costs of up to U.S. $5,000.

A crucial prerequisite for these potential fuel-efficiency improvements to materialize is that the trend toward larger vehicle size, higher engine performance, and other consumer attractions can be stopped. Further reductions in fuel consumption are possible if the vehicle fleet is significantly downsized. Downsizing the average new automobile sold in the United States by one size class would offer an additional reduction in fuel consumption and CO_2 emissions by 5–10 percent.

Achieving still greater reductions in vehicle energy use requires moving from a petroleum-fueled hybrid-electric to a hydrogen fuel cell powertrain. If integrated into a low driving-resistance vehicle, this technology would allow a reduction in vehicle energy use by 75 percent compared with the average new midsize automobile in the United States of the early 2000s and a complete elimination of tailpipe CO_2 emissions. However, the production, transport, and storage of hydrogen fuel can release a larger amount of GHG emissions than the upstream emissions of petroleum-fueled vehicles. Yet, even if the GHG-emission problem could be resolved during the next two to three decades, other constraints are likely to remain. Such fuel-cycle issues are examined in chapter 6. Before we turn to these opportunities, however, we proceed with an analysis of reducing fuel consumption of the fastest growing transport mode, which is aircraft.

5

Aircraft Technology

Having examined the technology and fuel opportunities for reducing greenhouse gas (GHG) emissions from road vehicles, we now continue our analysis with respect to commercial passenger aircraft. In chapter 2 we conclude that the relative importance of air traffic rises strongly at mobility levels greater than 10,000 passenger kilometers traveled (PKT) per capita (6,200 passenger miles per capita). Currently, three regions of the world are in that mobility regime. Depending on the economic growth rates underlying our projection of future levels of travel demand, residents in nearly all regions may also experience mobility levels above that threshold value by 2050.

Although we focus here on energy use and carbon dioxide (CO_2) emissions, aircraft also have several other climate impacts. These additional impacts can be direct or indirect and can occur at different scales of space and time since they depend on the type, time of day, and location of emissions. At typical cruise altitudes of about 10 kilometers (nearly 33,000 feet), aircraft nitrogen oxide (NO_x) emissions affect the climate indirectly through production of tropospheric ozone (a warming effect) and through accelerating the removal of methane introduced into the atmosphere by other sources (a cooling effect). Due to the different lifetimes of ozone and methane, the warming effect is regional, occurring where aircraft are flown, whereas the cooling effect occurs uniformly around the world.

In addition, contrails from aircraft engines directly and indirectly increase cloud cover and therefore alter the atmosphere's radiation budget (tending to produce a net warming effect). This phenomenon also is a regional effect, occurring where aircraft are flown, and its overall impact has a high degree of scientific uncertainty.

Notably, these kinds of effects have very different time scales. The effects of contrails and cirrus clouds, ozone production and methane removal, take place over a period that could be hours or decades after a flight is made. In contrast, the impact of CO_2 occurs over a period of centuries. The different regional and temporal scales over which these effects occur make evaluating their impacts difficult—and this topic remains the subject of considerable scientific study. However, various studies suggest that the impacts of these non-CO_2 effects may be as significant as the impacts of CO_2 effects or even more so.[1]

This inherent complexity and the scientific uncertainty it gives rise to also make it difficult to judge the relative merits of reduction strategies. For example, it is not yet known with confidence if there would be net climate benefits associated with reducing contrails and cirrus by flying around regions of the atmosphere where these phenomena may form, but at the expense of burning additional fuel and therefore adding more CO_2. Given this uncertainty in the relative contribution of non-CO_2 emissions to climate change, the focus of our analysis remains on CO_2 while being mindful of the existence of non-CO_2 effects. We begin our analysis with a broad discussion of technology opportunities and constraints before looking more closely at the determinants of aircraft fuel burn and the opportunities alternative fuels may provide for CO_2 reduction.

Technology Fundamentals

For road vehicles, we derived energy use per PKT by combining the equation of motion with the first law of thermodynamics. A similar approach can be used to estimate aircraft fuel use. During cruise flight at constant altitude, the lift force (L) balances the airplane's weight. At constant cruise speed (V), engine thrust equals aerodynamic drag (D). Combining the equations for these two relationships with the definition of engine-specific fuel consumption (SFC, or fuel flow rate typically expressed in grams per second, divided by the thrust produced, given in thousands [kilo] of newtons), and then dividing this result by the number of passengers (PAX) and the distance traveled, gives us equation 5.1:

$$\frac{E}{PKT} = \frac{W \cdot Q \cdot SFC}{PAX \cdot V \cdot (L/D)} \tag{5.1}$$

Here E/PKT is energy use per (revenue) passenger kilometer traveled, Q is the energy content or lower heating value of the (jet) fuel, and W is the total weight of the aircraft (including fuel, passengers, cargo, and the aircraft structure). Thus, the lighter the plane's structure (the smaller W is), the more fuel-efficient its engines (the smaller the SFC), the greater the number of passengers, and the higher its aerodynamic efficiency (L/D), the less fuel needs to be burned per PKT. Note that an increase in flight speed does not necessarily lead to a reduction in energy intensity, since the higher speed is tied to a change in the fuel flow rate and will affect the aircraft's lift-to-drag characteristics.

Note also that this equation is valid for only one point in time during the aircraft cruise flight, since the aircraft's weight declines significantly as fuel is burned (as much as 25 percent of the aircraft's takeoff weight may be fuel). To estimate the energy used on a full trip, one must thus integrate equation 5.1 as a function of time (producing a relationship similar to the well-known Breguet range equation, which can be found in many aviation textbooks).[2]

Owing to three-dimensional movement in the atmosphere, where weight, volume, and safety constraints are more stringent and engineering trade-offs more complex than for road vehicles, realizing reductions in fuel use can be more challenging for airplanes. Many engineering trade-offs must be made because many variables are interconnected. For example, increasing wingspan to increase the lift-to-drag ratio also increases—all other factors being equal—aircraft weight. Similarly, more fuel-efficient, higher-bypass-ratio engines are heavier and experience higher drag, and thus lose as much as half of their isolated efficiency gain once they are installed on an aircraft.[3] Additionally, requirements for noise reduction tend to favor higher-bypass-ratio engines, thereby exacerbating negative effects on aerodynamics and weight—sometimes at the expense of overall aircraft fuel efficiency.

Many such engineering trade-offs are not static and can change during flight. At takeoff, for example, SFC and L/D may deviate from cruise values by as much as 50 percent. And while there are many other possible engineering trade-offs with regard to cost penalties, many may violate environmental constraints. For example, raising temperatures and pressures in the engine to increase thermal efficiency may also increase complexity, manufacturing costs, and NO_x emissions.

Realizing significant reductions in fuel use is also challenging because of safety constraints. For example, flight regulations specify that a 30–45-minute fuel reserve must be loaded onto commercial aircraft. These reserves are typically not used, but they add weight to the airplane, thereby increasing the amount of fuel burned over a flight. Similarly, public safety agencies require aircraft to be designed with redundant systems so that failures of critical systems do not lead to catastrophic outcomes. In particular, twin-engine airplanes must still be able to operate, take off, and land safely if one engine fails. Thus, the amount of thrust designed into the engine—and thus the size and weight of the engine—is typically greater than would be required if this safety requirement were not mandated. These particular examples represent only the tip of an iceberg; there are thousands of other goals and constraints that must be balanced against fuel use.

Because of these and other complexities, the development of new aircraft bears significant economic risk. In several recent instances, billions of dollars have been lost because of design problems even very late in the development process. Development delays before the Boeing 747 was introduced in 1970 and, more recently, before the introduction of the Airbus A380, threatened the survival of their respective manufacturers.

To mitigate this risk, new aircraft often are produced from proven designs, with only a limited number of modifications made from one model to the next. Thus, as in the case of automobiles, many of the possible fuel-saving technologies that are discussed today have a long history. Take the flying wing, an air vehicle without a distinct fuselage or tail surfaces: inspired by the idea of removing all vehicle components that don't contribute to lift, the design of such tailless aircraft began in the late nineteenth century. One of the numerous patents filed for wing-only flying machines was one in 1910 by Hugo Junkers, the prominent German aircraft pioneer. His early hollow wing design accommodated pilots, engines, and other loads.

Over the past century, flying-wing designs converged to increasingly resemble Junker's early vision of a relatively uniform, blended entity. Among early models of nearly perfectly blended wing bodies were several designs by Jack Northrop in the United States and by the Horton brothers in Germany, which were tested during the 1940s. However, it took another fifty years before flying-wing aircraft were put into routine service in the United States in the form of the B-2 bomber. All the same, flying wings are still considered too risky and expensive for general commercial service and are not anticipated to be introduced into the commercial airline fleet for at least another ten to twenty years.

Yet, examples of more rapid development of a dramatically new technology exist. One such example is the jet engine. Sir Frank Whittle first speculated on the possibilities of jet engines in 1928 in England—Whittle was one of several people around this period with similar ideas. By the mid 1930s, both he and Hans von Ohain in Germany, who are credited with independently developing the first working gas turbine engines for jet propulsion, started development of prototype engines and produced working versions by the end of the decade. The first operational military uses followed shortly in the early 1940s (with the Gloster Meteor, the Arado AR 234, the Bachem Ba 349, the Heinkel He 162, the Messerschmitt Me 163, and the Messerschmitt Me 262 all seeing service in World War II). The first jet-powered commercial airliner was the De Havilland 106 Comet, which was introduced into service in 1952, only twenty years after the first working gas turbine engine designs were drawn.[4]

Thus, as with automobiles, technology development times for commercial aviation are typically measured in decades. However, the average lifetime of an aircraft—the time between the manufacturer year of an aircraft cohort and the retirement of half these airplanes—is around thirty years, about twice that of automobiles. Hence, all other factors equal, the time until new technology impacts the passenger aircraft fleet characteristics would be longer than that of automobiles. However, because a growing fraction of passenger aircraft are converted to freighters as their age approaches twenty years, and because the higher projected growth in air travel demand may lead to more rapid introduction of new aircraft, we also assume that passenger aircraft introduced during the 2020s will be representative of the 2050 fleet characteristics.

In addition to design complexities, many of the consumer acceptance issues discussed for automobiles also apply to aircraft. Indeed, for short flights at lower speeds, turboprop-powered aircraft would be more energy efficient than jet-powered aircraft because of the turboprop's higher propulsive efficiency.[5] However, passengers find turboprops less comfortable to ride, because they generate more noise and vibration. Since consumers are less willing to pay the same price for tickets on turboprop-powered aircraft as for jet-powered aircraft, airlines are less willing to purchase them. Only recently, the rise in oil price during the early twenty-first century has resulted in increasing orders for turboprop-powered aircraft.

Or take another example: the fuselage. It would be more fuel efficient to design aircraft without windows (since a pressurized hull without structural discontinuities can typically be made lighter), or without video screens in every seat, or without expansive first-class seating. Passengers, however, value these amenities, particularly for longer flights. Further, the passengers themselves have gained weight. Between the early 1960s and the early 2000s, the average person weight has increased by 15 percent in the United States, which, according to the equation in endnote 2, has increased aircraft fuel burn by two to three percent.[6] Nevertheless, despite the many technological constraints, and the influence of passenger preferences, a range of options does exist for reducing aircraft GHG emissions.

Reducing Aircraft Greenhouse Gas Emissions

While purchasers of light-duty road vehicles can choose models that maximize their overall satisfaction, selecting from among a wide range of available attributes, airlines choose an aircraft to maximize their profits, selecting a vehicle that offers passenger comfort at the lowest possible ownership and operating costs. As discussed in chapter 3, in 2007, fuel accounted for about half of the total direct operating costs and purchase price (the direct operating costs and investments, or DOC+I) of large aircraft, making fuel expenditures the single most important operating expense.

Thus, airlines are inherently interested in low-fuel-consumption aircraft, but only to the extent that they minimize overall operating costs. It is not surprising, then, that aircraft have experienced a strong decline in energy intensity during the past decades. According to our analysis at MIT, between 1959 and 1995, average new aircraft energy intensity declined by nearly two-thirds. Of that decline, 57 percent was attributed to improvements in engine efficiency, 22 percent resulted from increases in aerodynamic efficiency, 17 percent was because of more efficient use of aircraft capacity through higher load factors, and 4 percent resulted from other changes, such as increased aircraft size.

The historical trends in energy intensity of the U.S. fleet and for individual aircraft by year of introduction are shown in figure 5.1.[7] Year-to-year variations in energy intensity for each aircraft type, caused by different operating conditions—such as stage length, passenger load factor, flight speed, altitude, and routing controlled by different operators—can be significant and are on the order of 30 percent, as represented by the vertical extent of the data symbols. Despite the strong historical decline in energy intensity, further reductions are possible.

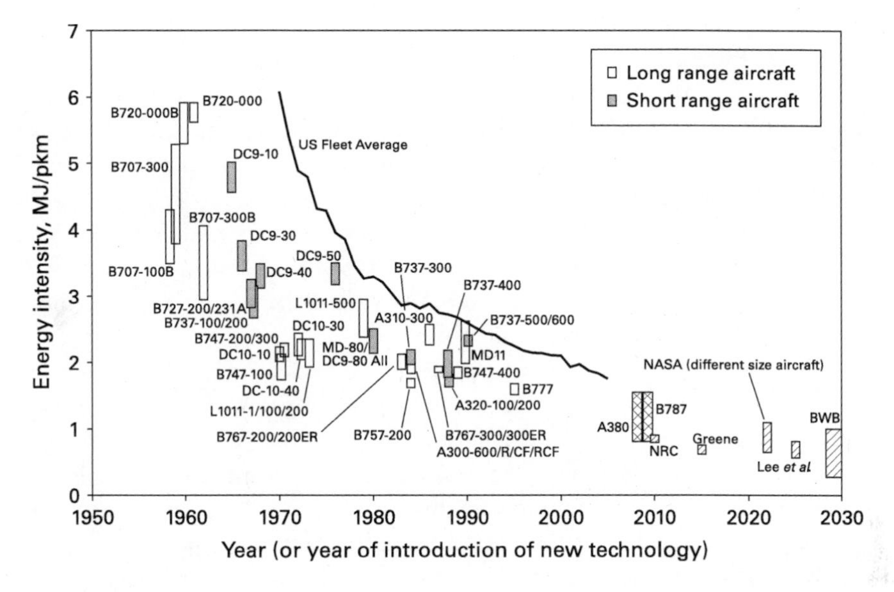

Figure 5.1
Evolution of aircraft energy intensity, new aircraft and U.S. fleet average. Sources: Lee, J.J., S.P. Lukachko, I.A. Waitz, A. Schäfer, 2001. Historical and Future Trends in Aircraft Performance, Cost, and Emissions, *Annual Review of Energy and the Environment 2001*, 26: 167–200. These figures were updated using U.S. Department of Transportation, *Air Carrier Summary Data (Form 41 and 298C Summary Data)—Air Carrier Summary: T2: U.S. Air Carrier Traffic and Capacity Statistics by Aircraft Type*, www.transtats.bts.gov/. The A380 and B787 energy intensity range was estimated using performance announcements on the Airbus and Boeing websites. The BWB data is derived from Liebeck, R.H., 2004. Design of the Blended Wing Body Subsonic Transport, *AIAA Journal of Aircraft*, 41(1): 10–25. References for other projections include Greene, D.L., 1992. Energy-Efficiency Improvement Potential of Commercial Aircraft. *Annual Review of Energy and the Environment*, 17: 537–573. Greene, D.L., 1995. Commercial Air Transport Energy Use and Emissions: Is Technology Enough? Presented at 1995 Conference on Sustainable Transportation Energy Strategies, Pacific Grove, CA. National Research Council and Aeronautical and Space Engineering Board, 1992. *Aeronautical Technologies for the Twenty-First Century*, National Academy Press, Washington, DC.

Improving the Efficiency of Aircraft Propulsion

The development of the gas turbine engine in the middle of the twentieth century was crucial for the evolution of high-speed commercial flight. Early aircraft were powered by piston engine–driven propellers. A key limitation on the cruise speed of these aircraft was the tip speed of the rotating propellers. As that speed approached the speed of sound, propeller efficiency declined rapidly. This limitation was overcome in gas turbine engines by the geometric design of the air intake, which slows down the speed of the airflow. Compared with piston engines, jet engines are also lighter per unit of power produced.

The last generation of large propeller-driven aircraft illustrate the advance. Each of the four eighteen-cylinder, 2,500 kW engines of the Lockheed 1049G Super Constellation had a power-to-weight ratio of 1.5 kW/kg. Given that these engines incorporated a number of sophisticated energy recovery technologies, this value of specific power was close to the technological limit that could be reached at that time. (Note that these levels are higher than the power-to-weight ratio for average automobile engines today and in 2020–30, as can be seen from table 4.1). The immediately following generation of aircraft, starting with the De Havilland Comet, put into commercial service in the early 1950s, was powered by turbojet engines. Each of the Comet's four engines had a power-to-weight ratio of nearly 5 kW/kg, more than three times that of propeller engines. That ratio has grown to 17–18 kW/kg for today's advanced turbofan engines. This greater power per unit of engine weight has made it possible to fly larger aircraft, reduce fuel consumption, operate at higher altitudes, and cruise at higher speed, thus greatly increasing an aircraft's productivity (the number of seats transported per hour).

All of today's aero-engines follow the same basic principles of jet propulsion: they generate thrust or force by increasing the speed of the airflow between the inlet and outlet of an aero-engine. Increasing that speed relies on a thermodynamic cycle that converts the chemical energy in the fuel into thermal and kinetic energy in the exhaust jet. It is convenient to express the total efficiency of an aero-engine as the product of its thermal efficiency and its propulsive efficiency. Thermal efficiency measures the percentage of fuel chemical energy transformed into the kinetic energy of the airflow moving through the engine. Propulsive efficiency then expresses the contribution of the airflow's kinetic energy to the propulsive power produced.[8] Much like an automobile engine, thermal efficiency is largely related to the *inner* workings of the engine—although those familiar with aircraft engines will recognize this as a simplification. Generally, the higher the temperatures and pressures in the engine, the higher its thermal efficiency.

And although this too is a simplification, the propulsive efficiency can be regarded as relating to the *outer* workings of the engine—specifically to the mass flow of air that composes the propulsive jet and the speed of that jet relative to the flight speed. For two jets with the same exhaust kinetic energy (the product of mass and the square of velocity, divided by two), the jet with the higher mass flow and smaller change in velocity is more energy-efficient. This is one reason why commercial aircraft use ever-larger engines; larger engines accelerate a larger mass flow, although with a smaller change in velocity.

In modern gas turbine engines, the bypass ratio—the amount of air that bypasses the combustor in relation to the amount of air passing through the engine core combustor—is five to ten. (Thus, propeller-powered airplanes can be more energy efficient than jet-powered airplanes in some instances—especially at lower cruise speeds—because they give a propulsive impulse to a larger mass of flow.) Today's most efficient aircraft gas turbine engines have a thermal efficiency of around 55–60 percent and a propulsive efficiency of approximately 65–70 percent, so that overall efficiency is around 40 percent.

Figure 5.2a shows the historical reduction in engine-specific fuel consumption. Most of that improvement was realized prior to 1970 through increases in propulsive efficiency, that is, the introduction of high-bypass engines. Over the past fifty years, the cruise SFC (again, fuel flow rate per unit of thrust) of newly introduced engines has decreased by approximately 40 percent, averaging an annual reduction of 1 percent. However, as bypass ratios have increased, engine diameters have also become larger, leading to an increase in engine weight and aerodynamic drag, and thus offsetting some of the improvement in total aircraft fuel use.

The efficiency of aircraft propulsion can be further improved through increasing the two underlying efficiencies. Propulsive efficiency can be further increased by using turboprops and unducted fan engines—turbofan engines without the nacelle and thus without its weight and drag penalties. However, as noted earlier, these types of engines are most efficient at lower flight speeds and often have noise and vibration challenges to overcome. Given the current state of engine technology, a 10–20 percent reduction in SFC seems to be possible with the development of advanced

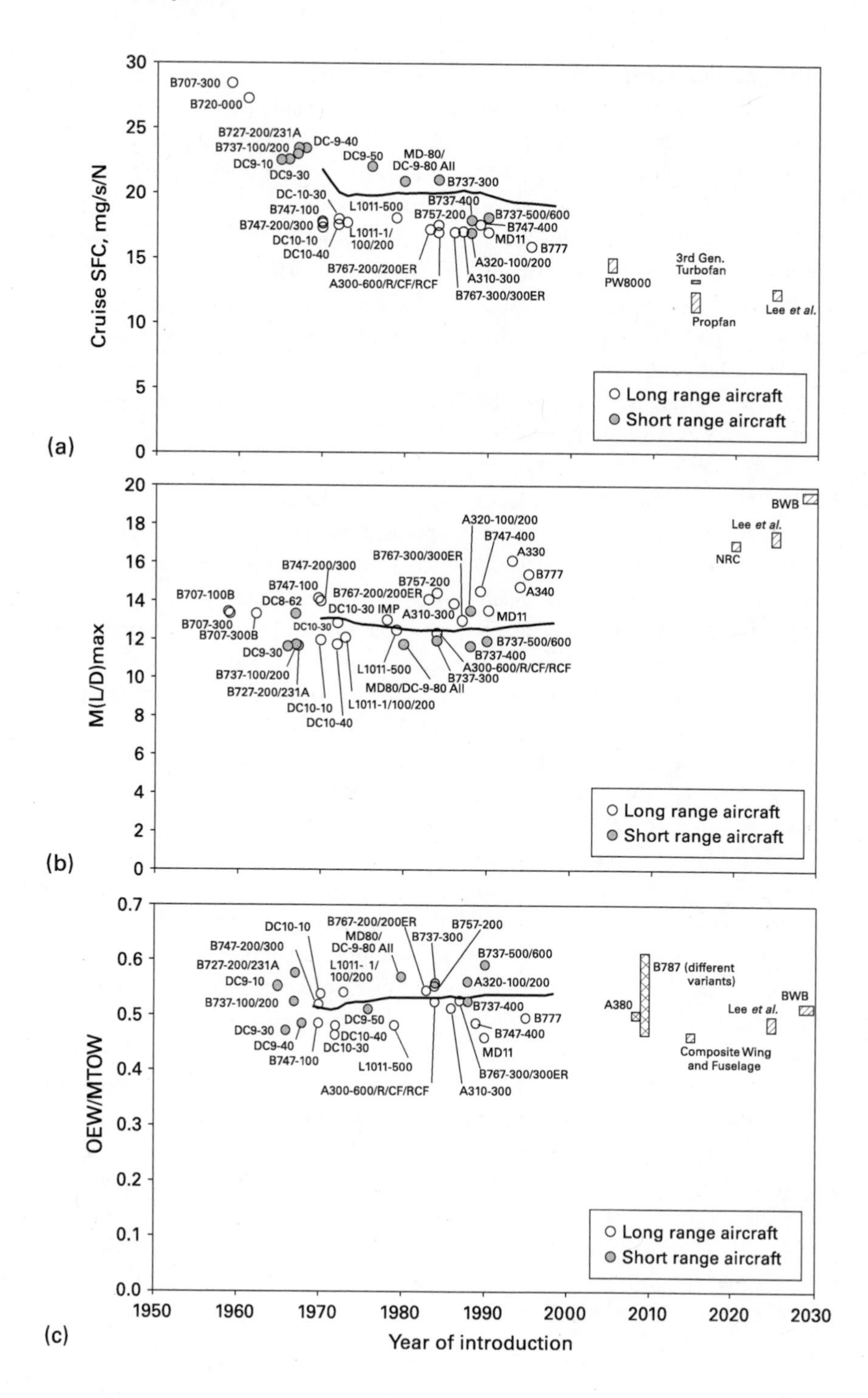
(a)
Cruise SFC, mg/s/N
B707-300
B720-000
B727-200/231A
B737-100/200
DC9-10
DC9-30
DC-9-40
DC9-50
MD-80/
DC-9-80 All
B737-300
DC-10-30
B747-100
B747-200/300
DC10-10
DC10-40
L1011-500
L1011-1/
100/200
B737-400
B757-200
B737-500/600
B747-400
MD11
B777
A320-100/200
A310-300
B767-200/200ER
A300-600/R/CF/RCF
B767-300/300ER
PW8000
3rd Gen.
Turbofan
Propfan
Lee et al.
Long range aircraft
Short range aircraft
(b)
M(L/D)max
BWB
Lee et al.
NRC
A320-100/200
B747-400
A330
B777
A340
B767-300/300ER
B747-200/300
B747-100
B707-100B
DC8-62
B767-200/200ER
B757-200
DC10-30 IMP
A310-300
MD11
B707-300
B707-300B
DC10-30
DC9-30
B737-100/200
B727-200/231A
DC10-10
DC10-40
L1011-500
L1011-1/100/200
MD80/DC-9-80 All
B737-300
A300-600/R/CF/RCF
B737-400
B737-500/600
Long range aircraft
Short range aircraft
(c)
OEW/MTOW
DC10-10
B747-200/300
B727-200/231A
DC9-10
B737-100/200
DC9-30
DC9-40
B747-100
B767-200/200ER
MD80/
DC-9-80 All
L1011- 1/
100/200
B737-300
B757-200
B737-500/600
A320-100/200
B737-400
B777
B747-400
MD11
B767-300/300ER
A310-300
A300-600/R/CF/RCF
L1011-500
DC9-50
DC10-40
DC10-30
A380
B787 (different
variants)
BWB
Lee et al.
Composite Wing
and Fuselage
Long range aircraft
Short range aircraft
Year of introduction

unducted turbofan engines. Since these technologies would require slightly slower flight speeds, they would probably be used mainly in shorter-distance traffic.

In addition to potential increases in propulsive efficiency, the overall efficiency of an aero-engine can be increased by improving its thermal efficiency. Several opportunities for doing this exist, but most require an increase in the complexity of the engine—and thus of costs. Much like the indicated efficiency of piston engines, the thermal efficiency of aero-engines can be increased by optimizing the underlying thermodynamic cycle. This can be accomplished by employing higher pressure ratios and temperatures, which in turn, however, would lead to materials challenges and higher NO_x emissions—unless effective cooling devices are developed and implemented.

Higher thermal efficiencies can also be achieved through the use of heat exchangers and intercoolers, as is commonly done for gas turbines used for land-based power. While the demanding weight and volume constraints of aircraft have thus far presented too much of a challenge for such technologies, overcoming these restrictions is an important area of research.

Finally, another way thermal efficiency can be increased is by improving component efficiencies within the engine so that the cycle performance more closely matches that of an ideal cycle for a given pressure ratio. However, as with the opportunities above, challenges of complexity, costs, and weight must be overcome. Overall, based on recent historical trends and possible improvements to existing engines, our recent study at MIT estimates that a 15–25 percent reduction in SFC is likely by the mid 2020s.[9]

Reducing Aircraft Drag

Commercial aircraft are typically designed to maximize the distance that a certain amount of payload can be moved per unit of fuel. As shown in equation 5.1, this requires minimizing the drag force on the airframe per unit of lift force, or maximizing the lift-to-drag ratio. Overall aircraft

Figure 5.2
Determinants of aircraft energy intensity: (a) cruise specific fuel consumption, (b) lift-to-drag ratio, and (c) weight ratio. Source: Lee, J.J., S.P. Lukachko, I.A. Waitz, A. Schäfer, 2001. Historical and Future Trends in Aircraft Performance, Cost, and Emissions, *Annual Review of Energy and the Environment 2001*, 26: 167–200.

drag consists of contributions primarily from two sources: skin friction on the surfaces of the airplane and induced drag, or drag due to lift. Induced drag occurs because all bodies that produce lift also produce vortices. The vortices influence the flow of air around the vehicle, which in turn produces a force with one component that is perpendicular to the lift force and opposite to the flight direction—drag.

For a typical commercial aircraft, skin friction drag accounts for nearly half of total aircraft drag, induced drag for slightly more than one-third, and a number of other drag components for the remainder. The lift-to-drag ratio then determines the quality of an airplane's aerodynamic design. That number is between 15 and 19 for modern commercial aircraft. Figure 5.2b shows that aerodynamic efficiency has increased by approximately 15 percent historically, averaging 0.4 percent per year. Better wing design and improved propulsion and airframe integration, enabled by improved computational and experimental design tools, have been the primary drivers of these improvements.

There are several ways to further improve the aerodynamic efficiency of aircraft. Some options are evolutionary, such as improving the design of the airfoils or the way the wing is integrated with the fuselage. Other options are more radical. Among these are dramatic changes in aircraft configuration, such as a blended-wing-body (BWB) aircraft. Like the B-2 Stealth Bomber, such an aircraft essentially corresponds to a flying wing—but one that would offer enough space for several hundred passengers. Another option is active flow control on the wing surfaces, where a small amount of air would be sucked through a porous wing to generate laminar flow over the entire chord (that is the length of the wing section measured parallel to the direction of flight, as compared with the span, which is measured perpendicular to the direction of flight) and thus create better overall aerodynamic performance. Combining these options leads to a laminar flying wing, and such a flying wing, with active flow control, could reduce fuel use by more than 50 percent.[10]

The menu of radical changes also includes wing shapes that change depending on the flight speed, such as variable sweep or variable camber wings that attempt to emulate some of the behavior seen in nature, such as when birds' wing shapes change to accommodate flight conditions. Of all these options, perhaps the one with the greatest potential is an integrated blended-wing body similar to a flying wing. The other options are less attractive because they present more significant challenges in cost, complexity, and operating constraints.

The blended-wing-body aircraft concept integrates the fuselage with the wings while eliminating the tail from the design. Such a configuration has reduced wetted area compared with a conventional tube-wing configuration ("wetted area" is the external surface area of an aircraft and is related to skin friction drag). The reduced wetted area results in an improved lift-to-drag ratio, an improvement that increases with aircraft size since the internal volume of the configuration can be used to hold passengers and fuel more efficiently. A recent Boeing study estimated that a blended-wing-body configuration could achieve a lift-to-drag ratio of twenty-three, allowing an aircraft to fly 480 passengers a distance of 16,700 kilometers or 10,400 miles (a mission similar to that of the Airbus A380), with nearly 30 percent less fuel than would be required by a tube-wing configuration aircraft of today.[11]

A study conducted by MIT and Cambridge University suggests that further improvements in the lift-to-drag of a blended-wing-body design (to an estimated value of twenty-five) can be obtained by aerodynamic shaping of the center body.[12] However, such radical changes in aircraft architecture would require considerable research and development efforts and thus entail financial risks that go significantly beyond those related to the introduction of a new tube-and-wing configuration (and we have noted that even these carry significant development risks). It is therefore likely that aircraft will continue to have a tube-wing configuration for at least the next ten to twenty years and that by the mid-2020s, improvements in L/D are likely to be limited to about 10 percent.[13]

Reducing Aircraft Weight

The efficiency of an aircraft's structural design can be assessed by comparing its operating empty weight to its maximum takeoff weight. The operating empty weight of an aircraft is the unloaded weight of a vehicle that is ready to fly—that is, without payload and fuel. This includes the weight of the aircraft structure, of safety and entertainment systems, and the crew and their luggage, food and beverages. Adding the maximum allowed weight of passengers, cargo, and fuel produces the maximum takeoff weight. Airlines are inherently interested in a low operating empty weight in relation to the maximum takeoff weight, since a lower structural weight ratio allows them to carry a relatively larger amount of payload and fuel. This structural weight ratio has been about 0.5 for today's aircraft. It will vary for the same type of aircraft depending on configuration, such as the number of seats in business class versus coach, and the type of market it is intended to fly, such as short or long range.

Figure 5.2c suggests that historical reductions in the structural weight ratio have been modest. The stability of that value is the result of two largely compensating trends. Over much of the history of aviation, large commercial aircraft have been constructed almost exclusively of aluminum compounds. Composites—fiber-reinforced plastics with high strength and low weight characteristics—were used only for a limited and slowly growing number of components, largely because of higher manufacturing costs and concerns about increased maintenance requirements. In this evolutionary manner, the range of composite-based material components spread mainly from one model to a later one. Overall, composite materials increased from about 1 percent of the total structural weight for the early Boeing 747s in the early 1970s to about 12 percent for new aircraft in the mid 1990s, to 25 percent and 50 percent, respectively, for the recently introduced Airbus A380 and Boeing 787.

All the same, the structural weight ratio has remained level because reductions in aircraft structural weight have been traded for other technological improvements and passenger comfort. For example, despite composite materials accounting for half of its structural weight, the structural weight ratio of the Boeing 787 is still about 0.5. This suggests that, as in the past, at least part of the weight benefit is being traded for improvements in other areas, including improved aerodynamic designs and items for improved passenger comfort (higher humidity, higher pressure cabin environment, larger windows, wider seats, etc.).[14] However, if a still larger share of lighter-weight, high-strength materials is substituted for the remaining metallic aircraft structures, some reduction in the structural weight ratio could be realized, provided the weight reductions are not offset by incorporation of other technologies (e.g., larger engines) or further increases in passenger comfort. Our recent study at MIT thus estimates that by the mid 2020s, a range of only 0–10 percent reduction in the ratio of aircraft operating empty weight to maximum takeoff weight is likely through enhanced use of composite materials.[15]

Given that the average life of a large commercial aircraft is about thirty years, it takes several decades until new aircraft have a significant impact on the fleet characteristics. Faster opportunities for reducing GHG emissions exist for changes in aircraft operation.

Operational Opportunities

There are numerous opportunities for improving the efficiency of air transport operations through better air traffic management, flight crew operating procedures, and new aircraft- and ground-based technologies. Historically, major air traffic control and management improvements were precipitated by safety concerns. As the safety of the system improved to present-day levels, however, new barriers to system change have emerged, but new challenges of capacity limitation and environmental impacts are providing renewed impetus to improve the system.[16] Some recent efforts at change have provided increases in airspace capacity and thus have introduced more flexibility for aircraft to fly closer to their optimum routes and altitudes. For example, the recently introduced Reduced Vertical Separation Minima standards allow suitably equipped aircraft to fly closer to one another at the most efficient cruise altitudes, reducing fleet-average fuel burn by approximately 1 percent.[17]

But despite these advances, it has been estimated that deviations from optimal routing because of inefficient air traffic control, restricted airspace, and air traffic congestion contribute to as much as 10 percent additional energy use.[18] Owing to the extra flight time—and thus airline expenditures for crew and fuel—the associated extra costs to the airlines are large. According to one estimate, these costs are about 1 billion U.S. dollars per year, both in the United States and in the EU.[19]

Initiatives are under way to address some of these shortcomings through major infrastructure upgrades. In the United States, the Next Generation Air Transportation System initiative (NextGen) is focusing on ways new technologies (such as satellite-based navigation, surveillance, and networking systems) can be leveraged to transform today's air traffic control system to reduce delays in the air and on the ground.[20]

In Europe, the Single European Sky Air Traffic Management Research initiative (SESAR) aims to unite European airspace, which is still largely structured along the national boundaries of its members and some of its neighboring countries. (Since the charge for using the airspace can differ significantly across countries, there is evidence that airlines make deliberate fuel-wasting detours through lower-cost airspace should this strategy lead toward lower operating costs.[21] A united European sky, however, would eliminate such price differences and thus the incentive for detours.) Both NextGen and SESAR rely on the deployment of advanced technologies that allow better surveillance and air traffic coordination, including more direct routing and earlier adjustment of aircraft trajectories. These capabilities will allow airlines to avoid long holding and excess fuel use in the terminal airspace of the arrival airport.

In addition to these infrastructure and technology evolutions, shorter-term operational improvements can also significantly enhance aircraft efficiency. For example, continuous descent approach trajectories keep aircraft higher and at lower thrust for longer periods of time compared with conventional step-down procedures, thereby reducing noise, fuel burn, and associated emissions. U.S. and European trials of advanced forms of continuous descent approach that utilize the flight management system incorporated into modern aircraft have indicated that fuel burn and emissions can typically be reduced by 10–20 percent during the descent and approach phases of flight, depending on the aircraft type and local airport conditions.[22] Fuel efficiency can also be given priority in modern aircraft through the "cost index" parameter in the flight management system that permits operators to balance fuel-related costs with time-related costs (such as flight crew salaries).

Additionally, as discussed in chapter 3, design of the air transportation system network can significantly influence fuel usage. A hub-and-spoke system, for example, will use fuel differently than a point-to-point system. Although the hub-and-spoke system allows many more markets to be connected with a given number of flights, a point-to-point model permits a better matching of aircraft size to demand in a given market pair.

There are also proposals to decrease fuel use by breaking up long routes into multisector journeys flown by more fuel-efficient medium-range aircraft (which do not have to carry excess fuel over long distances).[23] However, there are significant crew cost and passenger acceptance concerns associated with doing this (because of the longer trip times). After all, as outlined in chapter 2, the quest by travelers for ever-higher speed comes as they try to satisfy their rising travel demand within a given travel time budget. Further, there would be increased emissions that affect surface air quality and noise at intermediate airports.

Another fuel-saving strategy would be to use larger aircraft, an approach that would reduce airspace congestion and the associated extra fuel use. However, as described in chapter 3, airlines compete on frequency to prevent long waiting times for their customers. Yet another way for saving fuel and thus reducing GHG emissions would be to substitute some air traffic with high-speed rail. One of our studies at MIT found that replacing two-thirds of all aircraft-based PKT by high-speed ground transportation within the ten planned U.S. high-speed rail corridors would result in a reduction in CO_2 emissions by total high-speed transportation (air plus high-speed rail traffic) by 1–2 percent. At the same time, however, the generation of electricity, required for operating high-speed rail services, could cause an increase in sulfur dioxide emissions from coal power plants.[24] Since it is unlikely that the potential of any of these measures can be fully exploited, the combined impact of these individual measures might realistically reduce aircraft energy intensity and CO_2 emissions per PKT by 5–10 percent by the mid 2020s.

Characteristics of Low Greenhouse Gas–Emission Aircraft

Table 5.1 summarizes the evolution of major historical and current determinants of aircraft energy use and the associated aircraft energy intensity. In addition, the likely future technology parameters, the passenger load factor, and the fuel burn benefit of air traffic management are shown.

We chose the Boeing 777 as our baseline aircraft. Introduced in 1995, it typically carries three hundred to four hundred passengers. Applying our projections from table 5.1 and the structural weight ratio to the Breguet equation, we determined that new aircraft energy use per seat kilometer is expected to decline by roughly 25–45 percent by the mid 2020s.[25] When the projected improvements in air traffic management

Table 5.1
Determinants of the energy intensity of new aircraft, past, present, and possible future

	1960	1970–1989	1990–2005	2020–2030
Cruise SFC, mg/s/N	25–30	17–22	15–17	12–14
M $(L/D)_{max}$	12–14	11–14	12–16	17–20
OEW/MTOW	0.45–0.6	0.45–0.6	0.45–0.6	0.45–0.5
Passenger load factor (pkm/skm), %	0.6	0.5–0.6	0.6–0.8	0.85
Net benefit of ATM, %	≈ 0	≈ 0	≈ 0	≈ 5
Aircraft energy intensity, MJ/pkm	5–6	1.5–3	1.5–2	0.8–1.4

Notes: Cruise SFC: cruise specific fuel consumption; $(L/D)_{max}$: maximum lift-to-drag ratio; OEW/MTOW: the ratio of operating empty weight and maximum takeoff weight; skm: seat kilometer; ATM: air traffic management.

are included, aircraft energy intensity declines by roughly 30–50 percent by the mid 2020s. This is equivalent to a 1.2–2.3 percent per year reduction in energy intensity, compared with an average rate of 3.2 percent per year over the past thirty years.

In addition to suggesting general historical trends, figure 5.1 shows projections for a typical future aircraft derived by comparing several factors: historical trends, technology assessments, and near-term technology introductions for engine SFC, L/D, structural weight ratio, passenger load factor, and operational efficiencies. For comparison, included in the figure are recently published projections by other researchers of future aircraft energy intensity. (NASA has also projected the fuel efficiency of 100-, 150-, 225-, 300-, and 600-seat aircraft individually for internal studies; the entire range in energy intensity of these projected aircraft is included here.) The figure also shows that the projected energy use per PKT of a blended-wing-body aircraft, the new Airbus A380 and the new Boeing 787, are in line with the various projections.

Aircraft fuel efficiency and retail price are closely related. As discussed in chapter 3, airlines have a choice between less expensive aircraft that have higher direct operating costs (and thus fuel burn) and more expensive aircraft that have lower operating costs and are more fuel efficient. The direct relationship between aircraft fuel efficiency and price can be observed from historical data. As shown in figure 5.3, the price per seat of new aircraft has increased from an average of about $200,000 per seat in 1960 to about $350,000 in the mid 1990s.[26] Compared with short-range aircraft, the price per seat of long-range aircraft is slightly higher, with list prices for the new Airbus A380 and Boeing 787 roughly matching the historical trend of long-range aircraft. (Note that list prices are almost always higher than the sales prices, which include discounts.) Extrapolating this historical trend leads to a price of roughly $500,000 per seat for a new aircraft in the mid-2020s and a 25–45 percent reduction in energy intensity. If history is a reliable guide for the future, projected aircraft energy intensity and price characterstics will satisfy market conditions—that is, discounted savings in operating costs will more than compensate for an increase in aircraft price.

Summary

Aircraft emit a variety of GHGes, of which the non-CO_2 emissions may account for a similar climate impact as CO_2 alone. Given the scientific

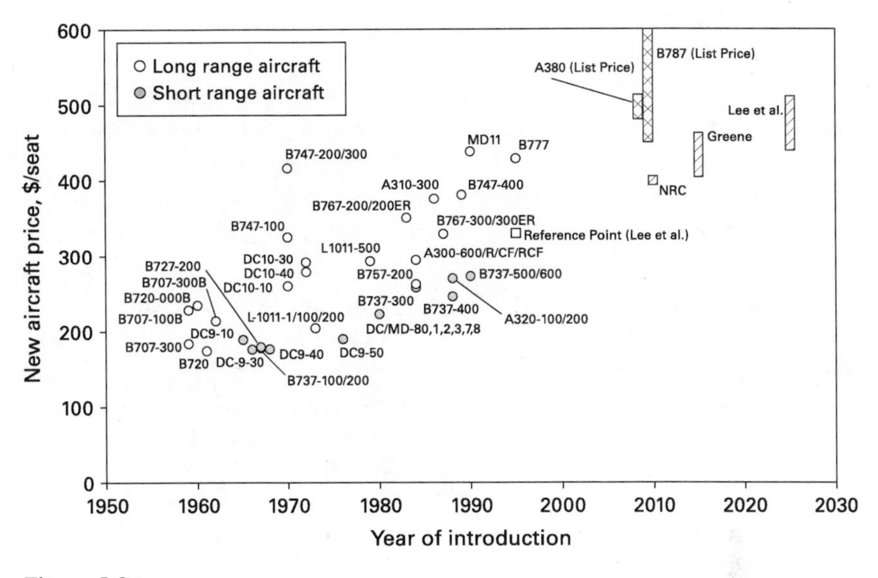

Figure 5.3
New aircraft price per seat, historical development and projections. Source: Lee, J.J., S.P. Lukachko, I.A. Waitz, A. Schäfer, 2001. Historical and Future Trends in Aircraft Performance, Cost, and Emissions, *Annual Review of Energy and the Environment 2001*, 26: 167–200.

uncertainty of these gases' climate impact, our focus was on CO_2 emissions.

Since airlines strive for minimum direct operating costs, of which fuel costs account for a significant share, the experience of the industry has been that the energy intensity and thus CO_2 emissions per PKT of commercial passenger aircraft have declined strongly over time. Between 1970 and 2005, energy intensity and CO_2 emissions per PKT of the U.S. aircraft fleet declined by 70 percent, an average reduction of 3.5 percent per year.

Over the subsequent three decades, new reductions in energy intensity and CO_2 emissions per PKT can be expected. However, because the required more sophisticated technology measures generate increasingly stringent technology, cost, and environmental trade-offs, we project the future decline to be at about half the historical rate. By the mid 2020s, therefore, CO_2 emissions per PKT of the average new commercial passenger aircraft could be 30–50 percent below current (early 2000) fleet emissions. Given market growth, the withdrawal of older aircraft for freight conversion, and natural fleet turnover, this reduction could be experienced on a fleet level by 2050.

In contrast to the historical decline, which was dominated by improvements in engine efficiency, future reductions are expected to be more balanced among all determinants of aircraft energy use. Additional significant reductions in CO_2 emissions per PKT would be possible, for example, through a departure from conventional tube-wing designs, in particular through blended-wing-body aircraft. However, given the high development costs and significant risk aircraft manufacturers already experience when developing new models from proven designs, radically different designs are unlikely to be introduced in the next decades unless substantially supported by government policy. Alternative fuels could provide an additional degree of freedom for reducing CO_2 emissions from aircraft. This option is studied in the following chapter.

6

Alternative Transportation Fuels

In chapters 4 and 5, we analyze the fuel-efficiency improvement potential for light-duty vehicles (LDVs) and aircraft. We concluded that over the next two decades, technologies could be developed that cut fuel consumption and carbon dioxide (CO_2) emissions for new ground and air vehicles by up to half of current levels, while keeping consumer attributes relatively unchanged.

In this chapter we examine an additional approach to reducing GHG emissions: the opportunities provided by alternative transportation fuels—fuels derived from nonpetroleum sources that have low carbon emissions. If such fuels were introduced on a very large scale, they could eventually contribute significantly to providing low carbon-emitting mobility.

Table 6.1 shows the main characteristics of petroleum-based transportation fuels and their potential substitutes. The fuels are ranked in a sequence of declining carbon-to-hydrogen ratios, determined by counting each fuel's carbon and hydrogen atoms. Together with a fuel's mass-specific heating value, the carbon-to-hydrogen ratio indicates how much CO_2 is formed per unit of fuel energy released when fully burned under stoichiometric conditions (when the amount of air provided to the combustion process is just enough to fully oxidize each of the fuel's carbon and hydrogen atoms under ideal conditions).

As shown by the fourth data column, the three oil-based fuels and the alcohols release about 70–75 grams of CO_2 per million joules (MJ). All other fuels in table 6.1 release a lesser amount. Burning one MJ of natural gas forms only three-quarters of the amount of CO_2 oil products release, while burning one MJ of hydrogen releases no CO_2. However, since hydrogen has to be produced from other fuels and substances, the CO_2 emissions of all energy inputs required to produce the hydrogen need to be accounted for.

Table 6.1
Transport fuel characteristics

	C/H ratio	Net energy content		Direct CO_2 emissions	Fuel density	Octane number	Stoichiometric air/fuel ratio	Flammability limits in air
	–	MJ/L	MJ/kg	gCO_2/MJ	g/L		kg_{Air}/kg_{Fuel}	Fuel/Air Vol (%)
Oil products								
Diesel	0.55	35.5	42.5	74.8	835	—	14.5	0.6–7.5
Jet fuel	0.55	34.7	42.8	74.3	811	—	14.5	0.7–8
Gasoline	0.54	32.2	43.5	73.0	740	87–93	14.8	0.6–8
Alcohol fuels								
Ethanol	0.33	21.3	26.8	71.4	795	103	9.0	3.5–16
Methanol	0.25	15.6	19.7	69.8	790	104	6.5	5.5–26
Methane								
CNG (248 bar)	0.25	10.4	50.0	55.0	208	120	17.2	5–15
LNG	0.25	21.0	50.0	55.0	420	120	17.2	5–15
Hydrogen								
CH_2 (345 bar)	0.00	2.8	120	0.0	23.0	60	34.3	4–77
LH_2 (20 K)	0.00	8.5	120	0.0	70.8	60	34.3	4–77
Electricity	—	0.6	0.3–0.5	0.0	—	—	—	—

Notes: The C/H ratio and calculation of CO_2 emissions in grams of carbon per MJ of energy are based on the following fuel composition: jet fuel and light diesel $C_{12}H_{22}$ and gasoline C_7H_{13}. The chemical composition of alcohols is C_2H_5OH (ethanol) and CH_3OH (methanol). Natural gas is assumed to be methane (CH_4). In reality, 80–95 percent by volume consists of methane with minor shares of heavier gaseous components such as ethane, propane, and butane. The energy content of electricity reflects the electrical energy storage characteristics of lithium ion batteries. The octane number describes a fuel's resistance to spontaneous ignition ("knock"); higher octane numbers allow higher engine compression ratios, which generally result in higher engine efficiencies. The stoichiometric air-fuel ratio indicates the mass of air relative to the mass of fuel to achieve complete combustion. Fuels with a smaller stoichiometric air-fuel ratio require more fuel for a given fuel-air mixture energy. They thus require a larger fuel tank for the same driving range. The stoichiometric air-fuel ratio is lowest for alcohols, because these fuels are already partially oxidized. The flammability limits in air indicate the range of the air-fuel ratio relative to the stoichiometric air-fuel ratio, within which the air-fuel mixture can be ignited and thus the engine operated. Wider flammability levels allow leaner and richer engine operation. However, low flammability limits also increase the explosion hazards of fuel leaks.

A similar situation applies to alcohols, such as ethanol and methanol. Although burning one MJ of these fuels releases about the same amount of CO_2 as oil products, they can be produced from renewable biomass, which—under ideal conditions—offsets their GHG emissions. An important implication of these tabulations is that a careful environmental assessment of alternative fuels can be done only on a lifecycle basis, in which all fuel-conversion stages are included, from resource extraction to consumption by the end user. In the following sections, we take these lifecycle aspects into account.

An important factor that helps explain our society's "addiction to oil" is that oil products are liquid at atmospheric temperature and pressure and thus are convenient: that is, they lend themselves to rapid and simple fueling processes and easy storage, and they have outstanding weight-volume characteristics necessary for a long vehicle range. As table 6.1 indicates, oil products store the largest amount of energy per unit weight of all liquid fuels, that is, more than 40 MJ/kg, thus imposing the lowest weight penalty on a vehicle. Oil products also have the highest energy content per unit of fuel volume—more than 30 MJ/L—and thus store a given amount of energy within the least amount of space. Their favorable weight-volume characteristics are especially significant for aircraft, because they translate into shorter refueling times and less fuel needing to be lifted. A fuel having a low energy content per volume would require larger storage tanks, which results in increased drag owing to the increased wetted area (surface area contributing to skin friction). A fuel having a low energy content per mass would require additional fuel weight to fly a given range, which results in a requirement for more lift, and thus additional structure, engines, landing gear, and so forth. And this in turn would lead to still more fuel weight being required to fly the mission. Hence, a strong requirement for aviation fuels is low volume and mass per unit of energy.

In contrast to liquids, all gaseous transportation fuels have a low volumetric energy density, even when compressed to high pressures. This attribute requires that these fuels take up storage volumes several times larger than oil products to contain the same amount of energy. As can be seen in the volumetric energy densities in table 6.1, fuel volume alone would be three times larger for CNG than for oil products, and twelve times larger for compressed hydrogen. In addition to the loss of space, the associated high storage pressures required by these fuels also imply a loss of energy to doing the necessary compression and a need for heavier and much more expensive fuel tanks. Gaseous fuels also require longer refueling times and constrain some common road vehicle usage patterns, such as parking in closed public spaces, where minor gas leaks can lead to fire and explosion hazards. The ease with which these fuels burn is especially apparent from hydrogen's wide flammability limits in air.

Although storage volumes could be significantly reduced if natural gas or hydrogen were liquefied, the associated energy requirements would be further increased and the required cryogenic storage tanks would add further complexity and costs. For commercial aircraft, the use of (liquefied) gaseous fuels would require complete redesign of the aircraft. Such designs are unlikely to enter the aircraft fleet on a significant scale at least over the next several decades, and we thus exclude them from our analysis.[1]

Because of oil products' favorable transportation-related characteristics, since the beginning of the twentieth century, scientists have tried to convert abundant, low-value hydrocarbon fuels into this "black gold." In 1913, Friedrich Bergius demonstrated a process that directly converts coal and hydrogen into oil products. A decade later, Franz Fischer and Hans Tropsch produced oil products from coal using an indirect process. Starting with synthesis gas (a mixture of carbon monoxide and hydrogen, here derived from gasified coal and steam), their process rearranges carbon and hydrogen atoms (using appropriate catalysts) to generate a range of hydrocarbon products, including naphtha (a gasoline feedstock), diesel fuel, and kerosene. The process presents various opportunities to direct the conversion process so that these synthetic fuels can be tailored to produce high-quality oil products. To overcome supply shortages of oil-derived fuels during World War II, chemical plants that either directly or indirectly converted coal to liquid fuels were operated by several nations, including England, Germany, and Japan. The conversion of coal to gasoline was later pursued in South Africa, originally as a way to use the country's large coal resources to reduce its dependence on oil imports, and later to overcome the trade embargo imposed on it during the period of apartheid.

The dominance of oil products in transportation has also resulted from their comparatively low cost. At a long-term average crude oil price of some $20 per barrel, nearly all alternative fuels are more expensive to produce than products from conventional crude.[2] Their higher costs make alternative fuels inherently less attractive to consumers. In many cases, only a drastic rise in oil price combined with substantial government subsidies for alternative fuels could convince automobile owners to purchase alternative fuel vehicles. Also, as shown by various national programs that have promoted compressed natural gas as a transportation fuel, vehicle owners start to shift back to oil products as soon as subsidies are removed or the price of oil drops.

Over the course of the last century, trillions of dollars have been invested in expanding the exploration, production, refining, transmission, distribution, and storage infrastructure of oil and its products. This effort has become ever more sophisticated and has resulted in a transportation fuel supply system of enormous scale. This very scale has several implications. Perhaps most important, any alternative fuel that is compatible with the existing infrastructure has a clear advantage over less complementary competitors. Thus, efforts to supplement conventional crude oil extraction with more unconventional carbon-intensive resources in order to produce synthetic light oil products and to ensure secure fuel supplies have a very strong driver. Not surprisingly, several countries have already begun to convert carbon-intensive unconventional oil reserves into oil products, despite significant challenges associated with their extraction, transport, and processing into fuel.

Another consequence of the large scale of oil's infrastructure is that change can be only gradual. Cesare Marchetti and Nebosja Nakićenović at the International Institute for Applied Systems Analysis have shown that in the past, it has taken about one century for a new energy carrier to reach its peak market share.[3] Thus, unless pushed in other directions by very effective policies, we will have to live with products from conventional and unconventional oil for a very long time.

Finally, the scale of the existing oil-based distribution and storage infrastructure gives a sense of the enormous supply challenge alternative fuels face if they are to replace oil products. This all-encompassing scale also emphasizes the need for lasting government policies that will induce and maintain significant changes in the fuel mix. As we discuss below, only a few of our current low-carbon fuels have the potential to substitute for oil products completely.

Despite these challenges, there are some success stories. The ethanol economy in Brazil is perhaps the best-known example of a continuous large-scale use of a petroleum alternative. Taking advantage of favorable climatic conditions, the Portuguese conquistadores set up the Brazilian sugar industry five hundred years ago. Sugarcane-derived ethanol was first used as a transportation fuel in 1903. (Henry Ford's Model T was capable of running on 100 percent ethanol fuel.) As a response to two problems, higher fuel prices in the aftermath of the first oil crisis and the declining world market price for sugar, the Brazilian government set up the ProÁlcool program in 1975. This program encouraged the sugar industry to convert sugarcane into ethanol on an unprecedented scale.[4] Over the next 15 years, total investments in the ethanol program amounted to more than $13 billion.[5] The ProÁlcool program gained momentum, and the number of vehicles operating on pure ethanol rose steadily, accounting for 95 percent of annual new vehicle registrations by the end of the 1980s.

However, the combination of declining oil prices after 1985, increasing sugar prices, and the removal of subsidies, caused ethanol production to stagnate. Thus, supplying the large fleet of neat ethanol vehicles proved to be impossible, and drivers lost confidence in the fuel. Nonetheless, the government remained committed to the ethanol program and subsequently mandated gasoline blends having a minimum percentage of ethanol, currently 25 percent by volume. As this book goes to press, ethanol consumption in Brazil is on the rise again, accounting for more than half of all LDV fuels.[6]

In addition to becoming less dependent on oil imports and reducing the country's vulnerability to oil price increases, indigenous ethanol production has improved Brazil's trade balance. Based on a cumulative consumption of nearly 300 billion liters (80 billion gallons) of ethanol through 2005, and gasoline price increases since 1975, the associated hard currency savings from displacing gasoline are in excess of $70 billion.[7]

Biofuel use is also on the rise in many other countries. Between 1980 and 2000, U.S. ethanol production rose tenfold, mainly as a result of fiscal and regulatory policies introduced by the federal government both to reduce oil dependence and to improve urban air quality.[8] More recently, because of renewed U.S. energy security concerns, ethanol production has increased strongly, tripling to 5 billion gallons between 2000 and 2006, causing the United States to surpass Brazil as the world's ethanol capital.[9]

Ethanol production is expected to increase further as a result of subsidies and the 2005 Renewable Fuel Standard, which requires increasing the use of biofuels to 28 billion liters (7.5 billion gallons) by 2012. This biofuel target will likely be exceeded, because production capacity was in excess of 23 billion liters (6 billion gallons) in 2006, existing plants are being expanded, and new plants are under construction. The added capacity of these new plants will be of about the same order as current production. Thus, while ethanol currently accounts for an average of some 3 percent of U.S. motor vehicle fuel use (by volume), satisfying the 2012 target would increase its volume share to 5–6 percent.

In Europe, the major producer of biodiesel, a directive adopted by the European Parliament and the European Council in 2003 recommends that biofuels supply at least 2 percent of the total volume of all transport fuels in 2005 and 5.75 percent by 2010.[10] Although the 2005 target has not been achieved, biofuel production is rising strongly.[11] In early 2007, EU ministers agreed on a biofuel target of 10 percent by 2020 as a way of reducing both GHG emissions and oil dependency.[12] However, increased concerns about the wider societal impacts of biofuels, including the growing competition for land resulting in rising food prices and the conversion of important ecosystems into biofuel plantations, has caused the EU and some European member state governments to rethink these targets and the underlying subsidy system.[13] Meanwhile, programs to introduce ethanol fuel are developing in many other parts of the world, including China, Colombia, India, and Thailand.

In the following sections, we evaluate most of the alternative fuels listed in table 6.1 according to the criteria discussed above. We begin by examining the fuel cycle, which includes all phases of fuel production and distribution, from resource extraction to vehicle fueling. In addition to fuel-cycle energy use and GHG emissions, we also examine the challenges of cost, convenience, infrastructure compatibility, resource availability, and "time to market" impact. Finally, to complete our life-cycle assessment, we combine the fuel-cycle and a well-to-vehicle tank assessment with the vehicle usage characteristics discussed in chapters 4 and 5 (the tank-to-wheel and tank-to-wake energy use, respectively).

Alternative Fuel Characteristics

While numerous pathways exist for producing a wide range of alternative transportation fuels, we focus on the most promising candidates: biomass-derived ethanol, biodiesel, biomass-derived synthetic oil products, compressed natural gas, hydrogen, and electricity. A discussion of these potentially lower-carbon alternatives, however, would not represent a complete picture without also considering synthetic oil products from fossil fuel feedstocks, including coal, unconventional oil, and natural gas. As a reference point for these alternative fuel cycles, we begin with light crude oil.

Products from Crude Oil

Crude oil consists of a large number of different hydrocarbon molecules. It is classified as light, medium, or heavy depending on the relative proportions of its components.[14] Light crude oil, owing to its lower carbon-to-hydrogen ratio and molecular weight, flows naturally. Thus, it can be pumped out of a well without any prior treatment and can be readily transported to a refinery via pipeline or ship.

At the refinery, light oil is first separated—refined—into its natural hydrocarbon components in a series of distillation steps. The specific distribution of the distilled product streams corresponds to the chemical characteristics of any particular light oil; but in general, light oil contains about a 20 percent share of naphtha, a 30–45 percent share of the middle distillates kerosene and diesel, and a 35–50 percent share of residue, which is a mixture of low-value carbon-intensive compounds that include heavy oil and bitumen (a carbon-rich substance that has a high resistance to flow).

Since the gasoline share of the distillates is too small by itself to supply the global fleet of gasoline-engine vehicles, and the proportion of residues is more than enough to serve other markets (marine shipping, road paving, and the steel industry), refiners change the natural mix of distillates toward lighter products. They do this through an upgrading process. The longer carbon chains of the heavy oil components are broken down into shorter ones (cracked), which are then rearranged to yield the desired mix of products. In a subsequent step, hydrogen is added to improve the quality of the shorter chains. In the United States, the product streams leaving these refinery "cracker" units typically contain a gasoline proportion of some 50 percent of total yield.

In this way, total gasoline production is about twice crude oil's natural share. Another 45 percent of refinery outputs are middle-distillates, which include diesel, jet fuel, heating oil, and similar products. The remaining share of the total output consists of residues. In a final step, the structure of the hydrocarbon molecules may be again reformed to satisfy the specifications of each final product—such as a high-octane number gasoline for spark-ignition engines—and each product is then purified. After this refinery processing, the products are usually shipped to distribution centers, from which they are trucked to service stations.

Following the entire path of this processing and distribution process, the total fuel-cycle energy use from resource extraction to delivery at retail stations amounts to about 0.18 $MJ/MJ_{Gasoline}$ and 0.13 MJ/MJ_{Diesel}.[15] (For the purpose of this study, the similar composition of diesel and jet fuel suggested we use diesel fuel as a surrogate for aircraft fuels.[16]) Extracting, refining, distributing, and storing 1 MJ of oil products typically has required an energy input of 13 percent of its energy content for diesel fuel and 18 percent for gasoline. (Because of its greater need for upgrading, gasoline's energy input is higher.) Fuel-cycle emissions associated with these energy inputs result in 9.5 gCO_2/MJ_{Diesel} and 13.2 $gCO_2/MJ_{Gasoline}$. When we also take into account the release of methane during oil extraction—about 0.05 gram per MJ of crude—total fuel-cycle GHG emissions increase to 10.7 grams CO_2 equivalent (gCO_2-eq) per MJ_{Diesel} (or per $MJ_{Jet\ Fuel}$) and 14.4 gCO_2-eq/$MJ_{Gasoline}$.[17] Nearly all other fuels require higher levels of fuel-cycle energy inputs, and only a few alternative fuels release lower levels of fuel-cycle GHG emissions.

Products from Unconventional Oil

While conventional crude accounts for nearly all oil products produced today, a small but growing fraction is derived synthetically from "unconventional oil." According to one estimate, unconventional oil accounted for about 12 percent of crude oil produced in 2000.[18] Some unconventional oil deposits have not yet been added to crude oil reserve estimates because, until recently, they have been unprofitable to exploit and process on a large scale, given extended periods of low oil prices over the past fifty years. However, in light of rising oil prices in excess of $100 per barrel in the early twenty-first century, many of these resources have become profitable and thus part of the oil supply equation.

Unconventional oil can be divided into two groups. Natural bitumen and heavy oils share the same formation history as conventional crude oil. However, these deposits were degraded through bacteria attacks and erosion, and thus lost the light components over the course of millions of years.[19] Thus, as with the upgrading process in a refinery, the long carbon chains of these substances are broken down into shorter ones.

In contrast to these "over-mature" oils, oil shale (a mixture of calcium carbonate and clay) contains a hydrocarbon compound, kerogen, which is a precursor of crude oil. Given favorable conditions, that is, sufficient reservoir depth to ensure high temperatures and pressures, the "immature" kerogen would convert naturally over the course of millions of years. Thus, all efforts to process kerogen into shale oil attempt to accelerate the geological clock by providing these conversion conditions. Currently, programs worldwide are investing in all of these hydrocarbon sources, such as Canada and Venezuela in bitumen, and Brazil and China in shale oil.

Common to all forms of unconventional oil is their resistance to flow and the associated effort of getting the feedstock out of the ground (a result of the high carbon-to-hydrogen ratio). At a temperature of 20°C, the flow characteristics of bitumen are between those of peanut butter and chocolate. Under these conditions, the top layers of bitumen reservoirs can be surface-mined like coal. However, in lower reservoir depths, the high viscosity requires more sophisticated methods of extraction, typically steam injection: just as hot, melted chocolate flows more easily than cold, so heated bitumen becomes less viscous. In addition to the more complex resource extraction, unconventional oils' high viscosities also require more expensive transport to refineries.

While temperatures are well below 20°C in the Canadian bitumen reservoirs, higher reservoir temperatures, in excess of 50°C, reduce the viscosity of bitumen in Venezuela and thus allow up to 10 percent of the resources to be extracted through "cold" production. (Because of higher reservoir temperatures and the associated lower resistance to flow, Venezuela's bitumen is generally referred to as "extra-heavy oil").

In contrast, solid oil shale rock is mined like coal, both underground and from the surface. Underground, or in situ, oil shale production methods have been tested, but they come with significant challenges. Similar in principle to the hot extraction of bitumen from deeper reservoirs, this strategy would try to provide a sufficient amount of process heat in a number of ways: through controlled nuclear underground explosions, underground combustion, or more recently, electric heating elements submersed in bore holes.[20] However, in situ retorting is still in the pilot stage—and it raises a range of additional environmental concerns, including groundwater contamination.

A major drawback associated with mined feedstocks is the need to separate hydrocarbon compounds from sands and clay or source rock and thus the enormous amounts of feedstock material required to produce the oil. To producing one barrel of synthetic crude, which weighs about 140 kilograms (about 300 lbs), requires roughly 2,000 kilograms (about 4,400 lbs) of oil sands or oil shale.[21] Thus, a refinery that produces 100,000 barrels of synthetic crude oil per day must move 200,000 (metric) tons of ore to the refinery per day—and almost that entire amount back from the refinery to landfills. The transport requirements away from the processing plant are even greater for oil shale, because the finely crushed source rock expands during retorting. In addition to the ore transport, the production process is water intensive, and total requirements are roughly three barrels of water per barrel of bitumen or shale oil produced.[22] Unconventional oils also contain comparatively large amounts of nitrogen, oxygen, sulfur, and heavy metals, which require special treatment, separation, and handling facilities at refineries.

Because of these undesirable characteristics, synthetic crude oil is more challenging, more energy consuming, and thus more costly and GHG-emission-intensive to produce than crude oil directly extracted from the well. According to field data from the Canadian oil sand program, the average energy use for extracting and upgrading the bitumen corresponds to one-quarter of the synthetic crude oil's energy content, or 0.25 MJ per MJ of synthetic crude oil.[23] Combining oil refining and product distribution, fuel-cycle energy use is roughly 0.40 MJ per MJ of oil product. (Because energy inputs are higher for upgrading the synthetic crude oil to gasoline, fuel-cycle energy use results in 0.45 MJ per $MJ_{Gasoline}$, compared with about 0.38 MJ per MJ_{Diesel} or $MJ_{Jet\ Fuel}$).

Thus, to unlock one unit of energy from oil sands, convert it into oil products, and transport the products to a retail station, an additional 40 percent of that unit of energy is required—an energy input that compares to only 13–18 percent for the fuel-cycle energy use of diesel and gasoline from light crude oil. Given that most of the energy input is supplied by natural gas (to deliver process heat, generate on-site electricity, and provide hydrogen atoms to increase the fuel quality), fuel-cycle GHG emissions result in about 26 $gCO_2\text{-eq}/MJ_{Diesel}$ (or per $MJ_{Jet\ Fuel}$) and 30 $gCO_2\text{-eq}/MJ_{Gasoline}$. These amounts, which include methane emissions from the oil sand operations, are twice those of the fuel cycle of conventional crude oil.[24]

Future levels of fuel-cycle GHG emissions are uncertain. Significant reductions have already been achieved through technological improvements. Between 1990 and 2005, GHG emissions per barrel of synthetic crude declined by 40 percent. Innovations in underground bitumen extraction, such as the use of solvents instead of steam, may continue that trend. However, the possible substitution of lower-cost but more carbon-intensive refining residues for natural gas, together with rising requirements for higher-quality synthetic fuels, may offset benefits from technological innovations in the future.

Fuel-cycle energy use and GHG emissions are higher for shale oil. During the retorting process, in which the shale oil is produced under elevated temperatures, the oil shale's calcium carbonates and magnesium carbonates are reduced to CO_2, which adds to the fuel-cycle GHG emissions. Based on the composition of Estonian shale oil, this CO_2 accounts for 18 percent of total fuel-cycle CO_2 emissions.[25] Assuming a retorting efficiency of 70 percent, which is representative of the U.S. shale oil program in the 1980s, fuel-cycle GHG emissions result in excess of 80 gCO_2/MJ—or six to eight times the level of petroleum-derived gasoline and diesel, respectively.[26] (If a conversion efficiency of 80 percent were assumed, fuel-cycle GHG emissions still would be four to six times those of gasoline and diesel.)

Given the difficulties in extraction, transportation, and processing of unconventional oil, what drives the oil companies to pursue its conversion? At least two reasons explain the companies' rising interest. One is location. While three out of four barrels of conventional crude oil reserves are controlled by the Organization of Petroleum Exporting Countries (OPEC)—and 93 percent of these reserves are located outside the industrialized world—most of the world's known unconventional oil reserves are located within the Western Hemisphere. As was shown in table 2.2, more than 50 percent of the world's known heavy oil deposits and around 85 percent of the world's natural bitumen deposits are concentrated in North and South America. In addition, about three out of four barrels of world oil shale resources are located in the United States.

A similarly important motivation that helps explain the push for unconventional oil is the recent rise in the world oil price, a condition that increases the profitability of synthetic crude oil. Although the price of oil may decline again as a result of mainly increasing production capacity, industry expectations are that the declining resource base of conventional oil necessarily will lead to rising prices over the long term, and thus the competitiveness and profitability of unconventional oil will increase. Already, producing synthetic crude from bitumen trapped in Canadian oil sands is claimed to be competitive at crude oil prices of about \$20–30 per barrel.[27]

In contrast, estimates of initial production costs for producing synthetic crude from oil shale are three times as high.[28] These costs probably would increase further, since such GHG-intensive facilities would likely need to operate with carbon capture and storage technology, at least within the industrialized world. Despite the high cost of converting oil shale to synthetic fuels, the abundance of oil shale resources has always been considered too large to ignore, driving ingenious efforts to improve the extraction of kerogen from the source rock. However, while the conversion of heavy oil and bitumen to synthetic crude oil is already under way and is likely to expand in scale, the near-to-medium-term future of shale oil conversion appears to be less promising. Over the long term, however, extraction and conversion technologies are likely to advance. The question is not whether, but at what rate and how far.

Oil Products from Coal and Natural Gas

By the end of World War II, Germany was operating nine indirect and eighteen direct coal liquefaction plants. They produced 4 million tons of oil products per year, satisfying 90 percent of the country's total fuel needs.[29] At the same time, coal-to-liquids (CTL) plants were also operating in England and Japan. South Africa developed CTL plants in the 1950s and intensified their development and operation through the 1970s and 1980s, during the period of the trade embargo. Currently, three plants in South Africa produce a maximum of 10 million tons of synthetic fuel per year, meeting more than 40 percent of total domestic fuel demand.[30] At Johannesburg's airport, blends of coal-derived and petroleum-based jet fuel have been supplied to international flights for many years, and in 2008 use of pure coal-based fuels was approved. Other countries, most notably China, are planning to invest in coal liquefaction to reduce oil dependence.

Compared with the long history of CTL conversion, the gas-to-liquids (GTL) experience is significantly shorter. The first commercial-scale GTL plant was built by Shell in Malaysia in 1993. Today, several plants are operating in the remote areas of the world where "stranded" natural gas, natural gas that is abandoned or burned off because there is no economical way to transport it, would require enormous investments to be moved via pipeline or to be liquefied and shipped to markets.

The reasons for converting coal and natural gas into oil products are identical to those for converting unconventional oil: an expected long-term increase in the price of conventional crude oil, the compatibility of the products to the existing oil infrastructure, and the abundance and concentration of large resources outside today's major oil-producing countries. Coal reserves in the United States alone account for more than 7,000 billion billion joules (or 7,000 exajoules, EJ), or 27 percent of the world's total reserves.[31] If converted into oil products, it would correspond to more than 3,500 EJ, or about half of proven light oil reserves (see table 2.2). Even if coal production were to continue at its current rate for the next one hundred years, the remaining reserves would still allow more than 2,000 EJ of liquid fuels to be produced. (This corresponds to 30 percent of today's global proven crude oil reserves.)

Although not as abundant as coal, the global proven reserves of natural gas are comparable to those of crude oil. However, stranded natural gas accounts for about half of all gas reserves.[32] Thus, converting these deposits into liquid fuels would increase proven oil reserves by one-quarter. The conversion of coal and natural gas into synthetic oil products would thus significantly extend the lifetime of conventional petroleum.

The conversion of coal, natural gas, or any other hydrocarbon feedstock to liquid fuels begins with breaking down the feedstocks' hydrocarbon molecules into hydrogen and carbon monoxide. This decomposition is accomplished by subjecting the feedstock to high temperatures and pressures in the presence of steam and a catalyst, and under a controlled supply of oxygen. The synthesis gas that results from this process is cleaned of impurities and then subjected to additional processing. This step, Fischer-Tropsch synthesis, rearranges the molecules to generate a spectrum of hydrocarbon fuels, including naphtha, kerosene, diesel fuel, and higher blending materials such as waxes.

In addition to being a proven technology, this indirect process produces a high-quality fuel. For example, due to the lower amount of aromatics, the synthetic oil product has a slightly higher heating value than the petroleum-derived counterpart (a characteristic that is not shown in table 6.1). Similarly, the removal of nitrogen, sulfur, and particulates from the synthesis gas makes it possible to generate cleaner-burning synthetic oil products—which in turn release lower levels of criteria pollutants (the six standard pollutants regulated by the U.S. Environmental Protection Agency).[33] This advantage is especially important for diesel engines, given their difficulties in controlling particulates (see chapter 4).

An alternative to the indirect process is converting coal to oil products directly, without prior gasification. In this direct liquefaction route, coal is mixed with a solvent to form a slurry. The slurry is exposed to hydrogen to reduce the carbon-to-hydrogen ratio and remove undesired substances at elevated temperatures and pressures. The product streams then need to be converted into lighter products in a refinery.

In contrast to the indirect conversion path, however, modern direct coal liquefaction can use only selected coals as feedstock, offers lower fuel design flexibility, and has not yet been proved at commercial scale. Yet, its chemical composition would make the product a suitable feedstock for gasoline. Current direct liquefaction technology converts coal to refined liquid fuels at higher efficiencies of 60–70 percent. Because the indirect technology path separates the feedstock into its molecular building blocks, efficiencies are only 50 percent for coal and about 60 percent for natural gas.[34]

The associated GHG emissions mainly depend on the respective feedstock's carbon content. Coal has an average carbon-emission factor of about 26 gC/MJ. Taking into consideration the conversion efficiency of the indirect process (and taking into account methane emissions from coal mining of 0.1 g per MJ of coal), the resulting fuel-cycle GHG emissions for coal- and gas-converted synthetic diesel and jet fuels are about 115 gCO_2-eq/MJ, an amount as much as ten times that from oil products produced from conventional crude. These emissions can be reduced by up to 90 percent—and thus would be comparable to those from conventional oil products—by adding a carbon capture and storage facility. However, higher levels of fuel-cycle GHG emissions are not the only environmental concern of increased coal use: additional impacts include groundwater contamination and land subsidence.

Because of natural gas's lower carbon content and the higher conversion efficiency of gas refineries, fuel-cycle GHG emissions of GTL processes are only 20 percent of those from coal-based processes without carbon capture and storage. Yet, GHG emissions of GTL processes are still roughly twice as high as those of a conventional crude oil fuel cycle.[35]

Production costs of synthetic crude oil are $20–$30 per barrel for GTL, are projected to be more than $30 per barrel for the direct coal liquefaction process, and are $50–$65 per barrel for the indirect process, depending on feedstock costs and other factors.[36] The nearly complete removal and disposal of CO_2 emissions is projected to add another $5 per barrel.[37] Although these costs do not appear to be especially high given an oil price in excess of $100 per barrel in the early twenty-first century, building a CTL facility bears substantial risk. Coal refineries are very capital intensive and have long lead times from their planning stage to beginning operations. Thus, an entire investment of hundreds of million dollars—the amount needed to build a conversion plant—would be at risk should the oil price drop substantially.

Such investment costs are roughly half for GTL. Given the enormous amount of stranded gas reserves and a crude oil price that may increase still more over the next thirty to fifty years, natural gas liquids are likely to play an increasingly strong role in the future—with the expense of higher fuel-cycle GHG emissions.

While the expansion of GTL production is expected to take place, the future contribution of coal liquefaction to the transportation fuel market is less certain. The most recent plant in South Africa began operation in 1982, and no further coal refinery has been built anywhere since. However, in China and other rapidly developing economies with little oil but large coal reserves, coal liquefaction may play an increasingly important role in the future. In the industrialized world, however, such a pathway would likely be prohibited because of the large GHG emissions that would be released—unless carbon capture and storage techniques are used.

Table 6.2 summarizes the fuel-cycle characteristics of gasoline and diesel/jet fuel for the hydrocarbon feedstocks discussed above. All synthetic gasoline processes require higher energy inputs and release larger amounts of GHG emissions than the crude oil–based process. The increase in fuel-cycle GHG emissions ranges from a doubling—associated with the production of gasoline from oil sands—to a roughly tenfold increase, which would result from coal liquefaction without emissions treatment. Considering these fuels on a life-cycle basis, which additionally takes into account CO_2 emissions released during fuel combustion, increases in fuel-cycle GHG emissions are mitigated. For oil sands–derived fuel, the increase is about 20 percent, while for fuel from coal (without emissions treatment), it is about 120 percent; and the increase for oil shale is about halfway between these two.

CO_2 capture and storage can reduce life-cycle GHG emissions dramatically. In the case of coal, GHG emissions can be reduced to about the level of emissions from crude oil–derived diesel; a similar mitigation potential applies to oil shale and oil sands. Table 6.2 also shows that the economic competitiveness of synthetic oil products strongly depends on the price of crude oil, which is assumed here to range between \$50 and \$100 per barrel.[38]

To provide a better understanding of the infrastructure requirements and land-use implications of large-scale production of liquid fuels from unconventional oil, coal, and gas, table 6.2 also reports the required

Table 6.2
Typical characteristics of synthetic oil products from various feedstocks

	Gasoline			Diesel/jet fuel			
	Crude oil	Oil sands	Oil shale	Crude oil	Coal w/o CCS[a]	Coal w CCS	[b]Natural gas
Fuel-cycle energy use, $MJ/MJ_{Product}$	0.18	0.45	0.65	0.13	1.02	1.10	0.68
Fuel-cycle GHG-Em., gCO_2-eq/$MJ_{Product}$	14	30	85	11	115	16	25
Life-cycle GHG-Em., gCO_2-eq/$MJ_{Product}$	87	103	158	86	190	91	100
Relative, oil product = 100	100	118	182	100	221	106	116
Costs of (synthetic) crude oil, $/bbl	50–100	20–30	60–85	50–100	50–65	55–70	20–30
Costs of supplied fuel (untaxed)							
$/L	0.28–0.78	0.28–0.31	0.53–0.68	0.28–0.78	0.34–0.46	0.38–0.49	0.18–0.24
$/gal	1.0–3.0	1.0–1.2	2.0–2.6	1.0–3.0	1.3–1.7	1.4–1.9	0.7–0.9
Scale impacts of supplying 25% LDV[c] fleet in 2005							
Increase in production over 2005 level, %	—	280	—	—	32	33	32
Increase in land use over 2005, mill. ha[d]/yr	—	0.004	0.001	—	0.015	0.016	—

Sources: Ansolabehere, S., J. Beer, J. Deutch, D. Ellerman, J. Friedmann, H. Herzog, D. Jacoby, G. McRae, R. Lester, J. Moniz, E. Steinfeld, J. Katzer, 2007. *The Future of Coal: An Interdisciplinary MIT Study*, Massachusetts Institute of Technology, Cambridge. Arro, H., A. Prikk, T. Pihu, 2006. Calculation of CO_2 Emission from CFB Boilers of Oil Shale Power Plants, *Oil Shale*, 23(4): 356–365. Bajura, R.A., E.M. Eyring, 2005. *Coal and Liquid Fuels*, GCEP Advanced Coal Workshop, March 15–16, Provo, Utah. Bartis, J.T., T. LaTourette, L. Dixon, 2005. *Oil Shale Development in the United States, Prospects and Policy Issues, RAND Corporation*, Santa Monica, CA. National Energy Board, 2004. Canada's Oil Sands: Opportunities and Challenges to 2015; www.neb.gc.ca/. Gibson, P., 2007. *Coal to Liquids at Sasol*, Kentucky Energy Security Summit, 11 October, Lexington, KY. National Energy Board, 2004. *Canada's Oil Sands: Opportunities and Challenges to 2015*; www.neb.gc.ca/. PriceWaterhouse-Coopers, 2003. *Shell Middle Distillate Synthesis (SMDS), Update to a Lifecycle Approach to Assess the Environmental Inputs and Outputs, and Associated Environmental Impacts, of Production and Use of Distillates from a Complex Refinery and SMDS Route*; www.shell.com/. Schäfer, A., J.B. Heywood, M.A. Weiss, 2006. Future Fuel Cell and Internal Combustion Engine Automobile Technologies: A 25 Year Lifecycle and Fleet Impact Assessment, *Energy—The International Journal*, 31(12): 1728–1751. UK Department of Trade and Industry (DTI), 1999. *Technology Status Report: Coal Liquefaction*, Technology Status Report 010, DTI, London. U.S. Department of Energy, 1983. *Energy Technology Characterizations Handbook: Environmental Pollution and Control Factors*, Third Edition, DOE/EP-0093, Washington, DC. Weiss, M.A., J.B. Heywood, E.M. Drake, A. Schäfer, F. AuYeung, 2000. *On the Road in 2020—A Lifecycle Analysis of New Automobile Technologies*, Energy Laboratory Report MIT EL 00-003, Energy Laboratory, Massachusetts Institute of Technology, Cambridge.

Notes: All dollars are U.S. 2000. Energy use and GHG emissions are our estimates and are based on several data sources. The most important ones are indicated in the endnotes in the text. The feedstock costs are $1.5–3.0/GJ (coal) and $0.5–1.0/GJ (remote natural gas). The costs of oil refining, distribution, and associated expenditures (but without taxes) are assumed to be $16.2/bbl (per barrel) for oil refining and profits and $7.7/bbl for distribution and marketing. Only marketing costs were added to the Fischer-Tropsch fuels. (The various components of gasoline supply costs were derived from the U.S. Energy Information Administration, *2005 Primer on Gasoline Prices*; www.eia.doe.gov/bookshelf/brochures/gasolinepricesprimer/.) The 2005 reference production level for oil sands is the Canadian synthetic crude, while that of coal and natural gas are U.S. production levels. The associated increase in land use is based on average geographic energy densities of 1.5 PJ/ha for Canadian oil sands, 10 PJ/ha for oil shale energy, and 0.5 PJ/ha for coal.

[a] CCS: Carbon capture and storage. [b] Remote natural gas. [c] For passenger transport only. [d] 1 ha = 0.01 km^2.

increase in fuel production relative to the 2005 level, together with the associated increase in land use necessary to fuel one-quarter of the 2005 LDV fleet supplying passenger travel in the United States. Because of the already large use of coal and of natural gas in sectors other than transportation, the percentage increase in coal or gas production compared with other feedstocks would be the smallest. Nonetheless, satisfying one-quarter of the 2005 U.S. LDV fleet gasoline demand would require a one-third increase in domestic coal or gas production.

Given an average energy density of coal mines of about 0.5 million billion joules (0.5 petajoules, or 0.5 PJ) per hectare (ha), the annual increase in land demand for coal mining would be about 15,000 ha—which corresponds to the *city* area of one Washington, DC, per year. (Geographic energy densities, however, differ widely; denser coal deposits would require less land to be given over to mining.) Owing to the greater concentration of oil sands, the increase in land use in this case would be smaller (3,600 ha)—corresponding to the size of Davis, California. The still higher density of oil shale deposits would result in the lowest increase in land use—although mining shale oil would cause a range of other non-GHG related problems as stated earlier.

Biofuels

Biomass-derived fuels include alcohols such as ethanol and methanol from plant matter, synthetic oil products from agricultural crops and residues, and diesel fuel from vegetable oils, used cooking oil, or animal fats. Biofuels are not confined to liquids. They also include gases, such as methane derived from municipal organic waste, and hydrogen, which could be obtained from various biomass sources. Thus, biomass can serve in principle as a feedstock for the entire range of fuels described in table 6.1. This wide range of possible feedstocks gives an inherent strategic advantage to biofuels: their widespread use could lead to the reduction of landfills, the mitigation of the recurring oversupply of agricultural products, and possible cuts in farm subsidies.[39]

The greater promise of biofuels, however, has been associated with their potentially carbon-neutral life cycle. If the production and distribution of biofuels were fueled chiefly by renewable energy sources, net CO_2 emissions would be low. In such a closed loop system, CO_2 emissions from fuel supply and from vehicle use would be naturally "recycled" into plant matter in the presence of sunlight—the photosynthesis process. As we show below, some feedstock-biofuel paths could come close to achieving this condition.

However, recent studies suggest that careful land management is an important prerequisite for realizing low life-cycle GHG emissions. If including the CO_2 emissions associated with the conversion of forests, peat lands, savannas, or grassland to new cropland, net CO_2 emissions may greatly *increase* because of the associated release of plant and soil carbon. This outcome can be avoided if producing biomass feedstocks more intensively on existing land or through converting degraded or abandoned agricultural into biomass plantations.[40] Similarly, the conversion of plant residues or organic waste would cause little or no extra CO_2 emissions.

The theoretical potential of a global biofuel supply is enormous. Given an annual production of about 220 billion tons of biomass, the comparable primary energy equivalent amounts to about 4,400 EJ, or ten times the 2005 world primary energy consumption.[41] However, the low energy density per unit of land area, the associated high costs for harvesting and transporting biomass, the competing demand for food, and the need to preserve areas of natural vegetation, including rainforest and wildlife habitats, reduce this potential significantly. Most studies estimate a global biomass potential of 150–300 EJ of primary energy per year by 2050, with a few analyses projecting a potential in excess of 1,000 EJ.[42] (However, the release of significant amounts of CO_2 associated with the required major land-use change makes the realization of such high estimates unlikely.)

Even the available amount of waste suitable for producing biofuels seems significant. Robert Perlack and his co-workers at the Oak Ridge National Laboratory conducted a detailed assessment of the biomass feedstocks in the United States. They found that unutilized residues from forest and agriculture alone of about 580–740 million tons could become available by midcentury.[43] Using the advanced fuel-conversion technology discussed below, around 5–6 EJ of liquid fuels could be produced, which would satisfy roughly one-third of the energy used by the total U.S. passenger transport in 2005. In practice, however, this potential may be significantly smaller because of the rising costs of collecting residues in more remote areas. Yet, even the exploitation of only one-third to one-half of that potential would provide a significant fuel resource.[44]

A favorable characteristic of liquid biofuels is their "blending flexibility" with petroleum fuels, which facilitates a transition toward using a greater share of biofuels. With the exception of biomass-derived synthetic oil products, higher shares of biofuels, however, require changes of the current oil and vehicle infrastructure to prevent fuel instability or material corrosion.

Ethanol is already being blended with gasoline in volume proportions of up to 10 percent. It is used to help engines achieve more complete combustion, thereby reducing emissions of the criteria pollutants, particularly in those metropolitan areas where local air pollution limits have been exceeded. This blending share can probably be increased up to about 20 percent by volume without having to change current oil and vehicle infrastructure. Tighter blending constraints of up to 5 percent by volume apply to biodiesel-diesel blends.[45] (As with ethanol-gasoline blends, biodiesel-diesel blends release lower amounts of most pollutants.[46]) Biodiesel–jet fuel blends are also being studied and tested as a short-term measure to reduce GHG emissions from aviation. Recently, a commercial airline has successfully tested one engine running with a 20 percent blend of vegetable oil during flight.[47]

Using higher ethanol shares would require significant adaptations in today's fuel and vehicle infrastructure. Flexible fuel vehicles are now being marketed that can operate using either pure gasoline or a gasoline-ethanol mixture in which ethanol makes up as much as 85 percent of the volume. An advanced version of such a dual-fuel vehicle is an ethanol-boosted engine recently developed at MIT. Based on a downsized gasoline engine concept, ethanol is injected directly into the vehicle's turbocharged, high-compression ratio combustion chambers during high power operation—when climbing a hill or accelerating rapidly—thus overcoming a gasoline engine's knock limit. For any given vehicle, such an engine would reduce fuel consumption efficiency by up to 20 percent.[48]

Ethanol can also be used as a stand-alone fuel in dedicated vehicles. In such an application, ethanol's lower energy density than that of oil products would have a sensible impact. As indicated in table 6.1, to store the same amount of energy as gasoline, ethanol requires about 60 percent more volume and adds a similar percentage of additional fuel weight. This increase is partly offset by higher engine efficiency (potentially up to 10 percent higher) in ethanol-dedicated spark-ignition engines than in gasoline engines (operating under similar conditions), because of ethanol's higher octane number. While the increase in fuel volume and weight does not impose major limitations on ground transportation, it would be a critical constraint if ethanol were used as an aircraft fuel (as discussed in chapter 5). Ethanol's lower energy density would deteriorate the performance of an ethanol-fueled aircraft by impacting size and range—a phenomenon similar to an electric vehicle with a low-energy density battery (see chapter 4). Analogous to today's battery-powered electric vehicles in road traffic, ethanol fuel would thus be suitable only for short-range aircraft. Most expectations of biomass-derived alternative aviation fuels are thus directed toward synthetic oil products. In contrast to ethanol, the physical and chemical characteristics of biodiesel are similar to those of their crude oil–derived counterparts.[49]

The production of ethanol by fermenting sugars is a well-proven process that has a history of several thousand years. Early fermenting processes used wild yeast from natural contaminants in flour or milk. In contrast, today's genetically engineered yeast is the subject of continuous research to achieve more efficient and effective fermentation. For today's ethanol production, the particular choice of sugar feedstock is determined mainly by its cost, ethanol yield, and experience in producing the feedstock. In Brazil, virtually all ethanol is derived from sugarcane because of its low cost and high conversion efficiency.[50] Sugarcane's raw sugar, which accounts for about one-third of the plant's weight, can be readily extracted and fermented to ethanol. The fiber left after the juice has been extracted is used to generate heat and electricity for the fermentation and distillation processes. This nearly complete independence from external energy sources results in very low GHG emissions for the sugarcane-based ethanol fuel cycle.[51] (The only fossil energy inputs occur in the agricultural stage and include the production of fertilizer and other chemicals and the use of diesel-fueled machinery for planting, harvesting, and transporting the biomass feedstock to the processing plant.) Since the next generation of biomass plants will absorb an amount of GHG emissions comparable to the emissions released during ethanol production and its combustion in a vehicle engine, sugarcane-produced ethanol life-cycle emissions are low.

In contrast, corn is currently used as the dominant ethanol feedstock in the United States. Using corn grain means that only half the weight of the feedstock plant can be utilized to produce ethanol (and several coproducts); the remainder—the stalk, leaf, cob, and husk—is left on the field to decay. The share of plant matter that can be used for ethanol production is further reduced to only one-third of the original plant mass, since starch accounts for only about two-thirds of the kernel mass. Other kernel components, including the germs and hulls, are used mainly for corn oil production and animal food nutrients. (Depending on the type of processing facility, these coproducts can account for a larger proportion of the facility's output than ethanol itself.[52]) After separating the starch from the other parts of the corn plant, enzymes are added to decompose it into simpler, more easily fermentable sugars.

Since the corn-based ethanol process requires significant inputs of natural gas or coal, the process results in only small reductions in GHG emissions compared with the gasoline life-cycle.[53] Growing corn also requires relatively good soil conditions, which make this feedstock more expensive to cultivate, because that land could otherwise be used for food production. In fact, the recent increase in the demand for ethanol in the United States led to a significant increase in corn prices in Mexico—which in turn resulted in an increase in tortilla prices, thereby adversely affecting the food supply of the low-income part of the population.[54] The increased use of agricultural land for corn production has also been identified as one cause contributing to the rise in wheat prices.[55]

An alternative to the sugar-to-ethanol conversion is the production of biodiesel fuel. The biofeedstock is usually a vegetable oil, such as palm oil, soybean oil (especially in the United States), rapeseed oil (especially in Europe), and others. Vegetable oils are triglycerides, compounds consisting of one molecule of glycerin combined with three molecules of a fatty acid. The fatty acids are straight chain compounds of about eight to eighteen carbon atoms (depending on the oil) with acid groups on one end. To make biodiesel, the fatty acids are chemically processed to be separated from the glycerin (which is sold as a valuable byproduct) and then reacted with an alcohol, usually methanol, to improve their properties for use as a fuel.

Table 6.3 summarizes typical biofuel characteristics for various feedstocks and regions of the world. The numbers indicated are averages of processes and take into account the size and type of energy inputs in each conversion step, including crop production (fertilizer and seed production, planting, watering, and harvesting), transportation of the feedstock to the fuel-processing plant, biofuel production, and distribution of the fuel to retail stations. The comparatively small fuel yield per unit of land area is quite apparent for corn-based ethanol—and especially for biodiesel fuel. This low yield is a consequence of being able to use only the starchy part of the corn plant and just the seeds of the biodiesel feedstock plants.

Table 6.3 also shows that all biofuels, independent of their feedstock, offer nearly complete displacement of petroleum fuel on an energy basis. That reduction in oil dependence would be achieved in different ways: by replacing petroleum with natural gas or coal in producing corn-based ethanol, or using an essentially self-sufficient process in the case of sugarcane-based ethanol. These different strategies affect the level of life-cycle GHG emissions differently. (For processes that also produce byproducts, such as the corn-to-ethanol conversion, the way the byproducts are accounted for also affects GHG emissions in table 6.3.[56]) The life-cycle impact of the corn-based ethanol process is roughly comparable to that of gasoline, while the life-cycle impact of sugarcane-derived ethanol is 85 percent lower. Recall, however, that these findings apply only to biofuels with feedstocks grown on existing or abandoned agricultural land. Otherwise, additional GHG emissions from land-use change could be released.

Table 6.3 also indicates the costs for producing and delivering various types of biofuels. Because of the costs involved in growing, cultivating, and harvesting the low energy-density biomass, together with subsequent fuel-processing cost, the production costs of biofuels are generally higher than those of their oil-based counterparts. The exception among the current generation of biofuels is sugarcane-based ethanol in Brazil. There, favorable climatic and soil conditions, and a long history of technological improvements allow the production and delivery of ethanol for slightly more than 30 cents per liter of gasoline equivalent. All other current-generation biofuels have two to three times higher costs in the United States and Europe. These differences are also expressed in the break-even oil price, the minimum price to make the respective biofuel cost effective. The break-even oil prices range from $35 per barrel in Brazil to $110 per barrel in the United States, to $160–$170 per barrel in Europe. Note that because of the large amount of coproducts produced by the corn-based ethanol process, a declining demand for coproducts would lead to higher ethanol costs.

The ethanol and biodiesel processes described above that are currently in operation represent only first-generation biofuels. Additional options for biomass conversion are currently being explored and show encouraging opportunities. To drastically reduce both the fossil energy

Table 6.3
Typical biofuel supply characteristics

	Current generation				Second generation	
	Ethanol			Biodiesel	Ethanol	BTL[a]
	Corn (U.S.)	Sugarcane (Brazil)	Sugar beet (Europe)	Rapeseed (Europe)	Wood	Wood
Fuel-energy yield						
GJ/ha[b]/yr	77	130	120	40	160–200	160–200
$L_{Gasoline\ Equivalent}$/ha/yr	2,300	3,900	3,600	1,200	4,800–6,000	4,800–6,000
Petroleum substitution potential, %	90–95	90–95	90–95	90–95	90–95	90–95
Life-cycle GHG-Em., oil-based gasoline = 100	95	15	60	60	13	13
Costs of supplied fuel (untaxed)						
$/$L_{Gasoline\ Equivalent}$	0.82	0.37	1.20	1.10	0.3–0.7	0.4–0.7
$/$gal_{Gasoline\ Equivalent}$	3.1	1.4	4.6	4.3	1.2–2.7	1.4–2.6
Break-even oil price, $/bbl	110	35	170	160	25–90	35–85
Scale impacts of supplying 25% LDV[c] fleet in 2005						
Increase in production over 2005 level, %	1,000	—	—	—	—	—
Required land area, mill. ha/yr	50	—	—	—	19–24	19–24
% U.S. land area, 48 states	6	—	—	—	2–3	2–3
Increase in land use over 2005, %	125	—	—	—	—	—

Sources: Farrell, A.E., R.J. Plevin, B.T. Turner, A.D. Jones, M. O'Hare, D.M. Kammen, 2006. Ethanol Can Contribute to Energy and Environmental Goals, *Science*, 27 January, 311 (5760): 506–508. De Carvalho Macedo, I., M.R.L. Verde Real, J.E.A. Ramos da Silva, 2004. *Assessment of Greenhouse Gas Emissions in the Production and Use of Ethanol in Brazil*, Government of the State of São Paulo, Brazil. Concawe, EUCAR, Joint Research Center, 2007. *Well-to-Wheels Analysis of Future Automotive Fuels and Powertrains in the European Context*, Version 2c; http://ies.jrc.ec.europa.eu/WTW. U.S. Department of Energy, 2006. *Annual Energy Outlook 2006—With Projections to 2030*, Washington, DC. U.S. Department of Energy, 2007. *Annual Energy Outlook 2007—With Projections to 2030*, Washington, DC. Shapouri, H., P. Gallagher, 2005. *USDA's 2002 Ethanol Cost-of-Production Survey*, U.S. Department of Agriculture, Washington, DC. Tijmensen, M.J.A., A.P.C. Faaij, C.N. Hamelinck, M.R.M. van Hardeveld, 2002. Exploration of the Possibilities for Production of Fischer-Tropsch Liquids and Power via Biomass Gasification, *Biomass & Bioenergy*, 23:129–152.

Notes: All dollars are U.S. 2000. The characteristics of current-generation processes are expected to improve gradually over time, while those of the second-generation processes are projections. The petroleum substitution potential is defined as 100 percent – petroleum energy input / biofuel-energy output × 100. Thus, 100 percent petroleum substitution potential corresponds to complete substitution. Life-cycle GHG emissions include nitrous oxide (N_2O) emissions from agriculture. The costs of supplied fuel are production costs plus $0.05/$L_{Biofuel}$ for distribution and marketing. The break-even oil price is based on the 2005 U.S. gasoline price structure and excludes fuel taxes. Energy use and GHG emissions are our estimates, based on several data sources.

[a] BTL: biomass-to-liquids.

[b] 1 ha = 0.01 km^2.

[c] For passenger transport only.

input and competition with food products (especially for the corn-to-ethanol process), much research is being dedicated to a range of high-cellulose and lower-cost plants that can be grown on marginal land. Crops such as switchgrass, willow, poplar, miscanthus, eucalyptus, and certain agricultural residues are composed of as much as 85 percent cellulose and hemicellulose. Their fuel yield would be significantly higher than that of corn-based ethanol. However, breaking down cellulose and hemicellulose, the tough cell material that gives a plant its structural strength, into fermentable sugars has been a critical barrier to producing cellulosic ethanol. Significant advances in a more efficient and cost-effective hydrolysis have been achieved only recently, and the conversion of cellulose feedstocks to fermentable sugars is only beginning to be commercialized.[57]

In the new cellulosic-ethanol process, as in the sugarcane-to-ethanol process, the major nonfermentable component of the feedstocks—in this case lignin—can be used as an energy source for producing process heat and electricity. Thus, cellulosic-ethanol production can be largely independent of external energy inputs, an attribute that also leads to drastically reduced GHG emissions. According to table 6.3, fuel-cycle GHG emissions for cellulosic ethanol are only 15 percent of those of the petroleum cycle, as opposed to more than 90 percent in the case of current corn-based ethanol approaches. (Note, however, that the energy and GHG-emission balance of the corn-based process would be significantly more favorable if corn stover and animal food were also used as fuel.) Cost projections of cellulosic ethanol delivered to fuel stations are uncertain and range from $0.3 to $0.7 per liter of gasoline equivalent ($1.2 to $2.7 per gallon of gasoline equivalent).

Another promising possibility is the conversion of biomass to synthetic oil products, including gasoline, diesel, and jet fuel. This requires, as in the case of indirect coal liquefaction, a multistage conversion process that includes biomass gasification and Fischer-Tropsch conversion—which would allow the production of clean synthetic fuels that have desired properties. Biomass-to-liquids (BTL) conversion uses the entire crop (whatever it might be) to provide both feedstock and a source of process energy. Its fuel-energy yield is thus similar to that of cellulosic ethanol.

The first commercial-scale BTL plant is about to start operation.[58] Current production costs are reported to be about 85 cents per liter ($3.2 per gallon) and cost projections range from $0.4 to $0.7 per liter of gasoline equivalent ($1.4 to $2.6 per gallon of gasoline equivalent) for the long term. Initial tests of the synthetic diesel fuel at Argonne National Laboratory resulted in reductions for all exhaust emissions, including an almost 10 percent lessening of CO_2 emissions per kilometer.[59] Another new technology opportunity is gasification of biomass to produce hydrogen—a possibility we discuss below.

Further promising longer-term opportunities may exist. Some fast-growing, high oil-yield, species of microalgae are receiving increasing attention (not shown in table 6.3). Cultivated in a sunny, aquatic environment, some microalgae are projected to produce an annual amount of oil that corresponds up to 90,000 liters (nearly 24,000 gallons) of biodiesel per hectare, a volume nearly twenty times higher than that of the future cellulosic biomass feedstocks projected for second-generation biofuels shown in table 6.3.[60] In the most favorable case, algae thus would require only about 5 percent of the land area for producing the same amount of fuel. They would neither compete for agricultural land nor induce any land-use change that releases CO_2 emissions. A prerequisite for algae to form a CO_2-neutral fuel is that the CO_2 necessary to feed the algae would need to be taken from coal power plants or from coal refineries nearby. Algae have been cultivated in a range of settings, including open ponds, seawater, and closed reactors, with each application posing different challenges. The main barrier, however, is that the cultivation of microalgae still needs to become economically competitive on a commercial scale.

Table 6.3 also illustrates the scale implications for the various biofuel feedstocks if they were to fuel one-quarter of the U.S. 2005 LDV fleet that provides passenger travel. For corn-based ethanol, current ethanol production would need to grow ten times. If the *entire* U.S. cornfield area were used as a base reference (some 15 percent of which has been dedicated to ethanol production in 2005), the total area of cornfield use would need to more than double. Although the up to twofold fuel-energy yield of cellulosic ethanol or synthetic oil products would reduce the land area needed, land requirements would remain significant—a clear disadvantage compared with fossil fuel–derived oil products. As can be seen from a comparison of tables 6.2 and 6.3, land requirements of a potentially low-carbon biofuel supply are more than a thousand times higher than those of carbon-intensive synthetic fuels produced from oil sands, oil shale, or coal.

The land-use implications for a biofuel-intensive road-passenger transport system are illustrated in figure 6.1 for the United States (top) and for Western Europe (bottom). The figures show the different percentages of

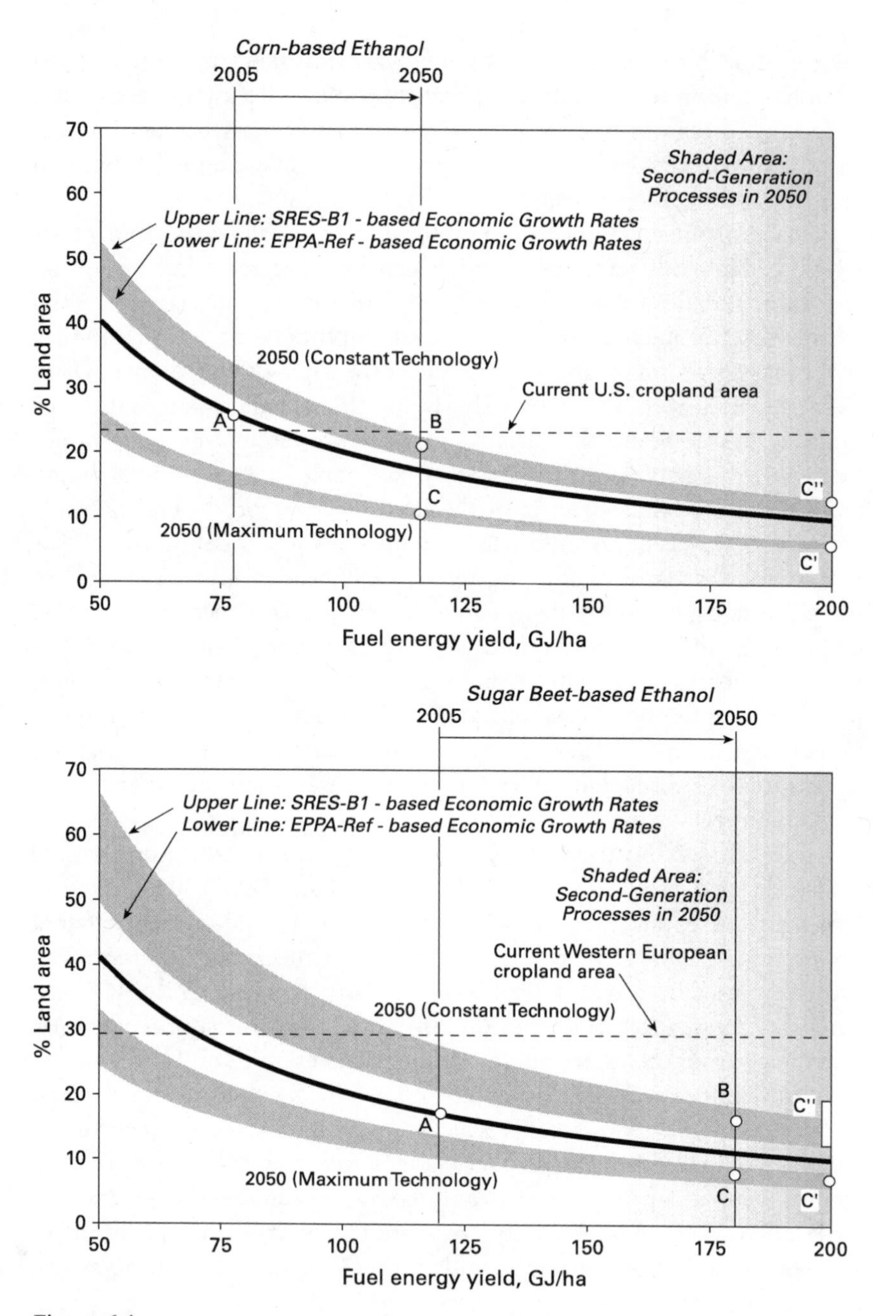

Figure 6.1
Required land area (in percentage of national territory) versus fuel-energy yield for the United States (top) and Western Europe (bottom).

total land area required to satisfy the current (2005) and future (2050) energy demand of the light-duty passenger vehicle and bus fleet. The boundaries of the 2050 projections result from the two economic projections that drive the travel demand, energy use, and CO_2 emissions scenarios used in chapters 2 and 3. The higher economic growth rates of the MIT model (EPPA-Ref) determine the lower limit (because of a higher relative importance of lower-energy-intensity air travel), while the lower growth rates of the IPCC SRES-B1 scenario define the upper boundary (because of a higher share of higher-energy-intensity LDV travel). These curves of constant biofuel demand are downward sloped—since an increase in fuel yield per land area reduces the land needed to grow biomass feedstocks that will satisfy a given fuel demand.[61] Also shown are two vertical lines, which indicate the current and projected fuel-energy yield of today's biomass feedstocks.

Figure 6.1 (top) illustrates the U.S. situation. To satisfy the 2005 energy demand of passenger-related LDV travel and bus transport of 15.4 EJ using corn-derived ethanol (which currently has fuel yields of 77 gigajoules per hectare [GJ/ha]) would require about 25 percent of the land area of the lower forty-eight states to be used for growing corn (point A). For comparison, less than 1 percent of the U.S. land area of the 48 states was dedicated to biofuel production in 2005. This amount of land is also larger than the current area of all U.S. croplands, the share of which is shown by the dashed horizontal line.[62]

Over the next decades, travel demand and transportation energy use are likely to increase. Depending on the economic growth rates in the EPPA-Ref or SRES-B1 scenarios, the 2050 road passenger travel energy demand would rise by 12 or 30 percent if vehicle fuel efficiencies remain unchanged. If assuming an increase in corn productivity by about 1 percent per year (about half the historical rate), the resulting 50 percent higher corn yield per unit of land area in 2050 would more than offset the land-use impact of an increase in energy demand. Thus, fueling the entire passenger travel–related LDV and bus fleet with 2005 fuel consumption characteristics would require a land area for growing corn slightly below the 2005 level (point B).[63] If, in addition, the growth in road passenger transport energy use could be reduced through a 50 percent reduction in the 2050 energy intensity of the vehicle fleet (by adopting the critical fuel-saving technologies identified in chapter 4), the land-use requirement would decline at a proportional rate, resulting in a need of slightly more than 10 percent of the U.S. land area (point C).

However, even in this limiting case, the land requirements for producing the corn feedstock would still correspond to about half the size of the current total cropland area.

Figure 6.1 also indicates the land-use implications for second-generation processes: cellulosic ethanol and synthetic oil products produced from biomass. Given the uncertain levels of fuel-energy yields in the future, we shaded the range to be between 160 and 200 GJ/ha/yr—in accordance with the data range given in table 6.3. Within that range, cellulosic biomass plantations would require about 6 percent of the U.S. land area (Point C′). If adding the projected energy requirements of commercial passenger air traffic (assuming for simplicity that the entire projected demand for high-speed transportation would be supplied by highly fuel-efficient aircraft), the energy requirements and thus the need for biomass plantations for supplying total passenger travel would approximately double (Point C″). The associated land requirements of about 13 percent of the U.S. land area would correspond to roughly 65 percent of the current agricultural land in the United States.

Importantly, even without such massive changes in land-use, biofuels could make useful contributions to future travel-related energy needs. Assuming that one-third to one-half of the projected technical potential of the biofuel equivalent of agricultural and forest residues of 5–6 EJ could be converted economically into synthetic fuels, and assuming also that fuel consumption of the future automobile and aircraft fleet would be cut by half, biofuels from residues could provide 5–8 percent of the entire projected 2050 passenger transport energy.

Figure 6.1 (bottom), which illustrates the Western European situation, reveals two main differences compared with the U.S. case. One difference is associated with the larger range in the projected future levels of energy use, which results from the larger difference in economic growth rates between the EPPA-Ref and SRES-B1 scenarios. In addition, biomass plantations in Western Europe require a larger percentage of land because Europe has a smaller overall land area available than the United States does, while projected levels of travel demand and energy use are similar. A comparable amount of biomass would need to be grown on an overall available land area that is only 57 percent of the U.S. area. (See chapter 2, note 2, for an exact definition of the Western Europe region.) Given a consistent location of Points A, B, C, C′, and C″, the interpretation of figure 6.1 (bottom) is identical to that of the U.S. case.

How feasible, then, could land requirements be for fueling *global* passenger travel? According to our projections in chapters 2 and 3, the 2050 fuel energy demand for global mobility would total about 150–250 EJ in a constant technology scenario, depending on the rate of economic growth. If we assume that the fuel consumption of the *global* fleet of air and ground vehicles can be reduced by 30 percent, the demand for fuel energy from biomass would decline to a more manageable level of 105–175 EJ. Using typical efficiencies of about 40 percent for biomass-to-biofuel-conversion from second-generation processes, the corresponding biomass feedstock energy of 260–440 EJ overlaps only slightly with the upper end of the range of the global biomass energy potential assumed in many studies of 150–300 EJ. The demand for travel and consequently for energy use, however, is likely to continue to increase, especially within the highly populated countries of the developing world. Thus, absent further vehicle efficiency improvements or significant technological advances in biofuel production, biofuels would be unable to fuel the entire global passenger mobility, unless significant changes in land use (and release of soil and plant carbon) were tolerated.

These projections of biofuel-energy yields include expected progress in genetically engineering microbes to better break down plant fibers but ignore a similar potential for significantly increasing land productivity through genetically modifying crops. The potential could be enormous. While current biomass plantations harvest only about 1 percent of solar energy input, their theoretical maximum is nearly 7 percent (with respect to an entire crop, and depending on the type of plant).[64] Thus, if it became possible to double photosynthetic efficiency by genetically modifying crops, the requirement for land devoted to growing biomass would drop by half. The associated step change in fuel yield would significantly increase the biofuel supply prospects.

This analysis suggests that biofuels almost certainly have the potential to become an important additional fuel stream. Under the condition of careful land management to keep CO_2 emissions from land-use changes low, biofuels could provide a significant reduction of GHG emissions even in the absence of vehicle fuel-efficiency improvements. Given sufficient technological progress in biofuel yields, new ways of growing feedstocks (e.g., algae), and major vehicle fuel-efficiency improvements, these fuels could become an important transportation fuel, especially in light of their potentially greatly reduced GHG emissions.

Compressed Natural Gas

According to table 6.1, burning one unit of natural gas releases about three-quarters the amount of CO_2 of an energy-equivalent amount of gasoline. Thus, a predominantly natural gas–based transport system could—in theory—significantly reduce GHG emissions. Proved global reserves of natural gas are comparable to those of crude oil, with about half the reserves being stranded in remote areas. The geographic distribution of gas reserves is similar to that of crude oil. More than three-quarters of the world's proven natural gas reserves are located in Russia and the OPEC countries, while the industrialized world accounts for less than 10 percent. The full potential resource base of natural gas, however, is vast and almost beyond imagination. If hydrates—natural gas trapped in crystals of frozen water—are included as possible reserves, current estimated natural gas reserves would be increased a hundredfold.[65]

Natural gas already accounts for one-quarter of total primary energy consumption in the United States. Its share worldwide is similar. Thus, a shift toward natural gas would be facilitated by the existence of a basic transmission, distribution, and storage infrastructure already in use in both the residential and the industrial sectors of the economy. Despite this, however, natural gas currently plays only a marginal role in the world transportation system. According to data collected by the International Association for Natural Gas Vehicles, the global fleet of natural gas–fueled automobiles, buses, and trucks accounted for some 6 million vehicles in 2005—less than 1 percent of the global road vehicle fleet.

Additionally, most of the natural gas vehicle fleets are concentrated in a few countries. Two-thirds operate in Argentina, Brazil, and Pakistan. In Argentina, natural gas–fueled automobiles currently account for slightly more than one-quarter of the country's fleet. In these countries, natural gas does account for a significant share of the road transportation fuel markets.[66] The motivation to use natural gas there is similar to the motivation in Canada and New Zealand during the 1980s, when increasing use of domestic natural gas in transportation was driven by rising oil prices and was promoted by government subsidies to reduce oil dependence. In contrast, the U.S. total of 150,000 natural gas–fueled vehicles translates into fewer than 1 out of 1,000 vehicles. (The main motivation in the United States has been reducing criteria pollutants emitted by buses.)

Compared with crude oil, natural gas requires very little treatment before it can be used as a (transportation) fuel. Natural gas is a mixture of hydrocarbons, chiefly methane.[67] After methane is separated from other gases at the extraction site, typically gas or oil wells, it is transmitted and distributed through a pipeline network to consumers, both industrial and residential. When used as a transportation fuel, natural gas needs to be compressed at the fuel station for storage on the vehicle. This requires additional energy inputs. Overall, the energy required for the extraction, transmission, distribution, and compression of natural gas corresponds to about 14 percent of the fuel's energy content, a level comparable to the fuel-cycle energy use of gasoline and diesel fuels.

Based on the amount of CO_2 released when methane is burned, shown in table 6.1, the associated energy-related fuel-cycle GHG emissions for natural gas are estimated to be 7.7 gCO_2/MJ_{CNG}. GHG emissions also include methane leaks during fuel extraction, transmission, and distribution. Related studies undertaken in the United States suggest a leakage rate of about 0.5 percent for natural gas production and 0.25 percent for both transmission and distribution, resulting in a total leakage of about 1 percent of the amount of natural gas transported from the wellhead to the end user.[68] Using a 1 percent leakage rate, a twenty-one-times stronger climate impact of natural gas compared with CO_2, and a fuel-cycle energy use of 0.14 MJ/MJ_{CNG}, the CO_2 equivalent of natural gas leaks then results in an additional 4.8 $gCO_2\text{-eq}/MJ_{CNG}$. Overall, fuel-cycle GHG emissions from natural gas result in about 12.5 gCO_2-eq/MJ. These numbers are shown in table 6.4. Note that the fuel cycle GHG emissions are comparable to those of petroleum fuels, but the composition differs. In the case of natural gas fuels, methane leaks have a significantly higher relative importance, nearly 40 percent of the total compared with less than 10 percent in the case of crude oil–derived fuels.

Assessing life-cycle energy use and GHG emissions requires also taking into account a natural gas–powered vehicle's operation. Natural gas's significantly higher octane number of 120 allows a higher compression ratio and thus higher engine efficiency. However, the higher volume occupied by natural gas in the fuel-air mixture in the cylinder compared with gasoline requires a larger engine, which partly offsets that gain. Overall, optimized dedicated natural gas engines are about 6 percent more energy efficient than gasoline engines under similar operating conditions. The higher vehicle weight, resulting from the larger engine and heavier fuel tank needed to store compressed natural gas, reduces this gain to about 3 percent.[69] In combination with natural gas's lower fuel-cycle GHG emissions, life-cycle GHG emissions from natural gas vehicles are about 25 percent lower per vehicle kilometer traveled (VKT) compared with similar gasoline vehicles.

However, compressed natural gas (CNG) vehicles illustrate the challenges that arise when gaseous fuels are introduced into the transportation sector. At a storage pressure of 250 bar (3,600 psi), natural gas's energy density is still only one-third that of gasoline. That translates into a travel range one-third that of a similar-sized gasoline vehicle, all other factors being equal. (The up to 6 percent higher energy efficiency of dedicated CNG vehicles compared to gasoline-fueled automobiles would not benefit vehicle range significantly.) To increase this range, natural gas vehicles require a larger, high-pressure fuel tank, which not only reduces trunk space, but also adds extra weight, refueling time, and several thousand dollars to the price of an equivalent gasoline vehicle. Extra costs for CNG-capable vehicles could in fact become even higher because of the need for more complex dual-fuel vehicles, vehicles capable of running on either gasoline or CNG (when or where natural gas is not widely available). Over what time horizon these extra costs can be amortized remains uncertain, especially in periods of volatile oil prices. But even if consumers were indifferent to the loss of carrying space and could amortize extra costs over an acceptable period, they would still face the obstacle of parking their natural gas vehicle in closed public spaces.[70]

This range of consumer concerns can be addressed realistically only through sustained subsidies for CNG vehicles—a policy measure that would be difficult to defend given the lack of large domestic reserves in both the United States and Europe. However, unless a shift to natural gas vehicles was major, the level of emission reduction is small. If we assume a natural gas vehicle penetration into the U.S. market to an extreme upper limit—say the current share of CNG-fueled automobiles in Argentina, about 25 percent—the degree of GHG emission reduction would be rather modest: roughly 25 percent of 25 percent, or some 6 percent. As discussed in chapter 4, that relatively low level of emission reduction can be achieved through a small number of other, low-cost, incremental measures that would affect neither carrying space nor parking restrictions.

An additional concern is the long time it would take to achieve such an impact, even though CNG vehicles are a proven technology. Retrofitting a significant fraction of the existing gasoline-fueled vehicles to natural gas would overwhelm the service industry. Therefore, most vehicles making up the 25 percent share would probably be new vehicles dedicated to

CNG. Given the rates of turnover in the replacement markets of the industrialized world, and assuming that 20 percent of all new LDVs would be powered by natural gas, it would take twenty years before the estimated 6 percent reduction in vehicle fleet GHG emissions could be achieved.

Thus, a significant use of natural gas vehicles faces serious hurdles: limited consumer acceptance, a need for sustained government subsidies to achieve a sizeable penetration share, and, in practice, a small emission-reduction potential. Taken together, these obstacles imply that the direct use of natural gas is likely to remain a niche-market fuel except in countries having large domestic natural gas reserves. More likely to happen is an increase in the indirect use of natural gas in the form of liquid synthetic fuel products (gas-to-liquids), a strategy, however, that would lead to an increase in life-cycle GHG emissions.

Electricity and Hydrogen

Electricity and hydrogen share several commonalities. Both are energy carriers and not "primary" fuels. Both offer a significant potential for low emissions at the point of consumption. In a climate-constrained world, nuclear energy and renewable sources could produce zero-GHG-emission electricity and hydrogen, thus potentially expanding the low environmental impact of vehicle use over the entire life-cycle.[71] And even when fossil fuels are converted into moving electrons or hydrogen, the fact that they are generated at a central location allows CO_2 to be captured and sequestered. The wide range of possible energy sources for electricity and hydrogen also increases energy security, an especially important objective of the United States, Europe, and other regions of the world that have comparatively low oil reserves.

Electricity and hydrogen could play complementary roles. Electricity produced from renewable energy or nuclear heat can be used to split water into its building blocks of hydrogen and oxygen. The recombination of these molecules either chemically in an internal combustion engine or electrochemically in a fuel cell completes the water cycle, thereby generating usable energy. Given the enormous potential of solar, wind, and nuclear fusion energy, this path—conceptually—represents the ultimate vision of a hydrogen economy. Our energy system, however, is still very far from this ideal end state. Today's commercial electricity and hydrogen use have evolved together to meet market-based needs.

From an energy-efficiency perspective, however, the use of electricity alone might be more advantageous. If hydrogen is produced by water electrolysis at a conversion efficiency of 65 percent, compressed to 350 bar (5,000 psi) for onboard storage at 90 percent efficiency, and then converted back to electricity in a fuel cell at an efficiency of 60 percent, overall efficiency is 35 percent. Thus, before electricity—via hydrogen—can drive a vehicle's motors, about two-thirds of its original energy would be consumed through conversion losses. In a hydrogen-fuel dominated transport system, this dissipation would add up to enormous losses of energy, requiring a huge increase in electricity-generation capacity. On the other hand, as we note in our analysis of life-cycle effects, many competing vehicle-fuel combinations perform even less efficiently, which—from an energy-efficiency perspective—makes the joint electricity-hydrogen option comparatively attractive.[72]

Most hydrogen today is derived from natural gas, which is a result mainly of the less complex process and the comparatively low cost. Additional amounts of hydrogen are produced from heavy oil and coal. Common to all hydrogen production from fossil fuels is the generation of a synthesis gas through reacting methane or gasified heavy oil or coal with steam in the presence of a catalyst under high temperatures—as with the first step of producing synthetic oil products. However, in contrast to the subsequent Fischer-Tropsch conversion for liquid fuels, in a second step, the carbon monoxide is subjected to steam at lower temperatures to form additional hydrogen and CO_2. Following this process, a significant amount of hydrogen is derived from steam, ranging from 50 percent in the case of natural gas to more than 60 percent for gasified heavy oil or coal. In addition to the fossil fuel-derived hydrogen, comparatively small amounts are produced from water electrolysis. Although water electrolysis is a state-of-the-art technology, its high electricity use constrains this process to locations with abundant and thus cheap electricity (such as at large hydropower stations). Several other, potentially low-carbon, hydrogen production paths exist, but they are either in the pilot or laboratory stage.

In addition to CNG, table 6.4 summarizes the performance characteristics of selected current and potential future hydrogen production paths. The shown *current* process routes include steam reforming of natural gas and water electrolysis. Following the natural gas–based route, splitting hydrocarbon and water molecules, transporting the resultant hydrogen to retail stations, and compressing the hydrogen to high storage pressures (about 345 bar [5,000 psi]) results in a fuel-cycle energy use of about 0.8 MJ/MJ of compressed hydrogen. This value is four to six times the 0.13–0.18 MJ/MJ for oil products, discussed earlier in this chapter. The associated fuel-cycle GHG emissions of the natural gas-based process amount to 107 gCO_2-eq/MJ of compressed hydrogen, eight to ten times above the amount from petroleum-derived fuels. However, since the use of hydrogen forms no CO_2 emissions, the greenhouse gases released during the fuel-cycle are also those of the life-cycle for all hydrogen paths. Nevertheless, per unit energy, GHG emissions are higher from natural gas-based hydrogen than those from petroleum-derived oil products on a life-cycle basis.

If the hydrogen is produced from water electrolysis, fuel-cycle energy use and GHG emissions strongly depend on the electricity fuel mix. The numbers in table 6.4 are based on the current U.S. electricity generation mix, which is likely to change only slowly over the next several decades. Based upon this assumption, fuel cycle energy use would be almost three times that of the natural gas-based process, while fuel-cycle and life-cycle GHG emissions would be almost 50 percent higher.[73] However, in the hypothetical case of a completely renewable or nuclear-based electricity supply, life-cycle GHG emissions would be very low.

The *future* options shown in table 6.4 include advanced processes based upon coal, biomass, or nuclear energy. Although hydrogen has been produced from coal at an industrial scale for a long time, advances in coal gasification and hydrogen separation are required to achieve the numbers shown in table 6.4. Similar challenges apply to the pathway for producing hydrogen from biomass, which follows that for fossil fuel-derived hydrogen described above. While several pilot plants that produce hydrogen from biomass gasification are already in operation, splitting water molecules using a thermochemical cycle driven by nuclear heat is still in the laboratory stage.[74]

The fuel-cycle energy use of these future options is between those of steam-reformed natural gas and water electrolysis using the 2005 U.S. electricity mix. The life-cycle GHG emissions from coal-based hydrogen show the significant impact of carbon capture and sequestration. If uncontrolled, life-cycle GHG emissions from coal-based hydrogen would be nearly 80 percent higher than those from steam-reformed natural gas. However, should CO_2 be sequestered at the conversion plant, life-cycle GHG emissions would be nearly 80 percent lower than those from the natural gas-based process without any CO_2 emission controls. In

Table 6.4
Typical supply characteristics of gaseous transportation fuels

	CNG	Compressed hydrogen					
	Current	Current		Future			
	Gas	Gas	Electrolysis	Coal	Coal-CCS[a]	Biomass	Nuclear
Fuel-cycle energy use, $MJ/MJ_{Product}$	0.14	0.80	[b]2.28	1.02	1.16	1.58	1.63
Life-cycle GHG-Em., gCO_2-eq/$MJ_{Product}$	13	107	[b]152	190	24	6	0
Relative, natural gas based hydrogen = 100	—	100	[b]142	178	22	5	0
Costs of supplied fuel (untaxed)							
$\$/L_{Gasoline\ Equivalent}$	0.34–0.54	0.47–0.71	0.67–1.13	0.38–0.44	0.39–0.46	0.76–0.85	0.61
$\$/gal_{Gasoline\ Equivalent}$	1.3–2.0	1.8–2.7	2.6–4.3	1.4–1.7	1.5–1.7	2.9–3.2	2.3
2X vehicle energy efficiency for H_2							
Break-even oil price, $/bbl	30–62	26–44	42–78	18–23	19–24	49–55	36
Scale impacts of 25% LDV[c] fleet fuel supply							
Increase in production over 2005 level, %	22	17	—	16	17	—	65
Increase in land use over 2005, mill. ha[d]/yr	—	—	—	0.015	0.016	19–24	—

Sources: CNG cost characteristics mainly based on Weiss, M.A., J.B. Heywood, E.M. Drake, Schäfer A., F. AuYeung, 2000. *On the Road in 2020—A Lifecycle Analysis of New Automobile Technologies*, Energy Laboratory Report MIT EL 00-003, Energy Laboratory, Massachusetts Institute of Technology, Cambridge. Cost relationships of hydrogen technologies are derived from National Research Council, 2004. *The Hydrogen Economy: Opportunities, Costs, Barriers, and R&D Needs*, National Academy of Engineering, Board on Energy and Environmental Systems, Washington, DC.

Notes: Capital and operation costs, except fuel costs, are averages of current technology or future projections. Ranges in costs and break-even oil price relate to lower and upper values of fuel and feedstock prices: \$6.0–12.0/GJ (natural gas), \$1.5–3.0 (coal and biomass), and 3–7 cents/kWh_{el} (electricity). Following the NRC study, the future costs for transporting and dispensing hydrogen were assumed to be \$5.6 per GJ. All dollars are U.S. 2000. Energy use and GHG emissions are our estimates based on several data sources. The performance characteristics of the indicated technologies can differ strongly across studies, because of different assumptions on scale, technology evolution, and other factors. Many of the characteristics shown are on the conservative side and could become typical for the performance of most plants put into operation during the 2020s.

[a] CCS: Carbon capture and storage.

[b] Number based on 2005 U.S. electricity generation.

[c] For passenger transport only.

[d] 1 ha = 0.01 km^2.

contrast, life-cycle GHG emissions from biomass or thermochemical processes are very small when the biomass feedstock or nuclear fuel supply and hydrogen distribution, compression, and storage are also powered with low-carbon fuels. (The low GHG emissions from biomass-derived hydrogen in table 6.4 result from cultivating the biomass feedstock.)

The fuel- and life-cycle characteristics per unit of energy shown in table 6.4 only indicate the comparative energy use and GHG emissions among the various hydrogen feedstock and process paths. Due to the different conversion efficiency of fuel to vehicle kilometers driven, it is important to compare life-cycle GHG emissions from different fuels on the basis of distance driven as opposed to on a fuel-energy basis. This is apparent from the comparative performance of the direct and indirect use of natural gas—that is, the performance of CNG compared with natural gas–derived hydrogen. While table 6.4 might at first suggest that the direct use of natural gas would be more energy efficient and less GHG intensive, the significantly higher energy efficiency of hydrogen fuel cell vehicles reverses the relative performance of the fuel pathways alone, as we show in the next section.

Table 6.4 also displays the cost characteristics of gaseous fuels. Assuming that the price of natural gas delivered to a hydrogen production facility is $6–$12/GJ or some 30 percent below the price of crude oil (assumed here to range from $50–$100 per barrel) per unit energy, and that the price of coal would be $1.50–$3.00 per GJ ($44–$88 per ton of coal), the unit energy costs of hydrogen from all production paths supplied to motorists would be within a range of $0.5 per liter of gasoline equivalent (natural gas–based process with lower end natural gas price) and $1.1 per liter of gasoline equivalent (water electrolysis with higher-end electricity price), corresponding to $1.9–$4.2 per gallon of gasoline equivalent. The electricity costs for water electrolysis were assumed to be between $0.03–$0.07 per kWh; higher electricity prices would increase the costs of supplied hydrogen significantly. However, as just noted, these costs need to be considered on a life-cycle basis, which does take into account the higher energy efficiency of hydrogen fuel cell vehicles and their higher cost compared with gasoline-fueled automobiles. We present this analysis in the next section.

As an approximate adjustment for such differences in the vehicle energy use, the break-even oil price and the scale impacts of fueling one-quarter of the U.S. LDV fleet assume the twofold energy efficiency of a hydrogen vehicle compared with its gasoline-fueled counterpart. It turns out that all hydrogen production pathways would be cost effective at oil prices above $50 per barrel, with the exception of electrolysis-derived hydrogen at electricity costs above $0.04/kWh. The scale impacts again illustrate the more than thousandfold difference in land-use requirements of biomass-derived hydrogen compared with the equivalent coal-based fuel.

Given the oil price rise in the early twenty-first century, all hydrogen pathways would be economically competitive, including the two proven ones shown in table 6.4. What then are the main barriers to a shift toward hydrogen fuel? As discussed in chapter 4, a large-scale shift to electricity has not been possible mainly because of the low energy density of batteries and their high cost. Also hydrogen storage onboard a vehicle is a major unresolved technical issue. Consider the highly fuel-efficient fuel cell vehicle projected in chapter 4, assumed to be available in the 2020s. When it achieves a range of about 650 kilometers (some 400 miles), it would need to store slightly more than 3 kilograms (nearly 7 lbs) of hydrogen. Several options for hydrogen storage exist, but none of them offers a satisfactory solution today.

Storing hydrogen as a compressed gas is the most obvious approach. This avenue requires high pressures to limit the volume of the hydrogen fuel to an acceptable portion of in-vehicle space. However, even at the high storage pressure of 345 bar (5,000 psi), the fuel volume of 3 kilograms of hydrogen corresponds to 130 liters (34 gallons) of gasoline, which is twice the size of the average fuel tank of today's midsize vehicles. The actual space requirement would be larger because of the cylindrical tank itself and the supporting infrastructure. In addition, onboard compressed hydrogen storage causes safety concerns.

Energy density can be increased significantly if liquid hydrogen is used. Because in its liquid state, one liter of hydrogen contains one-quarter of the energy of the same volume of gasoline, the vehicle under consideration would only need to carry a fuel volume of 42 liters (11 gallons). However, producing and then storing liquid hydrogen onboard a vehicle raises two concerns: the large amount of electricity required for hydrogen liquefaction, and safety (because of the unavoidable fuel evaporation and pressure buildup in a vehicle's cryogenic fuel tank).[75]

Because of these limitations that compressed gas and liquid hydrogen present, research and development in hydrogen storage technology is ongoing. One approach being pursued is to store hydrogen as a hydride, where a metal powder absorbs and releases the gas as the storage temperature and pressure change. Another approach aims to adsorb hydrogen within the pores of a charcoal-derived material that has a very large surface area. Although progress in these storage technologies has been reported, the consensus is that breakthroughs are still needed to establish commercial viability.[76]

Perhaps the greatest challenge, however, is the incompatibility of a new hydrogen-fuel regime with the existing fuel infrastructure. A shift to hydrogen involves the construction of numerous large hydrogen production facilities. Satisfying the energy demand for the 2005 U.S. LDV fleet used for passenger travel would require about 300 new chemical plants, each producing an average of nearly 1,200 tons of hydrogen per day. To supply this hydrogen to motorists, a dedicated transmission, storage, and refueling infrastructure would need to be created.

These infrastructure needs are intensified by the significant amount of energy necessary to produce hydrogen. Should hydrogen be produced from water electrolysis, the only proven potential zero-carbon pathway from today's perspective, the sheer scale of electricity generation would be enormous. Hypothetically satisfying the 2005 passenger travel energy demand by the U.S. LDV fleet with water electrolysis–derived hydrogen would require an additional electricity-generation capacity of about 900 GW_{el}—nearly doubling the total 2005 installed generation capacity of 1,000 GW_{el}.[77] It is highly unlikely that this extra capacity could be provided by renewably produced electricity before 2050, given these processes' low energy density and comparatively high costs. In fact, building that capacity with renewable plants before the end of the twenty-first century would still be a significant challenge.

Splitting atoms could be another potential source of carbon-free electricity. However, "going nuclear" would mean that the existing nuclear capacity of nearly 100 GW_{el} would have to increase ninefold. Given an average nuclear power plant capacity of about 1,000 MW_{el}, this strategy would require an additional nine hundred nuclear reactors. Even if public opposition to nuclear power declined, the acceptance of such an enormous expansion is difficult to imagine. Meanwhile, the hydrogen supply challenge would increase further, given the continuously rising travel demand projected in chapter 2.

In practice, the higher vehicle fuel efficiencies of hydrogen fuel cell vehicles could mitigate the electricity-generating capacity needed for a hydrogen fuel supply. If fuel-efficient hydrogen fuel cell vehicles were able to satisfy the projected 2050 travel demand (having the performance characteristics similar to those projected in table 4.3), the electric power needs would result in about 260 GW_{el}—or roughly one-quarter of the 2005 generation capacity. Based on today's average nuclear power plant capacity, the required electric power would translate into 260 additional nuclear reactors.

Recall that about 65 percent of the life-cycle electricity use of hydrogen fuel cell vehicles is consumed by the production of the hydrogen itself, its subsequent compression and conversion back to electricity. While the associated total electricity losses appear small on an individual vehicle basis, they accumulate to large totals for a fleet of several hundred million vehicles. In the U.S. context, 65 percent of the required extra electricity-generation capacity of 260 GW_{el} amounts to roughly 170 GW_{el} or about 170 nuclear reactors that would be needed just to supply electricity for the various conversion steps.

An alternative low-carbon hydrogen source would be coal, if carbon dioxide is captured and sequestered. Also in this path, the scale implications would be significant, albeit more manageable. As can be derived from table 6.4, the GHG emissions that would need to be captured and stored essentially correspond to the difference of the uncontrolled and controlled coal-based hydrogen fuel-cycle emissions, about 166 gCO_2-eq per MJ of hydrogen produced. Thus, supplying the 2050 LDV fleet with an energy-equivalent amount of hydrogen of 17–20 EJ would result in 0.7–0.8 billion tons of CO_2. A comparison with table 1.1 suggests that this amount corresponds to about 10 percent of the total 2005 U.S. energy-related GHG emissions. Given the annual large amount of CO_2 that would need to be stored, the coal-based process could plausibly only be a transition strategy to a truly low-carbon hydrogen production system.

Additionally, the disadvantages of using gaseous fuels (longer refueling times, banned parking in closed public spaces, and potentially the prohibited use of tunnels) would be even more severe for hydrogen than for natural gas.

These various concerns identify the major challenges that would need to be successfully addressed for hydrogen to become a large-scale transportation fuel: low-carbon hydrogen production, onboard vehicle storage, inconvenience of fueling and parking, lack of infrastructure compatibility, and the need for an unprecedented increase in electricity-generation capacity just for the sake of storing electricity in the form of hydrogen. Despite the large effort put into hydrogen research and devel-

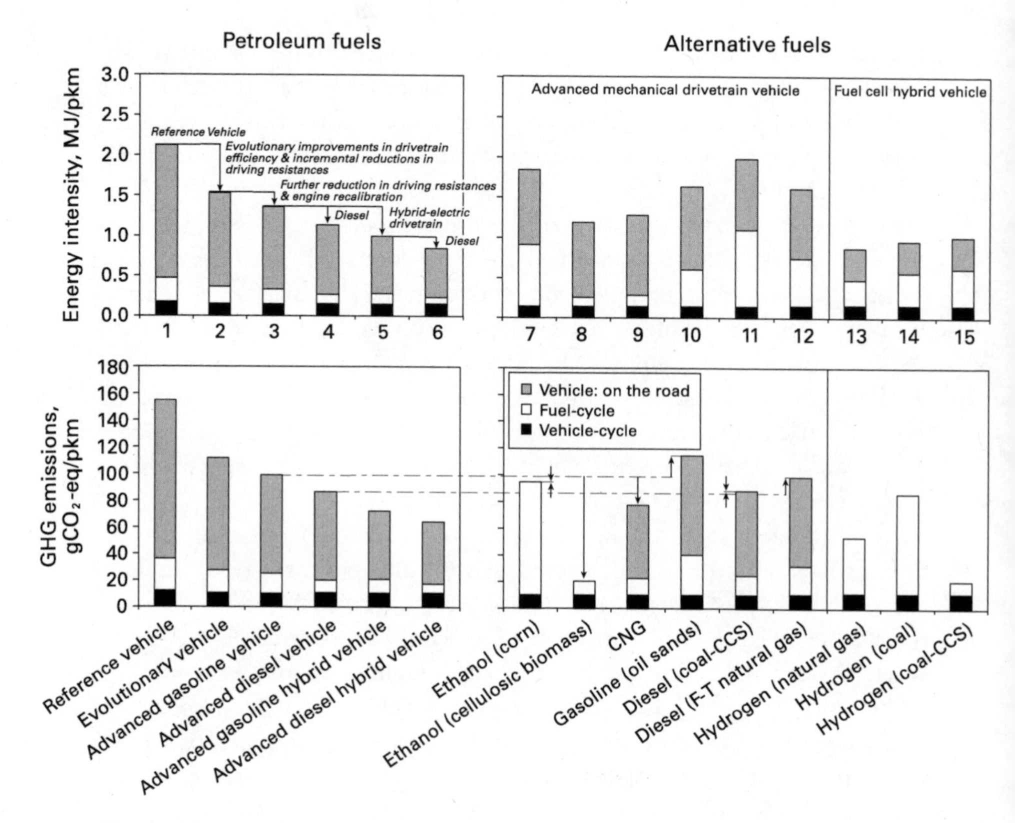

Figure 6.2
(a) Life-cycle characteristics of vehicle-fuel combinations: energy intensity (top) and greenhouse gas emissions (bottom). The vehicle occupancy rate was assumed to be 1.5 pkm/vkm. See table 4.3 for technology details of the shown vehicles. (b) Life-cycle characteristics of vehicle-fuel combinations: costs (top) and gasoline break-even costs (bottom). The vehicle occupancy rate was assumed to be 1.5 pkm/vkm. See table 4.3 for technology details of the shown vehicles.

opment, hydrogen's longer-term future remains uncertain. As discussed in chapter 4, energy density limitations significantly constrain the practicability of battery-only electric vehicles. Those limitations could be overcome, however, by hybrid-electric vehicles that incorporate batteries of sufficient storage capacity to allow more urban driving to be grid charged (plug-in hybrids). Thus, if the energy density of batteries can be further increased, the plug-in hybrid-electric vehicle could become preferable in light of all these hydrogen challenges.

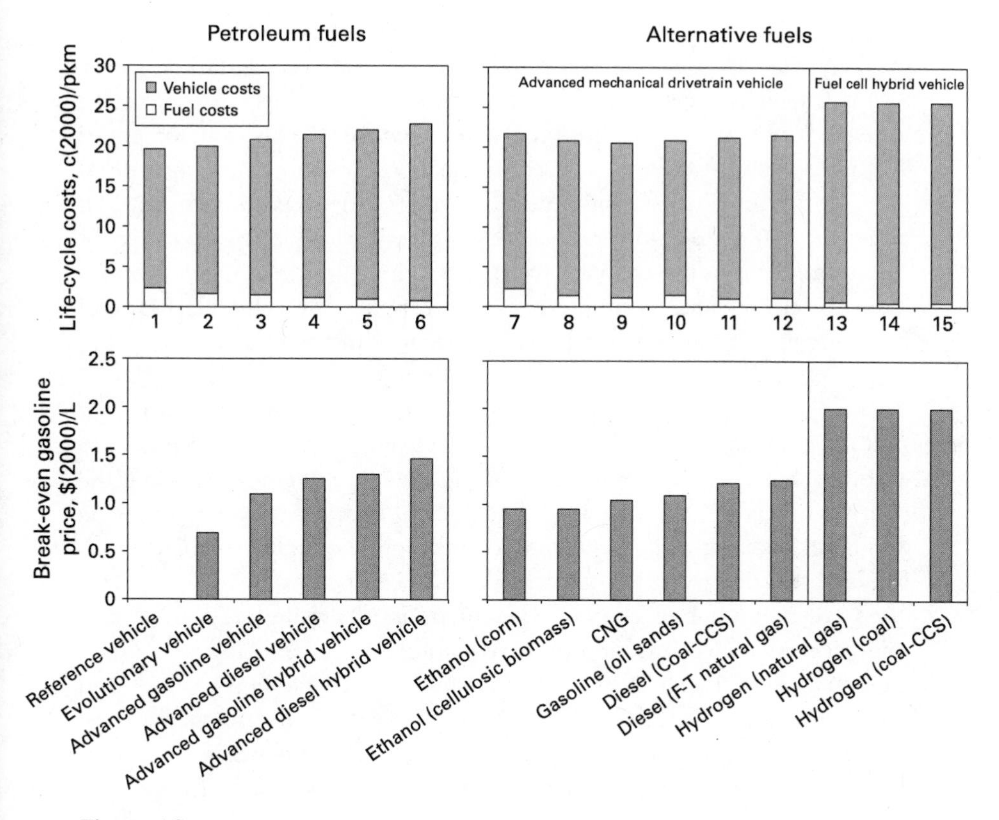

Figure 6.2
(continued)

Low GHG-Emission Light-Duty Vehicle–Fuel Combinations

To quantify the joint impact of fuel and vehicle technology, we combine the well-to-tank characteristics of the various fuel pathways discussed above with the tank-to-wheel characteristics of vehicle use summarized in table 4.3. We extend that analysis to include the energy use and GHG emissions associated with the production of the vehicles and their disposal at end of life, the so-called vehicle cradle-to-grave cycle.[78]

Figure 6.2a summarizes life-cycle energy intensity (top), and GHG emissions (bottom) for selected vehicle-fuel combinations, which are assumed to become available over the next 20–30 years. Figure 6.2b shows life-cycle costs and break-even crude oil prices. The reference vehicle for these comparisons is the typical new midsize automobile operating in the United States in the early 2000s. This vehicle's driving cycle-based, on-the-road fuel consumption is 7.7 liters per 100 vehicle-km (vkm) or 31 mpg, which is unadjusted for real-world driving. This corresponds to an energy use of 2.5 MJ/vkm and associated GHG emissions of about 180 gCO_2/vkm.[79] Using an average vehicle occupancy rate of 1.5 pkm/vkm, vehicle energy intensity and GHG emission translate into 1.7 MJ/pkm and 120 gCO_2/pkm, respectively. When fuel-cycle energy (for producing and providing gasoline from petroleum) and the energy requirement for producing, maintaining, and scrapping the vehicle at the end of its lifetime are added, total life-cycle energy intensity increases to 2.1 MJ/pkm, while GHG emissions rise to about 155 gCO_2-eq/pkm.

These characteristics are represented by bar 1 in the left top and bottom panels of figure 6.2a. As can be seen, on-the-road vehicle use accounts for about three-quarters of the life-cycle energy use, while the relative life-cycle proportion of producing and delivering gasoline is about 15 percent, with that of the vehicle cycle roughly 10 percent. A similar breakdown applies to life-cycle GHG emissions.

As we point out in chapter 4, reductions in driving resistances and improvements in propulsion systems can reduce energy use and GHG emissions significantly. We concluded that vehicle energy use and GHG emissions could be reduced by implementing several changes: incremental reductions in driving resistances, improvements in drivetrain components, and a reduction in engine size from six to four cylinders (reducing energy demand while maintaining the same power-to-weight ratio). Such an evolutionary approach would reduce vehicle energy use by about 30 percent—in this case from 2.5 MJ/vkm to 1.8 MJ/vkm—without changing the size and acceleration capability of the reference vehicle (1.7 MJ/pkm to 1.2 MJ/pkm, as shown by bar 2, evolutionary vehicle). Further reductions in vehicle energy use and GHG emissions are possible through more drastic changes in vehicle technology.

The strongest reduction in vehicle energy use (up to an additional 50 percent) is represented by a diesel-fueled hybrid-electric automobile (bar 6). A more fuel-efficient vehicle requires less fuel, which in turn translates into fewer energy inputs into the fuel cycle. Thus, fuel-cycle-related energy use and GHG emissions decline in proportion to less on-the-road energy use (with consequent fewer emissions). In contrast, *vehicle* cycle energy use and GHG emissions slightly increase for lighter weight vehicles, mainly because of the high share of aluminum used in them, which is more energy- and GHG-emission-intensive to produce. Bars

3–5 represent the life-cycle characteristics of vehicles from table 4.3 that have intermediate levels of energy use and GHG emissions.

Vehicle technology improvements represent only one avenue for reducing GHG emissions. Additional reductions can be achieved through deploying some of the alternative transportation fuels discussed in this chapter. Bars 7–12 in figures 6.2a and 6.2b represent selected fuel opportunities for an advanced mechanical drivetrain vehicle. The dashed and dotted lines shown in figure 6.2a are meant to facilitate making comparisons to vehicles fueled by petroleum-derived gasoline or diesel. Several of the vehicles using alternative fuels release lower levels of GHG emissions than does the corresponding oil-derived gasoline vehicle (bar 3). While currently produced corn-derived ethanol (bar 7) results in about a 5 percent reduction in life-cycle GHG emissions compared with the (petroleum-derived) gasoline-fueled alternative, that reduction would increase to more than 80 percent in the case of ethanol produced efficiently from cellulosic biomass material (bar 8). Note also that with these fuels, no net CO_2 emissions remain from vehicle operation, since the biomass-derived CO_2 is assumed to be reconverted to plant matter via the photosynthesis process; only the fossil energy inputs into the fuel cycle and the vehicle cycle show up in life-cycle emissions. Given the characteristics shown in table 6.3, life-cycle energy use, GHG emissions, and the costs of cellulosic ethanol would be roughly similar to a BTL life cycle.

As discussed earlier, CNG vehicles (bar 9) provide a roughly 25 percent reduction in life-cycle GHG emissions compared to the gasoline-fueled vehicle. The only technology-fuel combination showing higher life-cycle emissions is represented by bar 10, where gasoline is derived from oil sands. As can be seen by comparing bar 10 with bars 2 and 3, shifting toward oil sands would offset the preferential technology potential offered by shifting from bar 2 and bar 3. Bars 11 and 12 represent an advanced diesel vehicle powered with synthetic fuels from coal and natural gas, respectively. Bar 11 illustrates the potential for reducing GHG emissions at the synthetic fuel plant through carbon capture and sequestration. As a result, life-cycle GHG emissions of coal-based synthetic fuels are comparable to those from the same vehicle fueled with petroleum-derived diesel. In contrast, the uncontrolled natural gas-derived diesel fuel would result in about 15 percent higher life-cycle GHG emissions (bar 12).

Bars 13–15 show the life-cycle characteristics of a hydrogen-fueled hybrid fuel cell vehicle, for which the hydrogen is derived from various fossil sources. Life-cycle GHG emissions for an evolutionary vehicle (bar 2) are reduced by about half in the case of natural gas–derived hydrogen (bar 13), by about 10 percent if hydrogen is derived from coal (bar 14), or by 80 percent if the CO_2 emissions of the coal refinery are captured and sequestered (bar 15). Note that GHG emissions are released only during the vehicle and fuel cycles; the conversion of hydrogen to electricity onboard the vehicle releases no GHG emissions.

These examples attest to a key conclusion of the MIT study *The Future of Coal.*[80] Coal can play an important role in mitigating GHG emissions if, and only if, these emissions are captured and stored. However, as can be seen if we compare bar 11 to bar 4 and bar 15 to 14 in the upper panel, using coal as a feedstock to produce a low-carbon fuel comes at the cost of a higher life-cycle energy use. Additionally, not shown here are the associated nonenergy and non-GHG emission externalities, and the impacts of land subsidence and groundwater contamination that accompany increased coal mining.

Figure 6.2a also shows that the direct use of a fuel does not necessarily result in the least amount of GHG emissions. A comparison of bars 13 and 9 suggests that the significantly higher fuel efficiency of the fuel cell vehicle compared to the CNG vehicle more than compensates the higher fuel-cycle losses, which are associated with hydrogen production.

Figure 6.2b shows the consumer costs of owning and operating a vehicle per kilometer driven. Our cost estimates, which—for easier comparison—exclude fuel taxes, are based on a crude oil price of $50 per barrel. In keeping with our discussion of consumer behavior in chapter 3, we have employed a consumer discount rate of 25 percent to a vehicle life of fifteen years, which reflects an amortization period for fuel-saving technology of slightly less than four years. At that high discount rate, the current value of future fuel savings is too small to compensate the retail price increment of more fuel-efficient vehicle alternatives. As can be seen at the top of figure 6.2b, fuel costs decline from bar 1 to 6, but total ownership and operating costs increase. Thus, all other factors being equal, most consumers would not be willing to purchase a vehicle much different from the reference car.

Bars 7–15 show the life-cycle consumer costs of the various vehicle and alternative fuel combinations. At the underlying crude oil price of $50 per barrel, nearly all the synthetic oil products in figure 6.2 can be produced cost effectively (see table 6.2). Thus, consumers would likely experience a similar fuel price for these products as for the corresponding petroleum product. A direct implication of this assumption is that these vehicle–alternative fuel combinations impose life-cycle costs similar to those of petroleum product–fueled advanced gasoline and diesel vehicles.

Since CNG, ethanol, and hydrogen are not direct substitutes for gasoline, their future prices are likely to be different. Using the fuel prices from table 6.3 and 6.4, life-cycle costs of the corn-derived ethanol-powered vehicle (bar 7) would be slightly higher than those of an advanced gasoline vehicle (bar 3). Due to the projected lower fuel costs, the cellulosic biomass-derived ethanol-powered vehicle and the CNG vehicle would experience life-cycle costs similar to those of the advanced gasoline vehicle (bar 3).[81] The highest life-cycle costs would result for fuel cell hybrid vehicles (bars 13–15). These life-cycle costs are determined largely by the high retail price of hydrogen fuel cell vehicles, since these vehicles' high fuel efficiency causes fuel costs alone per kilometer driven to be very low.

Given their higher life-cycle costs, pushing vehicles that have fewer GHG emissions into the automobile market would require supporting government policy measures or incentives. For example, instituting higher gasoline prices would change the cost-competitiveness equation to favor more expensive, but more fuel-efficient, vehicles (see chapter 7). The bottom of figure 6.2b shows the break-even fuel prices for making several more fuel-efficient technologies cost competitive. For petroleum-derived fuels, the gasoline retail price would need to be at least $0.69 per liter ($2.6 per gallon) to initiate a technology shift away from the reference vehicle to what we've characterized as the evolutionary vehicle. The highest break-even gasoline cost of $1.46 per liter ($5.5 per gallon) is associated with the advanced diesel hybrid vehicle. The break-even gasoline prices for all liquid-alternative fuel vehicles shown here (bars 7–12) fall close to $1 per liter. Hydrogen fuel cell hybrid vehicles (bars 13–15) show break-even costs on the order of $2 per liter of gasoline ($7.6 per gallon).[82]

Fuel prices in excess of $1 per liter ($3.8 per gallon) have already existed in many European countries and more recently in the United States in the early 2000s. Should fuel prices remain at these levels, they would thus provide an incentive for vehicle purchasers to choose one of

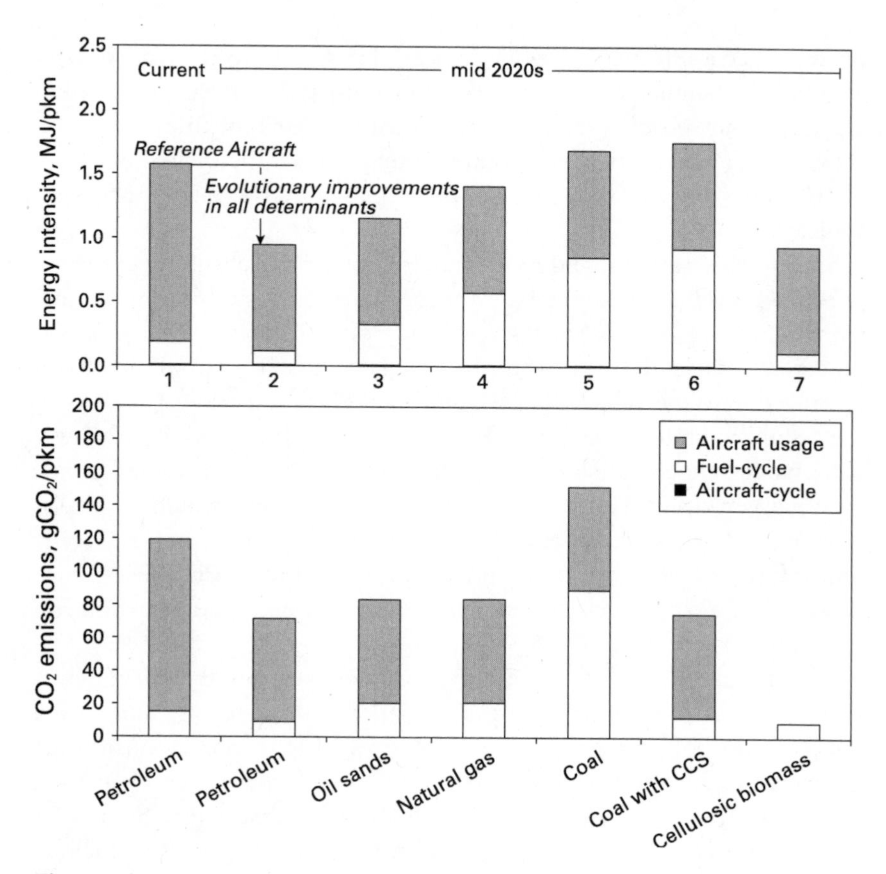

Figure 6.3
Life-cycle energy intensity (top) and CO_2 emissions (bottom) for several aircraft-fuel combinations. The reference aircraft is a Boeing 777 with energy intensity values as given in figure 5.1. The fuel-cycle energy use and CO_2 emissions were estimated using tables 6.2 and 6.3.

the projected lower GHG-emission vehicle fuel combinations in figure 6.2. These and other policy opportunities are discussed in chapter 7.

Low GHG-Emission Aircraft-Fuel Combinations

In analogy to figure 6.2a, figure 6.3 shows life-cycle energy intensity and GHG emissions for various aircraft-fuel combinations. Note that in the figure, the category "aircraft usage" includes only CO_2 emissions, while the category "fuel cycle" includes CO_2 and methane emissions. Life-cycle energy use and GHG emissions associated with aircraft usage and fuel production and delivery have roughly similar proportions to those associated with automobiles, as shown in figure 6.2a. Unlike automobiles, however, energy use and GHG emissions associated with aircraft production are negligible relative to those during use. This is partly because aircraft are operated an average of twelve to fifteen hours per day compared with only one hour per day typical for automobiles. Aircraft also carry 50–100 times more passengers at significantly higher speed during these hours.[83] Together, both factors imply that energy use and GHG emissions per PKT of aircraft production are roughly 1000 times lower than those of automobile production and are thus well within the uncertainty range of our life-cycle estimates.

As shown in figure 6.3, life-cycle energy intensity of today's "typical" commercial passenger aircraft (a B777) is about 1.6 MJ/pkm. The aircraft itself requires 87 percent of that amount, while the remaining 13 percent is expended in fuel production and delivery. The corresponding life-cycle CO_2 emissions (nearly 120 gCO_2/pkm) show a similar distribution between aircraft operation and fuel cycle. As discussed in chapter 5, by the mid-2020s, the energy intensity of the average new commercial airplane can be reduced by 30–50 percent. We assume a 40 percent reduction in aircraft energy intensity, which results in a proportional decline in life-cycle energy intensity and CO_2 emissions (bar 2).

Being more constrained in the choice of suitable transportation fuels than the automobile, further significant reductions in life-cycle CO_2 emissions—by almost 90 percent—can only be achieved if synthetic jet fuel produced from renewable biomass feedstocks is used (bar 7). As shown in figure 6.3, net CO_2 emissions from aircraft operation would be very small, since they are absorbed by later generations of biomass feedstocks. All other jet fuel feedstocks would result in similar or higher life-cycle CO_2 emissions than those associated with bar 2. Synthetic jet fuel from oil sands (bar 3) or natural gas (bar 4) would increase life-cycle CO_2 emissions by nearly 20 percent. Synthetic jet fuel from coal (bar 5) would even result in a doubling of life-cycle CO_2 emissions and thus more than offset the reductions achieved through aircraft fuel-efficiency improvements. However, capturing and storing CO_2 emissions at the coal refinery would result in life-cycle CO_2 emissions that are comparable to those from petroleum-derived jet fuel (bar 6).

Note that the fuel options shown in figure 6.3 are not exhaustive. As discussed earlier, other opportunities are currently being tested such as blends of petroleum-derived jet fuel with 20 percent vegetable oils or biodiesel by volume. These options, however, would result in reductions of lifecycle GHG emissions of less than 10 percent compared to the exclusive use of petroleum-derived jet fuel.

Comparing all Characteristics

In the preceding discussion, we identify several alternative fuels that have the potential for significantly reducing life-cycle GHG emissions. But how do these fuels compare with respect to other critical dimensions, such as environmental impacts other than greenhouse emissions, the existence of fuel resources outside the main oil-exporting countries, convenience of fuel use, compatibility with the existing fuel infrastructure, and the number of years it might take for the alternative fuel to have a substantive impact?

Table 6.5 summarizes the characteristics of alternative fuels relative to petroleum products according to these criteria. A 0 entry corresponds to characteristics similar to petroleum-derived fuels, whereas a plus (+) denotes superior characteristics; the number of pluses reflects competitive advantage. Similarly, the more minuses (–) a fuel has, the less attractive its characteristics or the worse its performance compared with petroleum products. These entries are qualitative and based on our judgment, so we have added a key at the bottom of the table to explain our choices.

Three columns contain most of the pluses: life-cycle GHG emissions (resulting from our desire to reduce GHG emissions), the degree of petroleum displacement (which results from alternative fuels' production processes being independent from oil), and fuel and feedstock resources within the Western Hemisphere (which is a consequence of the comparatively low occurrences of conventional oil and natural gas in this part of the world).

The largest block of zeros (no change relative to oil products) is associated with synthetic oil products from fossil fuels, because of their convenience as fuels and compatibility to the current fuel infrastructure. These fuel and infrastructure characteristics act as enablers, but the main driver for the growing interest in oil products produced from oil sands, natural gas, and coal has been concern about oil dependence. (The three pluses in the petroleum displacement column indicate those fuels' significant potential to replace oil.) These fuels' main disadvantage is their environmental performance. Unless carbon capture and sequestration technology is effectively employed during production, all synthetic petroleum fuels release a larger amount of life-cycle GHG emissions than do petroleum-derived fuels. Also, with the exception of GTL, synthetic oil products from fossil fuels generate additional environmental impacts.

The largest contrast between pluses and minuses exists for hydrogen produced from zero-carbon sources or from fossil fuels while employing effective carbon capture and storage. Under these circumstances, hydrogen's advantages of greatly reduced life-cycle GHG emissions, a nearly complete displacement of petroleum, and—with the exception of natural gas- and biomass-derived hydrogen—essentially unlimited resources in the Western Hemisphere, are compromised "only" by inconvenience issues: its very low density and its incompatibility with the current fuel infrastructure. For these reasons, hydrogen is unlikely to achieve a significant share of the fuel market, 25 percent, until after 2050—if at all.

The only column having only minuses describes scale-related challenges, here the time required to achieve a 25 percent market share. As discussed at the beginning of this chapter, changing transportation fuels takes time. If the objective is to rapidly lower GHG emissions through the use of alternative fuels, only biodiesel as a B20 blend, cellulosic ethanol as an E20 blend, or CNG could fuel one-quarter of the passenger travel–related vehicle fleet before 2030. Pursuing any one of these strategies, however, also requires thinking through a long-term strategic plan, because investments in new transportation fuels are significant and our earlier discussion questions, for example, the longer-term and large-scale feasibility of both B20 (a first-generation fuel) and CNG.

The distribution of pluses and minuses for biofuels is least extreme. In blends of up to 20 percent by volume, these fuels are largely compatible with the existing fuel and vehicle infrastructure. Given the already significant and rising production levels of first-generation biofuels, 20 percent volume blends with oil products could achieve a significant market share before 2030 (again, 25 percent). However, given the dependence of current biofuel profitability on coproduct markets, their further growth potential is likely to be limited.

The situation is different for second-generation biofuels. As the underlying biomass-to-fuel processes don't generate significant coproducts, the supply potential of second-generation biofuels is limited mainly by the amount of suitable land area available (and production costs). Since second-generation processes have only just begun to be commercialized, these fuels are unlikely to exceed one-quarter of U.S. road vehicle fuel use before 2030. Table 6.5 also shows that both first- and second-generation biofuels can significantly displace petroleum products.

Table 6.5
Characteristics of alternative transportation fuels relative to current petroleum products

	Life-cycle GHG emissions	Additional environ-mental impacts	Petroleum displace-ment potential	Oil break-even costs[a]	N. American/ W. European energy security improvement	Conve-nience of fuel use	Compati-bility with existing fuel infra-structure	Time to 25% market share
Petroleum diesel (LDV only)[b]	+	–	+	0	+/0	0	0	0
Synthetic oil products from fossil fuels								
Oil sands	–	– –	+++	+	+++/0	0	0	–
Gas-to-liquids (GTL)	–	0	+++	+	0/0	0	0	–
Coal-to-liquids (CTL)	– – –	– – –	+++	0	+++/+++	0	0	–
CTL with CCS	0	– – –	+++	0	+++/+++	0	0	–
Shale oil	– – –	– – –	+++	0	+++/+	0	0	–
First-generation biofuels								
E20 (LDV only)	0	0	+	0	+/+	0	0	–
Ethanol (LDV only)	+	0	+	–	+/+	–	–	–
B20	+	0	+	–	+/+	0	0	–

Second-generation biofuels								
E20 (cellulosic; LDV only)	+	0	+	0	+/+	0	0	–
Ethanol (cellulosic; LDV only)	+++	0	++	0	++/++	–	–	––
BTL	+++	0	++	0	++/++	0	0	––
Gaseous fuels (LDV only)								
CNG	+	0	+	0	+/0	––	––	–
H_2—CH_4	–	0	+	+	+/0	–––	–––	–––
H_2—coal	–––	–––	+++	+	+++/+++	–––	–––	–––
H_2—coal (CCS)	+++	–––	+++	+	+++/+++	–––	–––	–––
H_2—electrolysis[c]	+++	+	+++	0	+++/+++	–––	–––	–––
H_2—nuclear	+++	0	+++	+	+++/+++	–––	–––	–––
H_2—biomass	+++	0	++	0	++/++	–––	–––	–––
Electricity (LDV only)	––/+++	–––/+	+++	0	++(+)/++(+)[d]	–––[e]	–	––

Note: [a] Oil break-even costs of hydrogen fuels are based on 2X vehicle energy efficiency.
[b] Relative to gasoline
[c] Zero-carbon electricity
[d] Significant energy security improvement by 2030; major improvement by 2050
[e] Electric vehicle range limitation major negative

Key to table 6.5

	Life-cycle GHG emissions (%)	Additional environmental impacts (–)	Petroleum displacement potential (%)	Oil break-even costs[a] ($/bbl)	N. American/ W. European energy security improvement (–)	Convenience of fuel use (–)	Compatibility with existing fuel infrastructure (–)	Time to 25% market share (–)
+++	<–50	—	>50	—	Major	—	—	—
++	–25 to –50	—	25–50	—	Significant	—	—	—
+	–5 to –25	Decline in all emissions	5–25	<50	Moderate	—	—	—
0	±5	Largely unchanged	±5	50–100	Unchanged	No change	No change	Before 2020
–	5–25	Increase in non-CO_2 emissions	—	>100	—	More fuel stops	Modifications required	2020–2030
– –	25–50	plus high water use, possibly groundwater contamination	—	—	—	Parking constraints	New and partly existing	2030–2050
– – –	>50	plus land subsidence	—	—	—	Parking constraints and longer refueling time	Completely incompatible	Beyond 2050

As with other synthetic fuels, the amount of life-cycle GHG emissions for electricity strongly depend on the primary energy source. (To keep the size of the table manageable, we added three pluses and minuses instead of differentiating by fuel). Similar to the situation with gaseous fuels, convenience of fuel use, compatibility with the current fuel infrastructure, and the time needed to achieve a significant market share are all negatives for electricity. Without a breakthrough in battery technology or radical change in consumer expectations, battery-only electric vehicles are unlikely to enter the market on a large scale.

Electricity, however, can be an enabler of other fuels and technologies. Since no fuel satisfies every desired characteristic simultaneously, there could be a significant role for sensible combinations. We already have discussed the potential for plug-in hybrid-electric vehicles as a way to use grid electricity for most daily driving while using the internal combustion engine for long-distance trips. Following this logic, combining the rows in table 6.5 for electricity with any of the biofuels results in significantly improved overall performance.

Summary

The global fuel market is already in transition. Driven by rising oil prices and the increasing dependence on oil imports from politically less stable regions, the share of synthetic oil products from unconventional oil and natural gas is already growing. While coal-based feedstocks will likely join this trend, shale oil–derived fuels are unlikely to gain significant market share during the next several decades. All these pathways result, however, in increased levels of life-cycle GHG emissions, unless CO_2 emissions at the synthetic fuel refinery are sequestered.

At the same time, several governments are increasingly promoting biofuels through subsidies and regulation. Many factors underlie these actions, but they especially include concern over oil dependence, and, increasingly, concerns over climate change. Given current government targets, especially in the United States, this trend is likely to intensify and spread. In contrast to fossil fuel–based synfuels, these biofuel pathways offer promising opportunities for reducing GHG emissions, especially in the future. However, to avoid the release of significant amounts of soil and plant-related CO_2 emissions, careful land-use planning, intensification of existing plantations, and the use of forest and agriculture residues is critical.

Other low-carbon alternatives exist in addition to liquid biofuels, but their exploitation is constrained by numerous negative factors. The alternative low-carbon fuels we have discussed would lead to or come with some combination of these attributes: greater resource limitations, significantly less ease in fuel handling and storage, compatibility issues with the existing fuel infrastructure (thus making their introduction more challenging), or long lead times before being able to gain a significant market share. Thus, before 2030, the only lower-GHG-emission fuels that have the potential to gain significant market share are first-generation biofuels. These include corn-derived ethanol, biodiesel from vegetable oils blended with petroleum products at a volume ratio up to 20 percent, and biodiesel or vegetable oils blended with jet fuel. The resulting reduction in life-cycle GHG emissions, however, is below 10 percent. In addition, the economics of most first-generation biofuels depend on byproduct markers. Beyond 2030, second-generation biofuels, including cellulosic ethanol (for surface transport) and BTL (for surface and air transport), are expected to provide more significant reductions in GHG emissions. Provided that these fuels can be produced at acceptable cost, their main limitation becomes the amount of land available for growing biomass feedstocks.

Among the gaseous fuels, only hydrogen has potential for large-scale application. However, in addition to the fuel use and on-board vehicle storage challenges, and the inconvenience of gaseous fuels, a hydrogen transportation system would require a completely new production, distribution, and storage infrastructure. The generation of hydrogen from water electrolysis, the only proven potentially zero-carbon process, would also require an unprecedented increase in electricity-generation capacity. In road transportation, electric vehicles with advanced batteries, perhaps in hybrid mode with a cellulosic ethanol–fueled internal combustion engine for long-distance driving, could have a far smaller impact on the existing energy system and yet offer similar levels of low-carbon mobility.

7

Policy Measures for Greenhouse Gas Mitigation

Changes in vehicle technology have the potential to lower greenhouse gas (GHG) emissions from private automobiles and passenger air travel (chapters 4 and 5), and increased use of lower-GHG-emitting fuels could make an additional contribution to this effort (chapter 6). Unfortunately, without public policy to support these improvements, their cost will limit their contribution to controlling ever-rising GHG emissions (chapter 3). Here, therefore, we will explore government policies that have been proposed or are already being applied in various countries in an effort to bring technical advances and alternative fuels into use and to reduce vehicle kilometers traveled (VKT) on the road and in the air. Our focus is on the United States, which is the world's largest market for auto and air travel, but we will occasionally refer to the experiences of other countries.

As might be expected from the complexities laid out in earlier chapters, the development of an effective, economically efficient, and equitable emissions policy for passenger travel is no easy task. At issue are alternative modes of transportation that have different market structures and regulatory overlays as well as very different technical characteristics. Also important are concerns about interfirm and international competition and external effects like traffic congestion, urban air pollution, and the foreign policy implications of oil imports. Thus, we will first explore these special characteristics of passenger travel, and then review policy measures that can be used to reduce GHG emissions from this sector.

Characteristics of Passenger Travel

Perhaps most complex among the decisions that influence emissions from personal transportation is the individual consumer choice of an automobile. This decision involves trade-offs among many desired attributes: vehicle size and comfort, driving performance, safety, auxiliary devices, reliability, fuel consumption and other operating costs, convenience of fuel access, and initial price. As described in chapter 3, rising affordability has allowed consumers to value size and performance over fuel efficiency. Though more constrained by the technical and regulatory requirements of flying, aircraft present similar trade-offs among initial cost, fuel consumption, and expenditures on maintenance, air emissions, and noise pollution. The trade-offs among these auto and aircraft attributes, and decisions to scrap old equipment, are important targets of carbon dioxide (CO_2) mitigation policies, through changes in relative prices or restrictions of the range of consumer and industry choice.

A number of studies have attempted to determine how consumers choose among vehicle attributes, and most confirm that future fuel expenditures do not play a strong role. Purchasing patterns suggest that consumers expect investments in fuel-saving measures to pay off within three to four years, implying a discount rate of 25–30 percent.[1] In light of the long-term history of U.S. gasoline prices, this short amortization period means that, absent vigorous policy intervention, only a limited number of the fuel-saving technologies in chapter 4 would find their way into the on-road fleet. For example, at a gasoline price of 25 cents per liter (about $1 per gallon), the fuel-saving automobile technologies that consumers seem willing to buy would lead to a roughly 10 percent reduction in the average 2005 new vehicle fuel consumption.[2] In contrast, mid-2008 gasoline prices of $1 per liter (about $4 per gallon) could lower fuel consumption of future vehicles by an additional 30 percent.[3] However, it is questionable whether oil price–induced fuel prices will remain at this high level, and as this book goes to press fuel prices have declined. Moreover, these implicit consumer discount rates are sometimes augmented by rebate offers that favor heavier and thus less fuel-efficient alternatives.[4]

Airlines face similar choices, although they have a greater incentive than automobile buyers to pay close attention to fuel economy. As pointed out in chapter 3, the fuel bill can easily account for as much as 40–50 percent of the direct costs of owning and operating a commercial aircraft, so even minor fuel savings can have a significant effect on net revenue.[5]

Another factor influencing transportation emissions is the global movement away from public transport modes to the automobile in urban transport and toward high-speed transportation in intercity travel (see chapters 2 and 3). Reducing automobile dependence through land-use planning and increased use of low-speed public transport modes is both challenging and—because of the long turnover time of the infrastructure—long term. Even in intercity travel, where high-speed rail could displace short-to-medium-range air flights, these alternatives are commercially viable in only a few high-density corridors. And, unfortunately, there is not yet evidence that telecommunications will turn out to be a large-scale net substitute. Finally, as discussed in chapter 6, for cars and aircraft there is not yet any effective, large-scale substitute for fossil fuels. Biofuels and plug-in hybrid vehicles may offer possibilities for the future, but even given the needed technical advances, they will require several decades to achieve substantial market shares.

These factors—consumer and airline vehicle choice, the trend away from low-speed public transportation, and dependence on fossil fuels—combine to make the price elasticity of fuel use (and thus of CO_2 emissions) lower for personal transport than for some other sources of CO_2 emissions. For example, studies show that passenger transport is less responsive to fuel prices or emissions penalties than many household, commercial, and industrial end users.[6] Even when fuel substitution is more constrained, as in electricity demand for household appliances, price elasticities are still higher than for automobile fuel.[7] These different levels of consumer response have important implications for price-based GHG-emission-abatement policies.

The historically high rate of growth of auto travel, together with the relatively weak response of vehicle fuel efficiency to fuel price, has made auto travel a special target of mitigation measures. Increasingly, there is a call to do *something* to constrain emissions from this sector regardless of mitigation policies imposed elsewhere. This demand—the flip side of the supposed U.S. love affair with the automobile—raises the question of whether both consumer and corporate choices are somehow more flawed with respect to personal transport than for other consumption that uses energy and emits CO_2. Should a family be more heavily penalized for demanding interior room in its motor vehicle than in its home? Or for driving to Florida instead of turning up the thermostat? These are questions to which we return below.

The Existing Policy Context

Road vehicles, aircraft, and their operations are already subject to energy and environmental policies such as air pollution controls, fuel economy regulation, and subsidies to research and development (R&D) on fuel-saving technology. In considering additional interventions directed at GHG emissions, we need to take into account the potential interactions with all these measures—and their consequences. Three examples can illustrate the complexities involved: fuel taxes that differ both across and within countries and across modes of transport and fuel type, the regulation of non-GHG externalities, and the imposition of vehicle fuel economy standards.

Fuel Taxes

Taxes on transport fuel differ in many dimensions: among countries and between states within any one country, and by fuel and its use. In the United States, fuel taxes are imposed at both federal and state levels, and the current 11.1 cents per liter of gasoline (42 cents per gallon) is a combination of a federal tax of 4.9 cents (18.4 cents per gallon) and an average state tax of 6.2 cents (23.6 cents per gallon). The state component varies widely, ranging from as little as 2.1 cents per liter (8 cents per gallon) in Alaska to 9.3 cents per liter (35.1 cents per gallon) in Hawaii. The rates for diesel fuel are similar, and they similarly vary by state.

Fuel taxes are significantly higher in other developed countries, including European nations and Japan. In Germany, fuel taxes are currently 50.1 cents per liter of gasoline (about $1.90 per gallon) and 31.6 cents per liter of diesel. Both fuels are also subject to an ecological tax of 15.4 cents per liter and a tax to maintain strategic oil reserves of around 0.5 cents per liter of gasoline and 0.4 cents per liter of diesel fuel. On top of these taxes, which total 66 cents per liter of gasoline ($2.50 per gallon) and 56 cents per liter of diesel ($2.13 per gallon), is the value-added tax, currently set at 19 percent of the before-tax fuel price.

These patterns of taxation have long and varied histories. The U.S. taxes on motor fuel were implemented as road-user fees, with their proceeds applied to building and maintaining the nation's highway system. In the past, they financed the entire system, but they have not kept pace with the growth in costs and now cover only about one-third of expenses. The remainder of highway funding comes from real estate taxes, vehicle taxes and fees, tolls, and other sources.[8] In contrast, the relatively high motor fuel taxes in Germany, for example, originally were a response to fiscal and trade pressures. In the wake of World War II, demand for transport fuels grew rapidly, and there was a need to husband limited domestic refining capability and foreign exchange as well as to finance road construction. Over time motor fuel taxes became an important source of revenue and have remained in place ever since. Today, fuel taxes account for almost twice the total German government transport expenditures.

The taxation of aviation fuels has its own history and complexity. Since aircraft may cross several tax jurisdictions during a single flight, potentially creating multiple tax obligations, as early as 1951 the Council of the International Civil Aviation Organization adopted a tax exemption for fuel burned in international commercial air travel. On the other hand, several countries have imposed fuel taxes on domestic flights, including the United States, Japan, and the Netherlands. U.S. domestic flights are subject to a federal tax of 0.9 cents per liter (3.4 cents per gallon). Other countries are following suit, and the European Union will include air travel within its Emissions Trading Scheme (ETS) from 2012 on, in effect subjecting aircraft fuel to a CO_2 tax at the level of the ETS price.[9] This pattern of variation in tax rates among transport fuels in different places and uses, and the different rates of taxation of transport fuel compared with other commodities, can create price distortions. For example, fuel conservation or emissions reduction could be accomplished at less cost to the consumer by lowering the taxes on certain fuels and raising them on others. In this context, the imposition of a uniform carbon penalty could add to existing price distortions and increase the economic cost of the mitigation policy.[10]

Regulation of Non-GHG Externalities

In addition to CO_2 emissions, passenger transport creates other externalities, some of which are subject to regulation or price penalties, and some not. Consider LDVs. Several of their external effects are related to the amount of fuel consumed, such as macroeconomic costs associated with the volatility of world oil prices, national security effects of oil import dependence, and damage from oil spills. A U.S. National Research Council (NRC) report estimated that the aggregated external marginal costs from personal transport come to about 7 U.S. cents per liter of fuel (26 cents per gallon).[11] The estimate included the effects of oil price volatility, oil spills, and damage caused by climate change (quantified at $14 per ton of CO_2 or $50 per ton of carbon). Adding an estimate by the Oak Ridge National Laboratory of the cost associated with national security issues, which is between zero and 3.1 cents per liter ($0 to $5 per barrel), results in a total of up to 10 U.S. cents per liter (39 cents per gallon).[12]

Other external costs of automobile use derive not from fuel volume but from VKT. Principal among these is the value of the time lost because of congestion, followed by the cost of accidents. The range of estimates is wide, but according to a study by Randall Lutter and Troy Kravitz from the AEI–Brookings Joint Center, they center on a marginal congestion cost of 5 cents per kilometer (8 cents per mile) and a cost for accidents of 2 cents per kilometer (3 cents per mile). Additional external effects come from criteria air pollutants (carbon monoxide, lead, nitrogen dioxide, sulfur dioxide, particulate matter, and ground-level ozone) and automobile noise. Again, the range of estimates is wide, but the mean values are slightly below 2 cents per kilometer (around 3 cents per mile) for air pollutants and around 0.07 cents per kilometer (0.11 cents per mile) for automobile noise.[13] Public policies influence these externalities through road design and law enforcement, but with the exception of local ordinances and a few road-pricing systems (e.g., Singapore and the City of London) there are no measures at present to control VKT-related external effects directly.

Two shortcomings characterize attempts to control the effects of these externalities. First, vehicle emission standards ignore the trade-off between measures to limit criteria emissions and those directed at fuel consumption. Because of its higher particulate and NO_x emissions, a diesel engine requires more expensive emission-cleaning technology and sacrifices about 5 percent of the diesel energy-efficiency advantage over gasoline engines (see chapter 4).

Second, the economically most efficient way to correct for these VKT-related externalities would be to reduce travel. Absent widespread systems of road rationing or congestion tolls, the closest approximation would be a fuel tax calibrated to average vehicle use. For example, the 2005 fleet-average U.S. light-duty vehicle (LDV) operating at 10.7 liters per 100 vehicle-km (vkm) or 21.9 mpg has an externality cost (congestion, accidents, noise and air pollution) of 9 cents per kilometer (15 cents per mile) or 82 cents per liter ($3.10 cents per gallon). Adding the 7 cents per liter ($0.26 per gallon) estimated by the NRC for volume-related external effects (assuming national security costs to be effectively zero) would total about 89 cents per liter (nearly $3.40 per gallon). If these external costs were paid for by a fuel tax, the actual charge could be somewhat lower, since the tax should depress total driving and thus reduce the external costs. Still, U.S. fuel taxes are nowhere near these levels, although taxes in the EU and Japan are close.

The gap between non-GHG-related environmental costs and fuel taxes is even larger for aircraft, since jet fuel remains untaxed in many countries for domestic flights and for all international flights. Similar to the control of automobile criteria emissions, the regulation of aircraft pollutant and noise emissions often is in conflict with efforts to reduce CO_2 emissions. As discussed in chapter 5, NO_x emission standards and noise regulations impose design constraints on aircraft engines that reduce maximum fuel efficiency. Efforts to design an economically efficient program of GHG mitigation are thus greatly complicated by the patchwork regulations of other external effects.

Fuel Economy Regulations

In 1976 the U.S. Congress enacted a system of Corporate Average Fuel Economy (CAFE) standards for automobiles and light trucks. Although CAFE eventually reduced fleet fuel consumption per VKT, it also created undesired side effects. For example, lower vehicle fuel consumption reduces vehicle operating costs, creating a so-called rebound effect whereby the amount of driving increases. The effect has been estimated to offset 10–20 percent of the reduction in energy use brought about by the imposed standards.[14] In this way, controls can backfire because the costs of additional accidents and congestion can add up to more than the monetary value of the fuel savings and emissions reductions. These potential side effects require careful attention in the formulation of new controls directed at GHG emissions.

Potential Measures for Greenhouse Gas Emissions Control

GHG emissions from personal transport are the product of the chain of development, manufacturing, and use shown across the top of table 7.1. For both LDVs and aircraft, the process begins with the creation of a concept vehicle, continues through a several-year development cycle, and proceeds to incorporation into new vehicles and gradual diffusion through the in-use fleet. Emissions are then determined by the type of fuel used and the number of miles driven or hours spent in the air. Drawing on earlier efforts at MIT and elsewhere, the stub of table 7.1 shows the main groups of public R&D expenditures and regulatory and market-based measures to reduce emissions that are now in use or have been proposed in one place or another (or at several stages of development simultaneously).[15] In the body of the table are shown our estimates

Table 7.1
Positive (+) and negative (–) effects of policy measures on CO_2 emission reductions through action at various stages of the product cycle, and U.S. state and federal applications to CO_2 emissions

		Stages of product cycle				
	Applications	Vehicle concept	Product development cycle	Spread over new vehicle production	Fleet penetration and turnover	Vehicle use and fuel type
Market-based measures						
Price premium on fuel or CO_2 (tax, cap-&-trade)	Auto, aircraft	+	+	+	++	++
Modified vehicle price (tax, subsidy, feebate)	Auto	n/a	++	++	+	n/a
Accelerated retirement	Auto, aircraft	n/a	n/a	+	++	n/a
Subsidies for low-carbon fuels	Auto	n/a	n/a	n/a	n/a	++
Regulatory measures						
CAFE or CO_2 cap	Auto	+	++	++	+	–
Technology and fuel composition mandates	Auto	+	++	++	+	–
Research and development	Auto, aircraft	++	+	n/a	n/a	n/a

of the positive (mitigating) or negative (increasing) effects of these measures on emissions at each stage, and their relative strength.

Some policy instruments are designed to influence consumer and airline behavior at the time of purchase. One group of options includes market-based measures such as price penalties on CO_2 or fuel use as well as financial incentives for purchasing greener vehicles. As consumers and airlines rebalance their purchasing decisions in response, vehicle and aircraft manufacturers are given reason to develop and incorporate emissions-reducing technologies at a faster rate and on a larger scale than they would otherwise. Such broad market- or price-based measures have the advantage of achieving reductions at least cost because they tend to give the same incentive to all emitters. Unfortunately, price-increasing measures directly and perceptibly impact consumers' pocketbooks, and so generate opposition. An alternative effort, then, is to subsidize fuel-efficient vehicles or low-carbon fuels or both, also shown in the table.

Governments also can influence consumers' trade-offs among new vehicle features such as performance and fuel efficiency by enacting various forms of regulation. For the LDV, the most common of these is a tightening of CAFE-style regulations. No such restrictions have yet been applied to commercial aircraft, either as mandates or voluntary agreements, although measures to accelerate the retirement of high-polluting (or noisy) vehicles have been applied to both aircraft and autos. Perhaps because their impact on the consumer is not as immediately obvious as that of price penalties, these regulatory measures encounter less public (if not manufacturer) resistance, and thus far they have been the basis of most policies applied to auto and aircraft fuel use and emissions.

In addition to having an influence at the point of vehicle purchase, some measures are designed to reduce emissions from autos and aircraft while they are in use. A mandated increase in fuel prices, for example, would reduce emissions in two additional ways: by lowering the VKT of these modes of transportation, and by providing incentives to use less-emitting alternatives like public transport. Other measures are available that lower GHG emissions per VKT exclusively, while leaving energy use largely unchanged. The most prominent of these are a subsidy to biofuels and low-carbon fuel standards, but others (not shown in table 7.1) include regulation of the maintenance of in-use vehicles, limits on highway speed, and information programs to encourage less fuel-guzzling driving habits (see chapter 3). Finally, public investment in transport R&D can help speed technical improvement.

Several of these regulatory options can be implemented with a primary aim of reducing either fuel use or CO_2 emissions, and at first glance it may not seem to matter which goal is pursued since nearly all current transportation fuels are CO_2-emitting fossil hydrocarbons. In some cases, however, the effect on fuel vs. emissions can be very different. For example, a measure that aims to reduce GHG emissions but focuses only on gallons burned can retard a transition to the use of low-carbon fuels and lead to higher mitigation costs.[16]

We follow the table 7.1 outline in taking a closer look at the most prominent of these policy measures, first considering market-based measures and then taking up regulatory approaches and the role of R&D. Finally, we consider possibilities for seeking coherence among the various GHG policy approaches and for coordinating GHG mitigation and efforts to enhance energy security.

Price-Based Measures

Abating greenhouse gases is an economy-wide challenge, and an economically efficient policy will be important in sustaining a national effort. This goal is most effectively achieved by applying price-based measures equally to emissions across all economic sectors, including auto and air transport. Such policies might apply a penalty to the emissions themselves (although placing it on fuel) or on the energy-using device (such as a motor vehicle). Or they could subsidize fuel that is low in carbon or provide financial incentives to the retirement of high-emitting vehicles and other capital stock.

A Price Premium on CO_2 Emissions or Fuel

A uniform CO_2 penalty might be placed on all transport-sector emissions either by imposing a tax or by implementing a cap-and-trade system. A tax is simplest. At some place along the oil, natural gas, and coal supply system, an obligation is imposed to measure the carbon content of the fuel passing by and to pay a tax calibrated to the amount of CO_2 that will be released when the fuel is burned. For example, a tax of $15 per ton of CO_2 ($55 per ton of carbon) would raise the price of gasoline by 3.5 cents per liter (13 cents per gallon).

Such a CO_2 tax could be used to give incentives for marketing the advanced automobiles described in chapter 4. For example, assuming a gasoline price of 80 cents per liter (about $3 per gallon), to become cost effective with respect to the average new vehicle sold in the early 2000s (reference vehicle), the break-even costs of the advanced gasoline hybrid vehicle of about $1.3 per liter ($4.9 per gallon) shown in figure 6.2b would require a CO_2 tax of nearly $220 per ton of CO_2 ($790 per ton of carbon).

In addition to increasing the cost effectiveness of more fuel-efficient automobile and aircraft technologies, a CO_2 tax would raise the operating cost per VKT and lower total travel. It would also make aircraft detours through lower-priced airspace economically less attractive, a condition that can exist outside the harmonized U.S. airspace (see chapter 5).

A cap-and-trade system can give the same price incentive but is somewhat more complicated. First, a national cap is set on the total tons of CO_2 emissions that are to be permitted in some accounting period, say, a year. Then, emission *allowances* are created that add up to this amount. Points in the fuel supply system are identified where the carbon or CO_2 would be measured, and at the end of some accounting period, a number of allowances equal to this measured quantity have to be surrendered.[17] Whether allowances are auctioned or distributed without charge, a market will emerge and the resulting price will provide the incentive to reduce emissions, much the same as a tax would.

There are differences between a tax and a cap-and-trade system that are beyond the scope of our discussion, such as upstream vs. downstream implementation, likely price volatility, and distribution issues. All the same, the two approaches share important characteristics that are worth highlighting. First and foremost, if the penalty is determined in terms of CO_2 emissions, either approach would apply the same penalty on CO_2 wherever it is emitted. Importantly, this condition would apply across all elements of the stock of personal automobiles, whether old or new, and similarly to all components of the commercial aircraft fleet. The same penalty also would apply not only to all passenger transport modes but also to other sources of CO_2 emissions, including electric power plants, factories, and homes.

Moreover, the appropriate penalty would be applied to fuel blends that might consist of various percentages of fossil-based gasoline, diesel fuel, or a biomass-derived fuel. The price of a liter of fuel would incorporate the emissions penalty attributable to its gasoline or diesel component, and its possible biofuel component similarly would incorporate the penalties engendered by CO_2-emitting fuels used in growing, processing, and transporting the biomass feedstock. A similarly appropriate penalty would be levied on compressed natural gas or hydrogen. Even the emissions of plug-in hybrid cars would be properly accounted by these penalty procedures, because the bill for their electric input would include the price penalty paid for CO_2 released at the power plant.

As with a tax, such price penalties would apply immediately to the entire road vehicle and aircraft fleet, influencing decisions to drive or fly and lowering VKT. As discussed in chapter 3, for automobiles the short-term elasticity of demand for motor fuel (i.e., with the vehicle fleet essentially fixed) has variously been estimated to be between −0.2 and −0.3, meaning that a 10 percent increase in the gasoline retail price would lower demand by 2–3 percent. However, analysis of more recent decades indicates that by the first decade of the new century, the short-term elasticity may have fallen substantially below that of the 1970s and 1980s and is below −0.1.[18] A major portion of this adjustment is the reduction in VKT. Of course, as noted above, higher fuel prices would raise consumer demand for more fuel-efficient vehicles and give manufacturers an incentive to develop and produce them.

The same price incentive applies to aircraft. Most studies of the airline business have explored consumer reaction to higher ticket prices as they might result from higher fuel costs. Given a fuel cost share of about 22 percent of total (direct and indirect) operating costs for long-haul aircraft, the elasticity of air travel demand with regard to fuel price is between −0.1 and −0.3, depending on the nature of the flight (business or leisure). Since short-haul leisure flights can more easily be substituted by other modes of transport, a 10 percent increase in fuel prices would result in slightly larger reductions in travel demand (i.e., a 3 percent reduction in leisure travel versus a 1 percent reduction in international business travel).[19] Increased consumer and corporate interest in less fuel-consuming vehicles and aircraft would lead as well to the earlier retirement of fuel-intensive capital stock. And higher demand for fuel-efficient planes would, as in the case of automobiles, induce manufacturers to implement more fuel-saving technology and shift marketing efforts away from higher-consuming models.

The response to carbon penalties would grow stronger over the long term, with time for travel behavior to adjust and new vehicle characteristics to be implemented. The long-term price elasticity of motor fuel demand is variously estimated to be around −0.6 to −0.7, again with a portion of the adjustment to come from reductions in VKT.[20] The long-term fuel price elasticity for air travel is thought to be lower than that for LDVs, but we lack analysis of this aspect of the air travel sector.

How would the dynamics of technology change unfold if a carbon tax or equivalent cap-and-trade system were imposed on the U.S. economy? At MIT's Global Change Joint Program, we used a three-model construction to develop a scenario for a no-penalty reference run and scenarios for possible policy paths along which fuel-saving technology might be adopted.[21] The policy target applied here—in figure 7.1—is an aggressive one. It begins in 2010 with a reduction 7 percent below the 1990 level of national (all sector) GHG emissions. Emissions are then reduced by another 7 percent every five years after 2010. These individual steps sum to an emissions constraint 35 percent below the 1990 level by 2030. To meet that national CO_2 constraint, a uniform penalty per ton of CO_2 (which might be implemented through either a tax or a cap-and-trade system) is applied to all sectors and on the emissions from all fuels.

The key question explored in these simulations is how the technological options outlined in chapter 4 might achieve significant market shares under this type of cost-effective approach to GHG mitigation. We could have used the specific vehicle technologies shown in table 4.3 for this analysis, but these hypothetical vehicles are likely to become available only around 2030 given the underlying technology assumptions and evolution of costs. Since we are interested here in vehicle technologies that could become available during the next two decades, our simulation of the technology dynamics was based mainly on intermediate technology packages from other engineering economic studies.[22] The fuel consumption and cost characteristics of the technologies explored in these studies are consistent with those from our own analysis (as shown in the cost-curve results displayed in figure 4.3a).

The technologies assumed to be available in the coming years for reducing CO_2 emissions of the automobile portion of the LDV fleet are shown in panel A of figure 7.1. The figure shows the trade-off between fuel consumption and initial cost where each of the more fuel-efficient designs includes the fuel-saving features of the less-efficient ones. (An exception is the substitution of an aluminum-intensive vehicle body for a high-strength steel body with the transition from the "PowTrn2" to the "Alum" vehicle). A similar pattern holds for personal trucks, which were included in the study but are omitted from this brief overview.

The smallest reduction in fuel use and CO_2 emissions compared to the average new automobile sold in 1995 is offered by the "ZeroCost" technology. Its lower fuel use is attributable mainly to a slightly more compact design for a given interior size and reduced driving resistances. These incremental improvements, resulting in an 8 percent reduction in fuel consumption, have virtually no impact on retail price. In contrast, the largest reductions in fuel use and CO_2 emissions would be achieved by the advanced hybrid technology ("AdvHybrid"), an aluminum-intensive hybrid-electric vehicle. Compared with the average new automobile sold in 1995, this vehicle would cut fuel use and CO_2 emissions in half, at an additional initial cost of about $4,500.

The starting point of the scenarios is the automobile fleet in 1995, consisting of a large number of vehicles from previous model years. As new automobiles enter and older and damaged vehicles are withdrawn, the fleet characteristics (such as the fleet-average CO_2 emissions per VKT) change gradually over time. Under policy-induced increases in fuel price, consumers can move toward more fuel-efficient (and more expensive) car models, while largely holding on to all other desired vehicle attributes, such as acceleration and interior size.[23]

The reference case is applied globally in this scenario, but for the policy case, the emission targets are applied to the United States only. The base year for the analysis is 1995, and the gasoline price, which defines the reference level in the analysis, gradually rises from $0.32 per liter ($1.20 per gallon) in 1995 to $0.45 per liter ($1.70 per gallon) in 2030. This base price is substantially below the level of the 2007–2008 period, but the difference is not significant given that the purpose of the example is not to produce a forecast, but to illustrate the process and timing of fleet turnover in response to a policy of national restrictions on GHG emissions. The consumer discount rate used in this analysis was around 30 percent, somewhat higher than the rate that underlies figure 6.2b.

Panel B of figure 7.1 shows the U.S. passenger kilometers traveled (PKT) for the reference case. Here the gasoline price rises over time in response to an increasing world oil price, but no price penalty is imposed on CO_2 emissions. Over the period leading up to 2030, the 1995 vehicle ("1995 Fleet" in the panel) is displaced by designs incorporating the zero-cost features, and its PKT falls; the more costly improvements do not penetrate the market. By this simulation, it takes more than twenty years to displace the 1995 Fleet design.

In the policy case, the motor fuel price (refined oil price plus the national CO_2 penalty) doubles in 2010 when the emission target first becomes binding, and it ultimately achieves a level eight times the 1990 level (as shown in panel C). Panel D then illustrates how the on-the-road fleet of automobiles adjusts to changing fuel prices. (A similar pattern holds for light trucks.) The ZeroCost technology begins to displace the

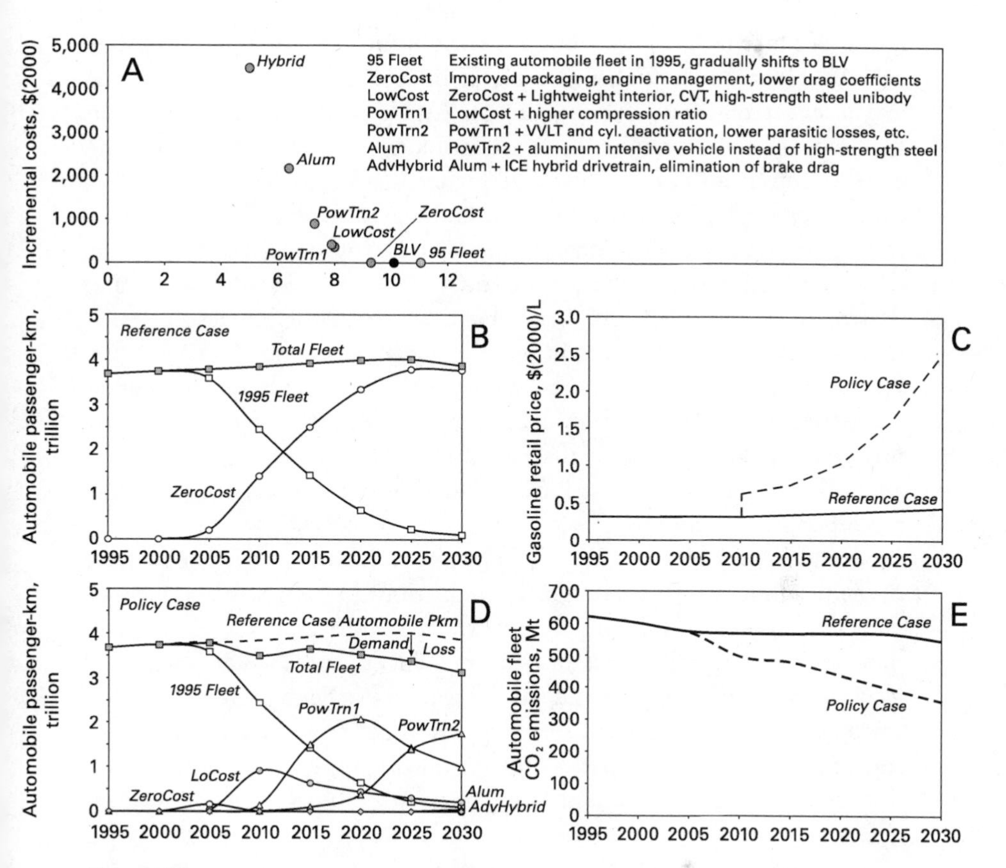

Figure 7.1
Technology options and technology dynamics under reference case and climate policy Conditions for automobiles in the United States. Source: Schäfer, A., H.D. Jacoby, 2005. Technology Detail in a Multi-Sector CGE Model: Transport under Climate Policy, *Energy Economics*, 27(1): 1–24.

1995 Fleet in 2005 but is itself soon displaced by the more fuel-efficient "LowCost" technology. Moreover, the LowCost design holds on to a market share only until around 2010, when it starts to be replaced by the next more fuel-efficient technology. This technology continues the LowCost features but incorporates the features we've designated "PowTrn1" (advanced power train 1). Beginning in 2015 the still more fuel-efficient "PowTrn2" technology begins to penetrate the market. Because less aggressive fuel saving technologies are more cost-effective over a wide range of fuel prices, only when consumer fuel prices reach U.S.

$2.50 per liter ($9.50 per gallon), according to this model, do the Alum and AdvHybrid vehicles begin to be introduced.

The effect of the increasing carbon penalty on U.S. CO_2 emissions from automobiles is shown in panel E of figure 7.1. The reduction indicates results from two factors. The higher fuel price makes auto travel more expensive whatever vehicle technology is in place and leads to a reduction in VKT indicated as "Demand Loss" in panel D. In addition, the rising fuel price stimulates a move to more fuel-efficient vehicle designs that emit less CO_2. As we have just seen, this movement takes time and is evolutionary: the initial ZeroCost vehicle never takes a substantial market share under this policy scenario and is quickly superseded by the LowCost technology and, after 2010, by the still more fuel-efficient PwrTrn1; later, the PwrTrn2 technology also enters service, and only after 2030 do the Alum and AdvHybrid vehicles become widespread consumer choices.

Several insights may be drawn from this simulation. Given the nature of the technology cost curve (greater reductions in fuel use being increasingly expensive) and the tightening trajectory of emission limits over time, the economically most efficient introduction of low-GHG-emission technologies is incremental and will be realized through gradual shifts toward ever more fuel-efficient technologies. Also, as shown in a comparison of alternative policy paths (not included here, however), the tighter the GHG emission constraint the greater the number of intermediate technologies that enter the market.

Also, and consistent with the discussion in chapters 3 and 4, our simulation shows that fuel prices substantially higher than today's will be necessary to introduce highly fuel-efficient vehicle technologies. A lightweight, low driving-resistance vehicle with a hybrid-electric propulsion system can of course find a niche market even at far lower fuel prices (and implicit consumer discount rates may turn out not to be as high as those that were assumed in these simulations). But in this analysis, for it to take a major share of the U.S. new vehicle market through fuel price–stimulated growth alone would require consumer costs as high as $2.50 per liter ($9.50 per gallon). As prices rose to these levels, of course, consumers would also move toward smaller vehicles, an option that is missing from the simulation shown here, further delaying the shift to the most expensive technologies.

The analysis also shows that reducing GHG emissions takes time. Even in our example of extreme emission reductions, it would take twenty-five years to reduce CO_2 emissions of the on-the-road automobile fleet by one-third relative to the projected reference case emissions, and this reduction includes a lessening of travel demand. (Achieving the same reduction by fuel economy standards takes even longer, as we discuss below.) While these simulations focus on U.S. conditions, they give insight into likely patterns in other parts of the world. Since gasoline prices are higher and vehicle size is smaller in Western Europe, the mix of technologies might differ. However, the same basic conclusions about the time path of change also apply there.

Attractive as these uniform price systems are, there may be problems in gaining public acceptance for their use as instruments of environmental control. The difficulties can be seen in the social unrest that attended the hikes in motor fuel prices in the UK in 2000 as a result of rising crude oil prices.[24] They can also be seen in the seemingly implacable resistance to increased motor fuel taxes in the United States.

There are also difficulties of coordination with pre-existing policies, as discussed at the beginning of this chapter. For example, the European Emissions Trading Scheme (ETS) leaves out surface transport and so excludes motor fuel, which already is subject to high taxes. Simply imposing an appropriately adjusted national emissions cap to include motor fuel could impose greater overall costs because of the distortions it would impose on the economy, as noted in the earlier discussion of fuel taxes. Joint adjustment of fuel tax and emissions price to achieve a uniform price pressure across different economic sectors would be difficult, particularly considering the importance of fuel tax revenues for some national and subnational fiscal systems.[25] Problems also arise with respect to aircraft fuel, as noted in a study done for the European Commission on including aviation in the ETS agreements, because of jurisdictional issues regarding fuel taxation.[26]

Also relevant are the arguments that the use of price as an emissions control instrument burdens lower-income consumers more than others. In fact, measured as a share of consumer expenditure, U.S. gasoline taxes have not been especially regressive.[27] However, low-income drivers in rural areas or people locked into long urban commutes would be differentially disadvantaged by an increase. Finally, many environmental advocates oppose the use of price penalties in principle, in favor of emissions constraints. They seek a guaranteed level of reduction, whereas consumer response to price is always uncertain.

A Premium or Subsidy on Vehicle or Aircraft Price

If it is not possible to impose a price penalty on emissions or fuel, other market-based measures can provide incentives to change the trade-off between vehicle performance and fuel efficiency. One approach is to raise the price of the auto or aircraft to reflect the expected CO_2 penalty over the life of the vehicle. Using automobiles as an example, an appropriate initial price premium could be based on an estimate of the total CO_2 to be emitted over a vehicle's lifetime, what we will call its "lifetime CO_2 burden," or LCB. Stated in tons of carbon, the LCB corresponds to roughly ten times the empty weight of a typical vehicle today, taking into consideration a normal lifetime VKT, a given level of fuel consumption, and the carbon content of gasoline. For example, a 2007 Toyota Camry that has a fuel consumption of 9 L/100 vkm (26.1 mpg) would release about 14 tons of carbon over its lifetime, an amount that is more than nine times its own weight (1.5 tons).[28]

Multiplying the LCB by an appropriate emission penalty per ton of CO_2, say $50 ($183 per ton of carbon), yields a tax that could be applied to the initial purchase price.[29] With this assumption, a buyer of our 2007 Toyota Camry would have to pay an emissions tax of $2,550. Similarly, a purchaser of a Ford Focus, which has a fuel consumption of 7.6 L/100 vkm (31 mpg), would have to pay a premium on the purchase price of $2,140. Alternatively, the present monetary amount of the tax could be calculated by applying a projection of expected prices and a discount rate (either the consumer rate or some social rate). Either approach could yield an appropriate tax, and more complex versions could deal with additional issues, such as the use of fuels other than gasoline and the expected change in the CO_2 penalty over the vehicle life.[30]

Another measure with similar incentive effects is the so-called feebate. In this approach, a pivot point (in L/100 vkm) is set such that vehicles that consume fuel over this point pay a tax (a fee), and those that consume below it are credited with a subsidy (a rebate). Based on a forecast of sales, the pivot point can be set to make the system revenue neutral, with fees just balancing rebates. Such a system would create an incentive for manufacturers to build fuel-saving technology into new vehicle models and for consumers to select more fuel-efficient vehicles. For example, with a pivot point of 8.1 L/100 vkm (29.0 mpg) and a feebate of $213 per additional L/100 vkm ($500 per gal/mile), a buyer of our 2007 9 L/100 vkm (26.1 mpg) Camry would have to pay an extra $192 at the time of purchase. In contrast, a purchaser of the 7.6 L/100 vkm (31.0 mpg) Ford Focus would enjoy a rebate of $109.[31]

A study by David Greene and colleagues at the Oak Ridge National Laboratory estimates that compared with a no-policy case, a feebate of $213 per additional L/100 vkm ($500 per gal/mile) would reduce new-vehicle fuel consumption by 13 percent.[32] Setting higher feebates and lower pivot points would further reduce the average new-vehicle fuel consumption. By Greene's analysis, these reductions would be achieved mainly by manufacturers integrating more fuel-saving technology into their vehicles; less than 10 percent of the reduction would come from consumers choosing smaller vehicles.

In its effect on emissions, an LCB-based purchase tax or feebate system differs from a penalty on fuel or tailpipe CO_2 emissions, even when the same penalty per ton of CO_2 is employed. On the one hand, policies directed to changing the new vehicle purchase prices provide no incentive *not* to drive, as does a premium applied to fuel, and they will have some rebound effect. Also, their effect only phases in with the turnover in the fleet, whereas the penalty of fuel applies immediately to all vehicles. On the other hand, in their effect on the choice of new vehicles, these measures differ by an amount that depends on the consumer's discount rate. As normally proposed, LCB-based approaches would sum up the penalty on expected future emissions at a zero discount rate. Since the average consumer discounts future operating expenditures at rates of 25 percent to 30 percent, the penalty on fuel would weigh much less heavily in the purchase decision, a condition that would lead to greater emissions impact from a LCB-based feebate.

Finally, subsidies can be used to influence vehicle purchase decisions. The prominent U.S. example is a tax credit for vehicles with a hybrid-electric propulsion system. The credit is not defined in terms of any particular cost of CO_2 emissions or the value of oil use avoided. Rather, the policy is intended to encourage the large-scale production of these vehicles in the hope of spurring cost reductions, and it is not likely to be permanent because of its potential budgetary cost. Besides encouraging hybrid vehicle growth, the U.S. program contains features that reflect two other objectives: to keep new technology devoted to fuel efficiency rather than to performance, and to protect domestic manufacturers. Thus, the tax credit for hybrid vehicles varies from $500 to $3,400 depending on a vehicle's fuel efficiency. And the subsidy to any import model is reduced once its sales exceed sixty thousand vehicles in any year.

Price Incentives for Retirement

Another approach to emissions reduction would withdraw the highest-emitting vehicles from the fleet. Accelerated retirement programs have already been implemented in various jurisdictions in the United States, Canada, and Europe as part of an integrated strategy to reduce air pollution. Although they made up only about 10 percent of the ground-vehicle fleet, these "gross emitters" generated more than half the emissions.[33] Various compensation schemes have been applied as incentives for turning in these polluting vehicles: a one-time payment, a bonus for a specific kind of replacement vehicle, tax reductions, and providing public transport passes.

A similar plan could also be applied to old and comparatively fuel-intensive vehicles. The impact on fleet energy use, however, would be significantly smaller compared with the reduction of vehicle emissions. While gross emitters can release ten times or more urban air pollutants per VKT than the average vehicle, differences in fuel consumption are significantly smaller. Accelerated retirement programs could be applied to aircraft as well, since, as shown in figure 5.1, older planes can consume up to three times the energy per seat kilometer than more modern aircraft.

Figure 7.2 shows the cumulative age distributions of the U.S. LDV fleet and the commercial aircraft fleet. Half of the LDVs on the road are less than seven years old, while half of the passenger aircraft fleet is below nine years of age. Similarly, 90 percent of both the motor vehicle and the aircraft fleet are less than twenty-two years of age. The survey was undertaken in the early 2000s, so a vehicle age of twenty-two years corresponds to a model year of about 1980. As can be seen from figure 3.2, automobiles older than the 1980 model year consume up to nearly twice the fuel per VKT than earlier models; and essentially the same rate of consumption applies to aircraft (figure 5.1).

It is uncertain how many owners of gross-consuming cars would participate in an early retirement scheme or what replacement vehicle they would buy, but in any case such a policy would be expensive. All pre-1981 model year vehicles on the road in 2001 made up 6 percent of the LDV fleet, but accounted for only 2 percent of total VKT, and produced 3 percent of the fleet CO_2 emissions.[34] Thus, even with the full participa-

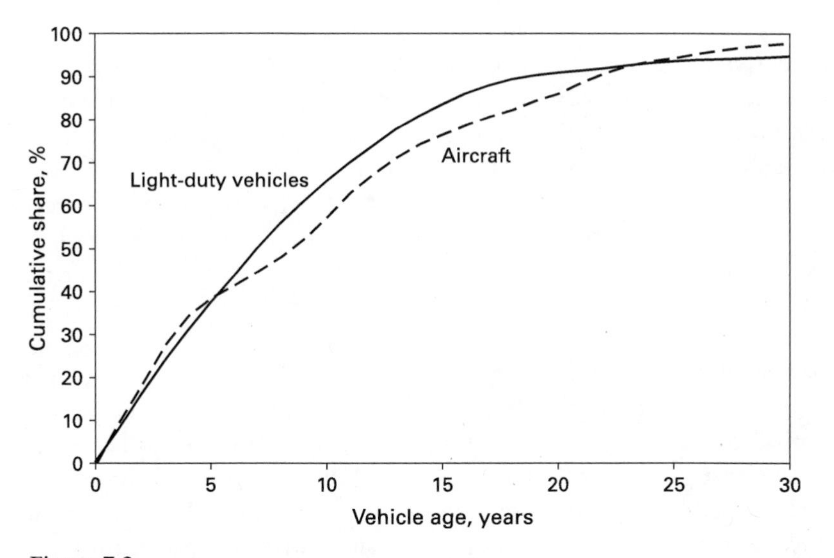

Figure 7.2
Cumulative age distribution of the U.S. light-duty vehicle and aircraft fleet in the early 2000s. Sources: Oak Ridge National Laboratory, 2004. U.S. 2001 National Household Travel Survey; http://nhts.ornl.gov/tools.shtml. BackAviation Solutions, 2007. *Fleet Worldwide Commercial Aircraft, Ownership & Transactions Database*, Version 4.9, June.

tion of owners of these gross consumers, the potential for success of an early retirement scheme is small. Depending on the compensation offered, which we've assumed to be \$500 to \$1,000 per vehicle, the compensation payments could total \$6–12 billion. Assuming that owners would buy a replacement vehicle with the average fuel consumption of the new vehicle year and use it as frequently as the retired vehicle, mitigation costs would be about \$650–1,300 per ton of CO_2 (roughly \$2,400 to \$4,800 per ton of carbon). The mitigation costs would drop to around \$200 per ton of CO_2 (about \$730 per ton of carbon) if the retired vehicle was not replaced.

These mitigation costs can be compared to an early retirement plan for passenger aircraft. According to figure 5.1, the energy intensity of aircraft introduced before 1970 was at least twice that of their modern counterparts. A rough estimate of the CO_2 reduction the retirement of these pre-1970 aircraft could bring about can be constructed using the age distribution of the U.S. aircraft fleet shown in figure 7.2. Replacing the 135 pre-1970 model year aircraft (which made up 2 percent of the

2005 U.S. commercial passenger aircraft fleet) with their modern counterparts would lead to a reduction in fleet-average fuel burn by less than 2 percent.[35] Correspondingly, replacing all pre-1980 aircraft, which would be about five hundred vehicles, or 7 percent of the 2005 U.S. commercial passenger aircraft fleet, would lead to a reduction in fleet-average fuel burn of about 5 percent. Because of the cost structure of direct operating costs, the CO_2 reduction costs are likely to be lower compared to those for automobiles, but still very high compared to other mitigation options.[36]

Subsidies for Low-Emitting Fuels

Among the market-based measures included in table 7.1 is the use of incentives to move to low-CO_2 fuels—already a widely used policy measure. Whether the ethanol economy in Brazil; the CNG programs in Canada, New Zealand, and elsewhere; or more recently the support of biofuels in the United States and the EU, alternative fuel programs have received some form of government subsidies.

Support for ethanol and biodiesel by the U.S. government has been critical to its recent increase in production. It has been estimated that the subsidy per unit of ethanol accounted for $0.26–0.34 per liter ($1.0–1.3 per gallon or $1.4–2.0 per gallon gasoline equivalent) in 2006.[37] Although U.S. ethanol subsidies were significantly higher in the mid-1980s, comparing current subsidies with the costs we have estimated for supplying ethanol ($0.82 per liter or $3.10 per gallon or $4.70 per gallon gasoline equivalent, which we presented in table 6.3) suggests that they still account for one-third of the ethanol supply costs.

These subsidies are justified by the potential of biofuels to reduce both CO_2 emissions and oil imports. According to table 6.3, corn-based ethanol reduces the amount of life-cycle GHG emissions per unit of energy by about 5 percent compared with petroleum-derived gasoline.[38] Given ethanol subsidies on the order of $1.0–$1.3/gal ($1.4–$2.0 per gallon gasoline equivalent) in 2006, the subsidies amount to $3,000 to $4,000 per ton of CO_2 reduced. A larger reduction in life-cycle GHG emissions compared with petroleum-derived gasoline would lead to lower overall costs, but the subsidies would remain high. Even a 30 percent reduction in life-cycle GHG emissions, an assumption that reflects the more optimistic life-cycle assessments, would still lead to ethanol subsidies in excess of $600 per ton of CO_2 reduced.

The need for future alternative fuel subsidies will critically depend on the development of the world oil price. As shown in table 6.3, the oil price would need to be at least between $25 and $90 per barrel to make second-generation biofuels cost competitive. Although the upper end of that range was exceeded in 2008, the oil price may well see long periods below this level (see the discussion in chapter 2). Thus, absent other policy instruments, and depending on the progress in reducing biofuel production costs, continuous fuel subsidies may be required in the future.

Regulatory Measures

Although there is strong interest in the United States, EU, and elsewhere in the use of market-based mechanisms, all but a few past environmental controls have been based on regulatory measures. Examples include controls on auto emissions of urban air pollutants, aircraft noise, and the fuel economy characteristics of LDVs. Such regulatory constraints have their strongest effect on product development and the spread of new technology through the production of new vehicles, as indicated in table 7.1. Significantly, they do not address vehicle use, and as noted earlier, they can foster additional driving through the rebound effect.

A Cap on Fuel Economy or CO_2 Emissions of New Vehicles

The United States first imposed Corporate Average Fuel Economy (CAFE) standards in 1978, applying them to all new LDVs. Separate standards were set for cars and light trucks, with those for automobiles being more stringent. Under laboratory test conditions (which underestimate fuel consumption under actual road conditions by about 20 percent), the initial maximum levels were 13.1 L/100 vkm (at least 18 mpg) for cars and 13.4 L/100 vkm (at least 17.5 mpg) for light trucks (pickup trucks, minivans, and SUVs). In subsequent years, these constraints were tightened, to 8.6 L/100 vkm (27.5 mpg) for cars, and to 11.2 L/100 vkm (21.0 mpg) for light trucks. Then, in 2008, the CAFE system was further tightened and restructured. A single minimum standard of 35 mpg (6.7 L/100 vkm) was set for the combined new car and light truck fleet sold by each manufacturer, to be achieved by 2020. And, though cars and light trucks are still treated differently, the standard is based on a mathematical function that avoids the sharp difference in treatment between them. Assuming an average vehicle lifetime of fifteen years, the 2020 CAFE target would be fully translated into the fleet until 2050.[39]

Over the history of the CAFE standards the imposition of one constraint on cars, and a less stringent one on light trucks, stimulated a major change in the composition of the LDV fleet. Consumers resisted the forced shift in trade-offs among car-model attributes, and manufacturers accommodated by developing user-friendly passenger vehicles that met the definition of a light truck. Compared to 1975, five times as many light trucks are used for passenger travel in the United States today. These passenger-friendly light trucks consume more fuel than other LDVs, and average fuel consumption of the combined fleet has therefore increased over this period (see chapter 3).

Several countries have followed the U.S. CAFE example, with either fuel economy regulations or design restrictions on GHG emission. A survey of these regulations by Feng An and Amanda Sauer found that they differ widely regarding the underlying driving cycle and whether they are mandatory or voluntary.[40] In addition, some countries, such as China and Japan, have attempted to achieve the same result through weight-based standards, while Taiwan set upper limits for fuel use based on engine size.

In some jurisdictions, such as the United States, constraints are mandatory and require each manufacturer to comply with the fuel economy regulations individually. In others, like the EU, voluntary targets have originally been applied to industry averages. In 1998, the European Automobile Manufacturers' Association voluntarily committed its members to collectively reducing the CO_2 emissions of the average new automobile. The target was 140 g/vkm by 2008 and was to be achieved through technological improvements and market changes. For petroleum-based fuels, this emission limit translates into fuel consumption levels of 5.8 liters of gasoline (or 5.25 liters of diesel) per 100 kilometers (44 miles per gallon and 48 miles per gallon). The same target has been set for new Japanese and Korean vehicles for a year later, 2009. The European target tightens to 120 gCO_2/vkm (51 mpg) by 2012. The EU Commission has threatened mandatory CAFE-type limits if it turns out that these voluntary targets are not being met. As these targets are now unlikely to be achieved within the required time frame, the commission recently adopted a proposal for a slightly relaxed target of 130 gCO_2/vkm by 2012 through vehicle technology alone. This reduction would need to be complemented by an additional 10 gCO_2/vkm through other technological improvements and enhanced use of biofuels.[41] Since manufacturers of larger vehicles would be disadvanatged in meeting the CO_2

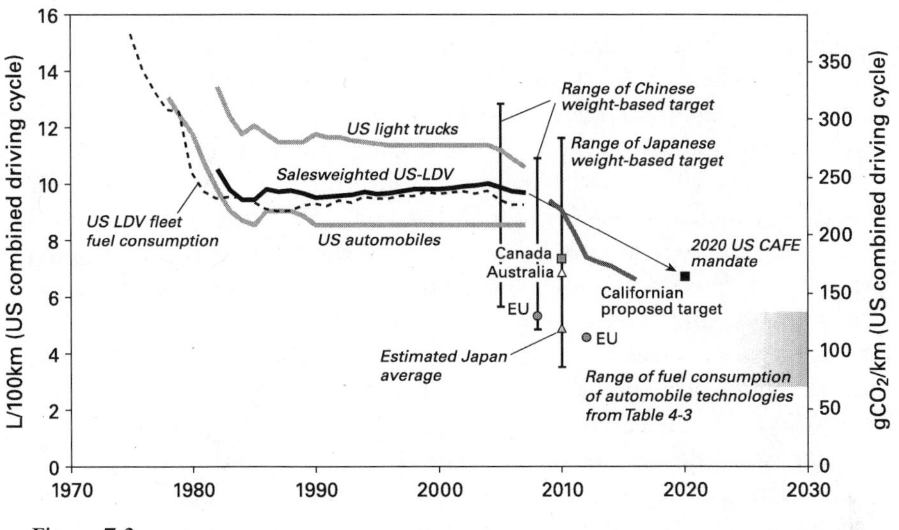

Figure 7.3
Fuel consumption and CO_2 emission targets for light-duty vehicles in the United States and abroad. Also shown is the weighted average U.S. light-duty vehicle fuel consumption (dotted line) and the range of achievable automobile fuel consumption levels from table 4.3. Data sources: An, F., A. Sauer, 2004. *Comparison of Passenger Vehicle Fuel Economy and Greenhouse Gas Emission Standards Around the World*, Pew Center on Global Climate Change, Washington, DC. U.S. Environmental Protection Agency, 2007. *Light-Duty Automotive Technology and Fuel Economy Trends: 1975 through 2007*, Compliance and Innovative Strategies Division and Transportation and Climate Division, Office of Transportation and Air Quality, Washington, DC.

target, the commission proposal includes vehicle weight–based standards; emissions over 130 gCO_2/vkm, however, would need to be offset by manufacturers of smaller and more fuel-efficient vehicles, such as through some form of emissions trading. Finally, some regulations distinguish between passenger cars and light trucks, as in the United States, while others apply the standard to an all passenger-vehicle aggregate.

An and Sauer converted the standards in these countries, imposed and contemplated, to a common yardstick. Figure 7.3 shows the historical development of fuel consumption in the United States and the projected 2020 target of 6.7 L/100 vkm (35 mpg) target in terms of this yardstick. The historical movement of consumers toward light trucks can be seen in the widening difference in fuel consumption between the sales-weighted LDV fleet and the automobile standards until the tightening of the light truck standards in 2005.

Figure 7.3 also shows future standards adopted by Australia, Canada, China, and Japan (with respect to maximum levels of fuel consumption and CO_2 emissions). Also included is a proposed California standard and the voluntary targets put forward by the European Automobile Manufacturers' Association. (These originally were expressed in terms of gCO_2 per kilometer driven. When converting the CO_2 into fuel consumption standards, we used the carbon content of gasoline.) These standards can be compared to our estimates of the future fuel consumption levels of midsize automobile technologies that are shown as gray areas in figure 7.3 and derived from table 4.3.

These various measures suggest that by the mid-2020s automobile fuel consumption regulations could be tightened to about 4 to 5 L/100 vkm (59–47 mpg) under laboratory conditions, or to between 4.8 and 6 L/100 vkm (49–39 mpg) under typical driving conditions—depending on whether diesel fuel or gasoline is used. Further tightening of the standards to about 3 L/100 vkm (78 mpg) would require the use of hybrid-electric propulsion systems.

So far, this discussion of technical potential has left out the issue of cost—not just the manufacturing cost of a vehicle that has every mandated characteristic, but also the full welfare cost—including the value of vehicle attributes that consumers would have paid for but can no longer be offered (the so-called consumer surplus lost). Effective as regulation may be in reducing automotive emissions, it has significant drawbacks on this score. First, the marginal costs of reducing emissions may differ substantially among manufacturers, and a uniform standard imposed independently across firms, such as the U.S. CAFE system, will create economic waste: total costs could be reduced by shifting control efforts from one firm to another. These losses could be reduced through a hybrid regulatory system that allowed manufacturers to trade emissions restrictions among themselves. A study by David Austin and Terry Dinan of the U.S. Congressional Budget Office (CBO) found that the cost of the CAFE system could be reduced by about 16 percent simply through the introduction of a system of trading CAFE debits and credits.[42] Since the further tightening of the standards in 2008 will disadvantage manufacturers of larger and more fuel-intensive vehicles, interfirm emissions trading could yield still stronger cost reductions.

Also, the CAFE system can be substantially more costly than a fuel tax while achieving the same emissions goal. The rebound effect is especially strong in countries that have lower income levels. It declines as income (and thus the opportunity cost of time) increases. It is argued that with rising income this effect is no more than 12 percent in the United States—and perhaps is even lower at fuel prices that are higher than those used in these studies.[43] Even at lower levels of rebound, however, the induced travel causes additional urban air pollution, accidents, noise, and congestion.

The fuel tax, of course, has the opposite effect on driving, lowering VKT as well as shifting new vehicle purchases to more efficient car models. The CBO study cited above found that, even without considering the costs of rebound, the overall cost of a 14–18 percent tightening of CAFE is substantially higher than that of a fuel tax that achieved the same reduction in fuel consumption.[44] This result highlights another characteristic of regulatory approaches like CAFE, which is that complex economic analysis of the type carried out by the CBO frequently is necessary to estimate the likely cost impact of a particular policy instrument. When such analysis is not undertaken, large variations in cost may be unknowingly imposed on different sectors of the economy—while opportunities for cost reduction go unexploited and even unrecognized.

A final disadvantage of regulatory standards is the long time required to design, test, produce, and sell low-GHG-emission vehicles and then to substitute them for older vehicles withdrawn from the fleet. Several decades may pass before the fuel consumption level of a new technology is fully represented in the vehicle fleet. Thus, if our hypothetical fuel-efficient vehicles from table 4.3 were forced into the market by regulation starting at around 2030, a substantive impact on fleet CO_2 emissions could be expected by midcentury at the earliest. In contrast, measures raising fuel price have an immediate impact on GHG emissions. In our example shown in figure 7.1, nearly the entire reduction in 2010 emissions would result from declining travel demand.

Despite their shortcomings, CAFE-type constraints have proved to be an attractive tool for limiting fuel use and emissions. Unlike fuel taxes, these regulations don't impose obvious, direct costs on consumers, which helps keep public opposition low. And although they fail on a criterion of cost transparency, they are comparatively simple and have low administration costs.

Technology and Fuel Composition Mandates

An alternative to the regulation of vehicle fuel consumption or GHG emissions is to mandate the use of specific technologies or fuels. Similar to the effect of instituting regulations on fuel economy or CO_2 emissions, mandating fuel-saving technology would influence the product development cycle and the spread of fuel-saving vehicles into new production. But they would not directly affect the existing vehicle fleet or its turnover rate, and so they would yield GHG-emission reductions only over several decades. Examples of mandated technologies include safety features for automobiles, and excess fuel reserves and various redundant systems for aircraft (see chapter 5).

In the most favorable circumstances, technology mandates can spur technology development and generate economies of scale. However, in picking specific technologies, there is also significant risk of misallocating resources—or even of complete failure. One example of a poorly designed technology mandate is the Zero-Emission Vehicle (ZEV) mandate as implemented in 1990 by the California Air Resources Board. The board originally required manufacturers selling in California to introduce ZEVs as a percentage of new vehicle sales, starting with 2 percent in 1998, increasing to 5 percent in 2001, and ultimately rising to 10 percent—124,000 vehicles—in 2003. No particular zero-emission technology was specified, but at the time the only technology that could meet the ZEV definition in the law was an electric vehicle. Thus, in effect the ZEV was a lead-acid-battery electric car mandate. (Also, the definition of "zero" did not include emissions from the electricity generation needed to power these vehicles, which, as discussed in chapter 4, can result in higher life-cycle emissions than crude oil–derived gasoline.)

The auto industry pursued a long series of lawsuits in which it agued that the ZEV mandate violated federal law. Several amendments rolled back the deadlines, so that today the ZEV mandate has become a complex formula that includes several technologies, among them both gasoline-electric hybrid and hydrogen-powered vehicles. The ZEV mandate did stimulate a substantial amount of research and development (R&D) on electric propulsion systems, however, and experimental vehicles were built and tested. But the battery technology could not meet the requirements of the consumer market.

In contrast to the technology mandates, fuel composition mandates could yield reductions in GHG emissions within a decade provided they are specified on a life-cycle basis (see chapter 6) and the vehicle infrastructure is compatible with the mandated fuel. Examples include the use of oxygenated fuels (gasoline with a minimum oxygen content) in areas that do not meet U.S. federal air quality standards. The Brazilian government also mandates minimum ethanol gasoline blends (see chapter 6). And the state of California has imposed a Low-Carbon Fuel Standard, which requires the carbon intensity of California's transportation fuels to be reduced by at least 10 percent by 2020.[45] Yet, prerequisites similar to those for the technology mandates apply here as well: for example, low-GHG fuel-conversion technologies adequate to satisfy the standards must be on the horizon.

Research and Development

Investment in research and development (R&D) is a prerequisite for producing many of the low-GHG-emission technologies and (low-carbon) fuels discussed in chapters 4, 5, and 6. In the past, R&D investments have funded mainly by industry, but more recently the role of the government has increased.

Auto and Aircraft Industry R&D

The automobile and aircraft industries typically invest 3–5 percent of their revenue in R&D, and given the size of the major firms, these percentages represent significant industrial activity. For example, in 2005 the combined global R&D expenditures of the big three U.S. manufacturers—DaimlerChrysler, Ford, and General Motors—totaled about $20 billion and were spread over a range of technological and cost initiatives, including performance, safety, reliability, comfort, fuel efficiency, and emission reduction.[46] Similarly, the Boeing commercial airplane group invested an average of slightly more than 3 percent of its revenue in R&D over the decade following 1995. That translates into about $900 million per year.[47] For both automobile and aircraft manufacturers, the focus of R&D effort has shifted in recent decades toward ever-greater emphasis on fuel saving and environmental issues, including GHG emissions.

Government R&D Funding

In many countries, governments also are active in this area. Until recently, however, environmental concerns have played only a small role. Even the most prominent U.S. R&D effort, the Partnership for a New Generation of Vehicles (PNGV), initiated in 1993, was motivated mainly by concerns over energy security, followed by industrial competitiveness. Total government support for this effort has been $200 to

$250 million per year, more than half of which was spent in government laboratories, while the remainder went to contractors, automotive suppliers, technology companies, and universities.[48] This funding was matched fourfold by industry, resulting in an overall budget of over $1 billion per year.

One PNGV mission was to develop (by 2004) vehicles with a gasoline consumption as low as 3 liters per 100 kilometers (up to 80 miles per gallon) while maintaining the fundamental characteristics of average 1994 sedans.[49] It is no surprise that this goal could not be realized. Figure 4.3a shows that reducing fuel consumption to this level, about 3 liters per 100 kilometers, would cause the vehicle retail price to increase by at least $5,000 if vehicle size and other attributes are maintained.

The Freedom Cooperative Automotive Research Program, a successor to PNGV initiated in 2002, seeks evolutionary technology while it maintains challenging long-term goals. It is more comprehensive than was PNGV in scale (it includes energy companies) and scope (it also includes light trucks). It also considers technology trajectories: improving the internal combustion engine, making a transition toward hybrid-electric vehicles, and identifying what is required to move toward a hydrogen economy. Collaborative government and interindustry R&D programs have also long existed in Europe and Japan. As with the United States, their primary motivation has not been environmental concerns but energy security and industrial competitiveness.[50]

Government support for aircraft R&D has been several times that for road vehicles, both in the United States and in Europe. However, no concerted PNGV-equivalent program exists. In the United States, the annual budget of NASA's Aeronautical Research and Technology section has been about $1 billion and has focused on fuel burn, noise, pollutant emissions, and increasing safety and capacity. Additional significant R&D funds come from the Department of Defense. The U.S. government's aerospace R&D has declined since the 1990s and currently roughly matches the industries' own funding.[51] In Europe, specific R&D targets funded by the European Union include reducing development costs and production times, developing higher fuel efficiency and environmental performance, and improving operational capabilities and safety. Funding was initially around $100 million per year, then was raised to $250 million by the 2002–2006 period.[52] Additionally, the EU-level R&D support is complemented by significantly greater funding from national and regional governments.

Government-funded R&D programs can be an important component of a comprehensive climate policy. Besides direct results in low-GHG-emission technology, expenditures in one country can also stimulate public investments elsewhere. For example, the U.S. PNGV program spurred government R&D in Europe, and similar knock-on effects can be expected in the future.[53] However, government support for industrial R&D can also create competition problems under existing trade agreements. The strategic nature of government R&D is perhaps most evident for aircraft. European governments dedicate a significant fraction of their aircraft R&D support to near-term technology development and to efforts to improve product marketability. So-called launch aids (low-interest loans repaid only once a sales target is reached), for example, allow manufacturers to assume more aggressive pricing and financing practices.[54] For the new Airbus A380 alone, these have totaled about one-quarter of its development costs, so it is a practice that draws strong criticism from U.S. competitors.[55] European firms, in turn, point to a range of nonrepayable tax breaks and federal subsidies that Boeing enjoys; the tax breaks for the new Boeing 787 apparently are of an amount similar to the launch aids for the A380.[56]

Even massive R&D investments do not, however, dependably lead to the development of new, improved products. As suggested by the steering committee for the PNGV program, "breakthrough ideas and talented people" can be more stringent constraints than insufficient R&D investments.[57] More important than the size of R&D investments is that they be sustained over long periods and coordinated across government departments. Also, it is questionable how truly effective R&D investments in near-term technology can be if the efforts they support are isolated from other emissions mitigation measures. And in following an R&D-only policy, a government can do nothing to advance fuel-saving technology that may be developed under its programs—and which undoubtedly will be expensive—but can only wait until market forces change the economics of the vehicle market. Thus, governmental R&D expenditures on low-GHG-emission technologies and fuels are an important but only complementary policy measure.

Multiple Objectives and Policy Coordination

Choosing policy measures from among the options in table 7.1 requires weighing a set of sometimes conflicting concerns, such as CO_2 emissions, other environmental effects, safety, oil security, tax revenue, competitiveness of domestic firms, and consumer impact. Moreover, proponents of one approach vs. another often disagree over the level of acceptance to be accorded to consumer behavior and corporate objectives. In policy design, for example, one dispute tends to focus on market-based vs. regulatory measures. It is frequently argued that a system of uniform, universal price incentives will tend to produce least-cost solutions for any level of emissions reduction. Historically, however, as discussed above, most environmental controls and fuel use policies have employed regulatory measures and selective subsidies. Examples include regulations on building construction and standards for appliances, and subsidies to solar power, insulation, and other conservation measures. Only in recent decades have charges on emissions come into play, most prominently in the regulation of sulfur dioxide (SO_2) and nitrogen oxides (NO_x) in the United States, and in the European ETS for GHGs.

How personal transport should be handled in evolving GHG policies in the United States, the EU, and many other countries is a continuing source of debate. By the logic of market-based measures, early CO_2 reductions should come more substantially from economic sectors other than passenger transport because of their relatively higher elasticity to emissions penalties. Nonetheless, strong arguments are advanced for special regulation of personal transport, without regard to price pressures or regulations imposed elsewhere, precisely *because* it has a low price elasticity and price measures won't achieve "enough" reduction. Underlying this primitive argument are more substantive concerns about market-based measures. Some argue that a set of "market failures" are to blame for high implicit consumer discount rates. Perhaps buyers lack the information, or the ability, to understand future operating costs and initial purchase incentives. Or maybe secondary markets do not appropriately value the fuel efficiency of used vehicles, so a purchaser who perceives some chance of selling the used vehicle before the end of its life has a reduced incentive to invest in lower fuel consumption.

Even more radical than the market-failure justification for regulatory solutions is a sweeping rejection of the vehicle attributes consumers apparently prefer. The appropriate level of GHG control is a social decision, it is argued, and the relevant way to add up additional vehicle costs and fuel savings is by using a social discount rate. The fact that consumers place so little value on future savings, or that they may experience an even higher cost in being forbidden to purchase a car with attributes they want, is simply irrelevant to social policy.[58] If the planet is at stake, we shouldn't worry about issues of economic inefficiency or excess costs imposed on firms or consumers, so the argument goes. (Even this position needs to be tempered by the understanding that there is a limit to the inefficiency and waste that can be tolerated, lest public support be lost.) The extreme version of this view is well captured in a campaign to attach bumper stickers to high-consuming vehicles that ask, "What would Jesus drive?"

Despite the moderate estimates of up to 3.1 cents per liter (11.7 cents per gallon), another important motivation for new policy initiatives is the reduction of perceived security costs of oil imports. If reducing oil imports is a problem, the best solution would be a penalty applied directly to the matter of concern: a tariff on imported oil at the estimated per barrel security premium. Nonetheless, energy security frequently is conflated with GHG control, and indeed many of the CO_2 measures considered above serve both objectives. More fuel-efficient autos and aircraft, or the introduction of especially second-generation biofuels, will tend to reduce imports as well as emissions. But proposed cures for these joint concerns also can conflict with one another. For example, the reduction of imported oil by the substitution of domestic (or continental) fuel from heavy oils, tar sands, or shale oil would involve substantial increases in CO_2 emissions because of the additional energy needed for extracting and upgrading these feedstocks (see chapter 6). Absent CO_2 capture and storage production of transport fuels from coal would double life-cycle GHG emissions per unit of transport energy provided.

Important to passenger travel as is the balancing of objectives and choices among policy approaches, these issues are likely to be resolved on a wider basis, whereby a national solution will involve all sectors of the economy. Even more likely, because there are several relevant objectives, and measures generally are adopted sequentially over time, many of the measures summarized above will ultimately be employed. This is, of course, already the case for the personal automobile. In the United States, for example, reductions in fuel use and CO_2 emissions are being pursued by CAFE standards with respect to design, by subsidies for the purchase of hybrid-electric vehicles, by subsidies for ethanol in motor fuel, and by establishing targets for the biofuels component of the national motor fuel pool, as well as through a number of government R&D programs.

Proposals for a national cap-and-trade system, which under most designs would place motor fuel under a cap, do not contain provisions to abandon other measures already in place. An issue, then, is the interaction between these different types of instruments. An example of lack of coordination is provided by the flexible fuel vehicle, which is designed to run on a fuel mix of up to 85 percent ethanol (see chapter 6). Each flexible fuel vehicle receives a CO_2 credit under the CAFE system as if it were running mainly on carbon-neutral ethanol whereas in fact this fuel is available only in very limited areas of the country, and even where it is available it may not be selected at the pump. Moreover, in crediting the contribution of the (imagined) biomass component of this fuel to the CAFE system goal of reducing oil consumption, the ethanol component is treated as if it were an exact substitute for energy or oil on the basis of the energy in the E85 fuel mix. The actual displacement of this mix is less, because of the petroleum used in growing the biomass and in the manufacture and transport of the alcohol fuel and the ethanol-gasoline mix (see chapter 6).

The issue of coordination among market-based and regulatory measures will arise in a still more important form if a cap-and-trade system is implemented in the United States. If a further tightening of CAFE also is a component of GHG policy, then the level chosen will imply some cost in terms of dollars per ton of CO_2 emissions avoided. Tighter CAFE standards might be calculated using the approach followed by Austin and Dinan for the CBO or by the Department of Transportation's procedures—the resulting estimate depending, of course, on the discount rate used. Concurrently, a price might be set on CO_2 emissions through a cap-and-trade system, and the CO_2 penalty in the two systems could be very different, implying that the same environmental gain could be had at less cost if the burden of reduction were moved among sectors.[59]

Whatever the level of public commitment to CO_2 mitigation, perhaps entangled with oil import concerns, there will be limits to the pressure that can be put on consumers, in terms of cost and constrained choice, and on manufacturers and airline companies, in terms of cost and competitive effects. Therefore, success in dealing with these challenges will require careful well-to-wheels analysis of options, and consideration of personal transport within the context of an integrated set of national policy developments.

Summary

A range of policy measures is available, applied at various stages of the vehicle and fuel cycle, that could lead to reductions in GHG emissions from passenger travel. Nearly all measures cause consumers or airlines to rebalance their preferences among attributes of new automobiles or aircraft, and they could also induce the vehicle industry to develop and offer less GHG-intensive vehicles and fuels at a faster rate and larger scale than they would do otherwise. In addition, some of these interventions would affect the usage of these transport systems once on the road or in airline fleets. A carefully combined set of actions, covering all stages of a vehicle and fuel cycle, could lead to a large-scale reduction in the growth of GHG emissions from passenger travel.

The task is not a simple one, however, for passenger travel is already the target of various policy concerns. Measures in place include safety and emissions regulations, fleet-average fuel economy standards and fuel taxes, subsidies to the producers and users of low-carbon fuels, and R&D expenditures. Some measures could be modified with the objective of reducing GHG emissions, but care in design would be required to avoid economic distortions caused by interaction among the various measures, or conflict with other objectives such as oil import security. For example, promotion of more fuel-efficient diesel engines would lead to increases in urban air pollution. Some biofuel substitutes for oil imports, and domestic petroleum derived from carbon-intensive sources (e.g., tars, oil sand or shale, and coal-to-liquids technology) may reduce oil dependence but increase CO_2 emissions.

The introduction of an economy-wide emissions penalty, by a cap-and-trade system or a CO_2 tax, could meet all of these objectives—reducing travel and biasing consumer choice to more efficient vehicles, inducing industry to focus on low-GHG-emission technologies and fuels, and at the same time reducing oil dependence and air pollution. Such an application of the same price penalty across all economic sectors would tend to produce the least-cost solution for any reduction target. Nonetheless, the relatively low price response of this sector leads to arguments that passenger travel, and particularly the automobile, is a special case that should be handled separately from other parts of the economy. So the appropriate role of personal transport in national climate policy remains a subject of contention.

In the final chapter, we thus put these issues of policy choice in the context of projected travel demand, energy intensity, and alternative fuels to explore the potential role of passenger travel in a climate-constrained world and the policy challenges ahead.

8

Future Prospects and Policy Choices

Over coming decades, nations will face challenging questions about the growth of passenger travel and its role in efforts to mitigate greenhouse gas (GHG) emissions. To close our analysis of this sector, therefore, we integrate insights from the previous chapters into a picture of the impact that technology and fuels currently under development, combined with adjustments in travel patterns, might have on GHG emissions. And we examine the likely contribution of passenger travel to announced goals for stabilizing the atmospheric concentration of GHGs. The discussion leads inevitably to a summary of the choices before us.

Opportunities for Controlling Transport Emissions

To facilitate our review of prospects and choices, we again turn to the relation we introduce in chapter 1 that identifies the components of passenger travel GHG emissions, denoted GGE:

$$GGE = \frac{GGE}{E} \cdot \frac{E}{PKT} \cdot PKT \tag{8.1}$$

Moving from right to left, a first contributing factor is passenger kilometers traveled (PKT) and the associated change in mode shares, discussed in chapter 2. There we argued that rising affordability will continue to increase travel demand in all parts of the world, and that this growth will be accompanied by a rising relative importance of faster transport modes. While the automobile is projected to experience the highest growth in the developing economies, high-speed transportation is likely to continue to grow fastest in the industrialized world.

Next in the identity is the ratio of total energy use E to PKT, or passenger travel energy intensity, which we discuss in chapters 3, 4, and 5.

For a given fleet of ground and air vehicles, E/PKT depends on consumer or airline behavior and characteristics of the prevailing technology. The consumer- and airline-related influences are treated in chapter 3, where factors are identified that lead toward an increase in energy intensity. Most important are a shift toward larger and more powerful light-duty vehicles (LDVs), an increase in the relative importance of more energy-intensive urban transport, and—outside the industrialized world—a decline in vehicle occupancy rates. Similar trends that lead toward higher levels of comfort, speed, and thus energy intensity can also be observed in aviation. Overall these trends result in an increase in passenger travel energy intensity. Thus, absent any technological progress, world average passenger travel energy intensity would increase through 2050.

That increase in passenger travel energy intensity, combined with the growth in PKT projected in chapter 1, yields a strong growth in mobility-related energy demand (assuming, perhaps naively, that this demand could in fact be satisfied with frozen early 2000 technologies). If oil products were to continuously fuel the world's transportation system (thus leaving the factor GGE/E largely unchanged) then, with early 2000 transport technology, global mobility-related carbon dioxide (CO_2) emissions could rise by a factor of three to five by 2050, depending on GDP growth between now and then.

To explore potential departures from such constant-technology projections, we discuss a range of road and air vehicle improvements that could reduce passenger travel energy intensity and thus GHG emissions. In chapter 4 we examine the technology opportunities for reducing the energy intensity of road vehicles, and the potential is large. Combined with a saturation of the long-term trend toward larger and more powerful vehicles, the fuel consumption of today's average new LDVs could be cut by 30 to 50 percent over the next twenty to thirty years. A percentage reduction of similar size and over a similar time horizon is also possible for new aircraft, which we analyze in chapter 5. Even larger reductions appear possible beyond 2030 for both autos and aircraft.

An additional opportunity for emissions reduction is available through the introduction of fuels with a low carbon-to-hydrogen ratio, a strategy that would reduce the ratio GGE/E. Our life-cycle analysis in chapter 6 suggests that all synthetic oil products derived from fossil sources (e.g., tar sands, shale oil, and coal) would result in significantly higher levels of GHG emissions compared with conventional petroleum-derived fuels, unless CO_2 emissions from fuel processing are captured and sequestered.

While nearly all current-generation biomass-derived fuels offer only marginal benefits in terms of life-cycle CO_2 reduction, second-generation fuels appear more promising, though they are still in a stage of R&D and early commercialization. Common to all biomass fuels, of course, are limits on land conversion. Over the long term, electricity and hydrogen may become important transport fuels, but both energy carriers still require technology breakthroughs in vehicle storage. The challenges are greatest for hydrogen because of the need for a completely new production, distribution, and storage infrastructure.

Because many of the potential technical means of reducing emissions are costly, market forces as experienced during most of the second half of the twentieth century will not be strong enough by themselves to induce motorists or airlines to adopt many of these improvements. Therefore, in chapter 7 we explore a range of policy options that could accelerate market penetration. Measures that raise the fuel price through a carbon tax or cap-and-trade system can reduce all three right-hand-side factors of equation 8.1. They would improve the economics of low-carbon fuels relative to petroleum products, reducing GGE/E, reduce E/PKT by affecting each stage of a technology's life-cycle from vehicle concept to fleet penetration and vehicle usage, and reduce PKT. Other potential policy interventions can influence various subsets of these factors, including regulations and subsidies that can be used to force technology adoption, and similar measures that can be applied to the carbon content of fuels. Road pricing remains at best a distant possibility, so an increase in fuel price is the most promising option for reducing PKT growth over the time horizon of this study.

Using these findings, we can address the prospects for reducing the GHG emissions from U.S. and global passenger travel and the likely implications for personal mobility, and discuss the unavoidable choices faced by government and industry as we seek the appropriate role for this sector in climate policy.

A View of the Technical Potential Alone

Even without changes in vehicle attributes, or policy measures that might lower PKT, technology currently under development could play a significant role in reducing the projected GHG emissions from passenger travel. Figure 8.1 shows the life-cycle GHG emissions from passenger travel for 1950 and 2005, and projections for 2050. Estimates are

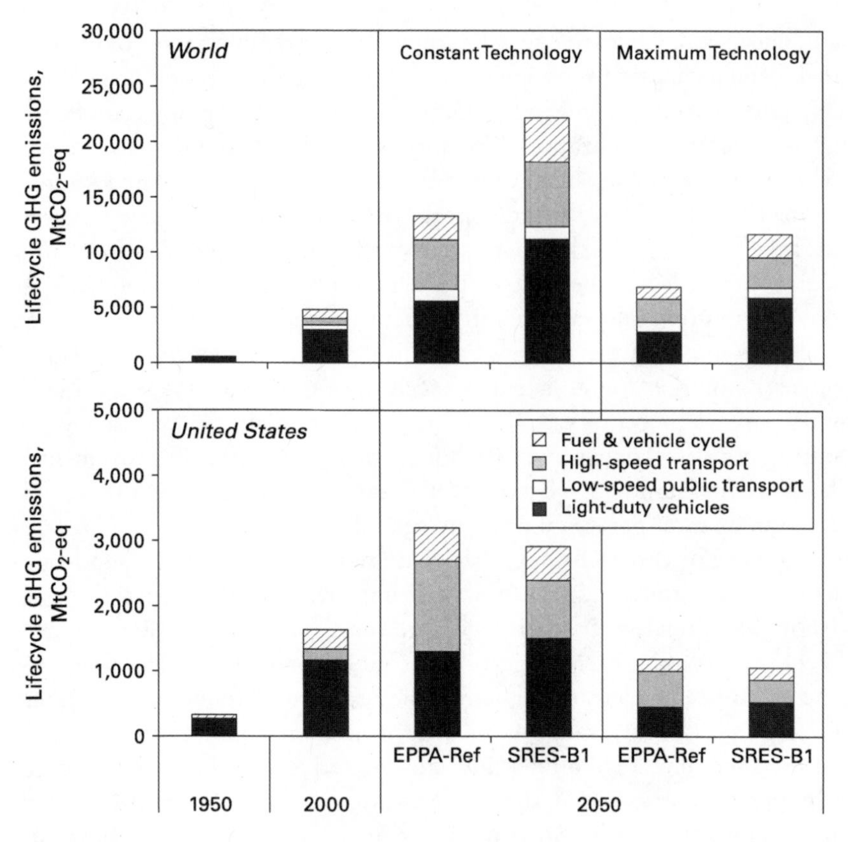

Figure 8.1
Passenger travel GHG emissions, past, present, and projections for 2050 for the world (top) and the United States (bottom).

provided for the world as a whole (top) and the United States (bottom). Two limiting cases are shown. The constant technology case assumes no improvements in fuel efficiency technology over the early 2000 level and that the penetration of biofuels is negligible (see chapter 3). In contrast, the maximum technology case exploits the full technology potential for reducing GHG emissions discussed in chapters 4 and 5 along with significant penetration of second-generation biofuels. Underlying such a scenario of technology penetration is a fuel price of at least \$1.3–1.5 per liter (\$4.9–5.7 per gallon) to make the most advanced fuel-saving technologies, which could become available by the end of the 2020s, cost effective. Also, the figure shows the changes if the full fleet is converted to this technical level. As in earlier chapters, the projections are generated for two sets of economic growth rates, those from the reference run of the MIT Emissions Prediction and Policy Analysis model (EPPA-Ref) and those from the Intergovernmental Panel on Climate Change's SRES-B1 scenario.

As shown in the top panel of figure 8.1, global life-cycle GHG emissions were slightly less than 0.5 billion tons of CO_2 equivalent in 1950. Over the subsequent fifty-five years, the factors GGE/E and E/PKT remained approximately constant, while PKT increased strongly. The overall effect of these changes has been a roughly tenfold increase in life-cycle GHG emissions, to some 4.7 billion tons of CO_2 equivalent in 2005. More than 80 percent of these emissions were released as CO_2 from automobiles, other passenger surface modes, and aircraft; the remaining GHG emissions mainly consisted of CO_2 and methane (CH_4) during the production and supply of vehicles and fuels. (See table 3.3 for an estimate of the directly released CO_2 emissions.)

Assuming the fuel consumption characteristics of early 2000 technology to prevail, life-cycle GHG emissions would nearly triple or quintuple by 2050, depending on the assumed rate of economic growth.[1] A maximum technology scenario as defined above would involve a decline in energy intensity of new early 2000 technology by about 35 percent for automobiles and 40 percent for high-speed transport (mainly air travel).[2] In addition, second-generation biofuels could account for one-fifth of world passenger travel energy consumed, so in sum, the 2050 global life-cycle GHG emissions could be reduced to a level between 50 percent and 150 percent above that in 2005 while accommodating the large projected increase in travel.

The bottom panel of figure 8.1 summarizes the results for the United States. In 1950, the U.S. fleet of automobiles, buses, railways, and passenger aircraft released about 250 million tons of CO_2 equivalent. Taking into account also the emissions associated with the production and supply of vehicles and fuels, life-cycle GHG emissions totaled around 300 million tons. The factor GGE/E remained approximately constant, E/PKT grew by about 10 percent, while PKT increased strongly because of economic growth, resulting in 1,600 million tons of CO_2 equivalent in 2005, roughly a quintupling of the 1950 emissions.

As with the global scale result, the constant technology case for the United States would result in a further increase in life-cycle GHG emissions to between 2.9 and 3.1 billion tons of CO_2 equivalent in 2050, about 80 percent to 100 percent over the 2005 level. However, in contrast to the global picture, the maximum implementation of technology under the assumptions above could mitigate GHG emissions to a level below 2005 even, as assumed here, with vehicle attributes held constant. For this case, our analysis shows that under full market penetration and under the price assumption above, LDV energy intensity would fall to half of that in the constant technology case, and second-generation biofuels could account for one-third of transportation energy in 2050.[3]

The Possible Contribution of Passenger Travel to Climate Goals

Figure 8.1 shows only limiting cases of the influence of technology alone on future life-cycle GHG emissions from passenger travel. In practice, the achieved reduction level will depend on the implemented policy measures and their effect on vehicle attributes and operations as well as on technology and fuel characteristics, and on the time necessary for these alterations to penetrate the vehicle fleet.

To illustrate the possible contribution of personal transport to a frequently discussed mitigation target, we adopt a set of projections developed for a study of scenarios of GHG emissions and atmospheric concentration targets prepared for the U.S. Climate Change Science Program (CCSP).[4] The study reported analysis by three U.S. integrated-assessment models, for a reference, or "no policy," projection and four levels of atmospheric GHG stabilization, and here we apply results from the MIT Integrated Global System Model.[5] For purposes of discussion to come, three aspects of this study should be highlighted. First, atmospheric stabilization was defined in terms of the combined effect of all anthropogenic greenhouse gases—CO_2, methane, nitrous oxide and the industrial gases (hydrofluorocarbons, perfluorocarbons, and sulfur hexafluoride)—but each stabilization level was related to a particular achievement in CO_2 concentrations. Our illustrations in table 8.1 are constructed using the case that leads to an atmospheric CO_2 level of 550 parts per million by volume (ppmv). Second, the study assumed participation by all nations from the outset. And finally, the calculations assume "what" and "where" flexibility, meaning that the same CO_2 penalty is applied to all sectors and nations in each period. This result might be achieved by application of a price penalty on CO_2 by a cap-and-trade system or a CO_2 tax, or by a set of regulations designed to impose a uniform opportunity cost across all emissions sources. Below we will consider the implications of these assumptions for judgments about transport policy. To view the personal transport sector in the context of these policy scenarios we adopt three data series from the MIT model

Table 8.1
Past, current, and projected future life-cycle GHG emissions from U.S. passenger travel

	Life-cycle GHG emissions ($MtCO_2$-eq)	Retail fuel price ($/L [$/gal])	Second-generation biofuel		Prevailing Technology
			Fuel share (% energy)	Land use (% U.S. cropland)	
1950	300	0.32 [1.20]	0	0	
2005	1,600	0.47 [1.77]	0	0	2005 fleet of LDVs and aircraft
2050					
Constant technology	3,130	0.41 [1.55]	0	0	Early 2000 LDV and aircraft technology
Reference case	1,450–1,730	0.72 [2.74]	25–10	30–10	Improved mainstream technologies: 30% reduction in LDV fuel use; 40% reduction in aircraft energy intensity
550 ppmv (CO_2-eq)	1,250–1,660	0.79 [3.00]	33–10	40–10	Improved mainstream technologies: one-third reduction in LDV fuel use; 40% reduction in aircraft energy intensity

Notes: Year 2050 PKT projections are based on the economic growth rates derived from the EPPA-Ref; all PKT projections but those in the reference case were adjusted using an elasticity of travel demand with regard to changes in 2050 fuel prices of –0.1, with the projected PKT in the constant technology case as the reference point. The prevailing LDV technology determining E/PKT is based on our cost-effectiveness calculations in figure 6.2b and determined by the 2030 fuel price, given the time required to penetrate the 2050 fleet. The range of secondary biofuel penetration was estimated based on the cost effectiveness of the numbers in table 6.3 with regard to the 2030 fuel price. The (constant) world oil price determining the fuel prices shown in this table underlying the 2050 constant technology scenario was $25/bbl—the long-term average between 1950 and 2005.

results: the U.S. economic growth rate (which is consistent with the EPPA-Ref used in figure 8.1), the world oil price, and the price penalty on CO_2 emissions.

Table 8.1 shows data for two historical periods (1950 and 2005) and three projections of passenger travel emissions in 2050. First is the estimate of 2050 emissions at constant early 2000 LDV and aircraft technology and a fuel price near that in 2005 (see chapter 3 for details on the constant technology scenario). This case yields a doubling of life-cycle GHG emissions by midcentury, reaching 3.1 billion tons of CO_2 equivalent in 2050, as noted above.

A more likely (if approximate) set of estimates of 2050 emissions can be made by assuming that all new fuel-saving technology that will be economic in 2030 (at the MIT-CCSP projected fuel price in that year) will be spread throughout the vehicle and aircraft fleet by 2050. This assumption defines E/PKT. Also, we assume that PKT adjusts to the estimated increase in fuel price over the period to 2050 with an elasticity of −0.1, a number that is consistent with our review in chapter 3. In the MIT-CCSP reference projection, the world oil price rises from $35 per barrel in 2005 to $75 in 2030 and $95 in 2050.[6] According to our literature review in chapter 6, an oil price of $75 per barrel is at the high end of the expected costs for producing and delivering second-generation biofuels. Thus, we assume these fuels to supply a share of 10 to 25 percent of transportation energy. As a result, shown as the reference case in the table, sector emissions are projected to be half what they might be under constant technology and fuel price and are roughly comparable to the 2005 emission level.

If a global policy is adopted as formulated in the CCSP study, with universal participation and "what" and "where" flexibility, the projection for 550 ppmv stabilization yields a CO_2 price of $112 per ton of carbon (around $30/ton CO_2) in 2030, rising to $245 per ton ($67/ton CO_2) in 2050. The rising emissions penalty gives incentives for slightly more efficient transport technology (here with a focus on the United States), with the retail fuel price rising from $2.74 to $3.00 per gallon. Because consumer and airline fuel prices are slightly above those in the reference case, we assume second-generation biofuels to supply 10 to 33 percent of total U.S. passenger travel energy. Under these assumptions, the projected emissions from personal transport are comparable to or below those in 2005, depending on the economic competitiveness and thus penetration of second-generation biofuels.[7]

As table 8.1 also shows, the relatively modest reduction from reference case GHG emissions is attributable to the penetration of automobiles with only moderately improved fuel efficiency. The result is a percentage reduction in this sector that is much smaller than for the economy as a whole, which is simply a reflection of the relatively low price elasticities in this sector, discussed earlier, when the effects of auto and aircraft technology and low-carbon fuel alternatives are combined.

The land-use requirements for growing and cultivating biomass to supply one-third of transportation energy in 2050 would be significant. Assuming a biofuel yield of 160–200 GJ/ha, and depending on the penetration of these fuels into the market, land requirements would correspond to as much as 40 percent of the current cultivated land area of the forty-eight lower U.S. states. However, this could be reduced to below 30 percent if also using agricultural and forestry residues as a fuel feedstock.

A similar analysis could be carried out for Western Europe and other world regions. Several conditions differ, however. The average Western European LDV is roughly two-thirds the size of that in the United States, as was shown in table 3.2. In addition, the performance escalation over the past decade has proceeded at a lower rate, meaning that new engine technology has been used predominantly to increase fuel economy rather than engine power. Hence the potential for reducing fuel consumption of Western European vehicles is lower. Because many of the lower-cost options have already been implemented, the costs for reducing vehicle fuel consumption are higher (the cost curve of figure 4.3 would be steeper). The imposition of the same (international) carbon price on top of the significantly more highly taxed fuel would likely result in the introduction of more advanced fuel-saving technology as observed for the United States in table 8.1. The penetration level would ultimately depend on the costs and opportunities for reducing GHG emissions from sectors other than passenger travel. Yet, the contribution of passenger travel to an economy-wide reduction effort would likely be comparatively low, given the effect of high fuel taxes on current fleet efficiency and the comparatively low-cost mitigation potential in other sectors.

The integration of the three determinants of GHG emissions, PKT, E/PKT, and GGE/E, as summarized in figure 8.1 and table 8.1, leads to some important conclusions. First, even in light of a continuously growing travel demand, industrialized countries like the United States could reduce mobility-related GHG emissions to current levels or below, provided policy measures that increase the cost effectiveness of the required fuel-saving technology are in place. In contrast, in developing countries strong growth in travel demand will lead to GHG emissions that will be difficult to control at least through midcentury. As a result, global mobility-related GHG emissions are likely to continue to increase, even under the most stringent GHG-emission-reduction policies that now seem likely, particularly in developing countries.

Second, CO_2 emissions mitigation would come mainly from changes in technology, not in travel behavior. In the medium term, the considerable potential for reducing GHG emissions offered by reductions in road and air vehicle fuel consumption could be further expanded using second-generation biofuels. These two options are interconnected, because a more fuel-efficient transport system would also help realize a larger benefit from a limited resource. Despite their comparatively small contribution to emissions reductions, losses in travel demand are an important element for reducing GHG emissions, since further improvement in the performance of already efficient technologies may come at high economic cost. But even under the most drastic policy measures (e.g., those that strongly affect fuel prices), rising incomes will likely lead to renewed growth in travel following adjustments in consumer driving habits and vehicle and technology preferences, and changes in airline operations. Should travel demand ever saturate, we do not expect it to be caused by CO_2 emission constraints.

And finally, with universal participation in an agreement to achieve a 550 ppmv target, assumed in table 8.1, the personal transport sector makes a relatively small contribution to the needed emissions mitigation even while there remains the technical potential for more, as illustrated in figure 8.1. If, however, it is not possible to achieve the emissions penalties assumed in the analysis, or (as is very likely) the assumption of universal participation is highly optimistic, then there remains the possibility of achieving reductions in this sector by regulation, independent of policies elsewhere. As discussed in chapter 7, reductions might be sought by more stringent regulation of vehicle design and limits on the fuel composition.

The Consumer Impact of GHG Mitigation Policies

The effect of emissions mitigation on consumers depends to some degree on the particular policy approach taken. If reductions are sought by higher fuel prices—perhaps by an economy-wide CO_2 tax or an emissions trading scheme (ETS)—then consumers would experience higher travel costs. Figure 2.1 shows that average U.S. travel costs have remained approximately constant since 1950, whereas costs per PKT have declined for all modes except the automobile. An increase in fuel costs would further increase automobile travel costs and eventually prevent air travel costs from a further decline. Such a fuel price policy thus could disrupt the long-term historical trend toward a greatly increasing affordability of travel, as discussed in chapter 2.

Even such a change would not, however, increase the fraction of consumer income spent on personal travel. According to our scenarios of economic growth, GDP per capita would be 25–40 percent higher in 2030 compared with 2005. In comparison, the cost-effective improvement in mainstream technologies from table 8.1 would add a comparatively modest 15 percent to the retail price of the average new automobile sold in the early 2000s. In addition, growing income in combination with the improved fuel efficiency of advanced vehicles would tend to compensate the oil price and carbon tax-induced increase in fuel prices shown in table 8.1. A similar compensating effect, with declining aircraft fuel burn and rising income balancing fuel cost increases, would be present for air travel as well.

If, on the other hand, emissions reductions were achieved by regulation, say by limits on technology or fleet fuel economy, then there would be costs nonetheless but of a different form. Taking the automobile fleet as an example, part of the consumer cost would hit the pocketbook as increases in vehicle price, for any set of attributes, as vehicle costs likely would rise somewhat to meet the new standard. But part of the cost would be borne in consumer surplus lost, because vehicles with certain attributes, which consumers would have preferred and would be willing to pay for, would be either forbidden or otherwise regulated out of the market.

Whatever the cost, the driving experience is not likely to change much even with strong mitigation measures. In countries with low fuel taxes, such as the United States, the internal combustion engine coupled to a mechanical drivetrain likely will remain the dominant road transportation technology at least over the next several decades. Although such vehicles will be far more fuel efficient than those sold today—through changes in drivetrain, fuels, and the vehicle itself—it is questionable whether consumers will notice any significant operational difference. In contrast, in countries with higher fuel taxes, such as in Europe, hybrid-electric vehicles could become cost effective and thus gain a stronger share in the vehicle fleet. Progress in battery technology could lead to the emergence of plug-in hybrid-electric vehicles, but such technology would not impose significant operational changes to consumer habits. Consumers would not be strongly affected by a rising proportion of second-generation biofuels partly blended with oil products; the lower energy density of alcohols would be compensated by higher fuel efficiencies yielding hardly any impact on the frequency of vehicle refueling.

Consumers are not likely to experience any significant change in the experience of air travel either. While aircraft technology is likely to continuously improve, especially in a setting with climate policies, the change will be evolutionary. Engines will have lower fuel burn; the density of aircraft materials will decline; and aerodynamic efficiency will further increase. But the basic tube-wing aircraft configuration is likely to remain the dominant configuration in the fleet for at least the next several decades. Although blended-wing-body aircraft would offer a significant reduction in fuel burn, problems associated with the financial risk to manufacturers and consumer acceptability are likely to delay its introduction at least for the next decade.

The Choices Ahead

Governments have important choices to make if they are to respond successfully to the threat of climate change. The transport sector presents special challenges. Choices will interact with existing energy and environmental policies; they can support or conflict with other policy objectives, such as the reduction of oil imports, and they will be constrained by efforts to manage interfirm and international competition. Also, there is the challenge of providing long-term stability in policy to take account of the long lifetime of the existing transport infrastructure and to provide a predictable environment for the capital-intensive investments that will be necessary for a significant mitigation effort.

One choice involves the mix of policy approaches to pursue. Either market-based measures or regulations, or both, could be used to change consumer and industry choice among the attributes of new ground or air vehicles. Influence on consumer choice is especially important in restraining the weight and power of LDVs, which have increased since the beginning of auto travel. Over sufficiently long time scales, either price or regulatory approaches could achieve a target level of GHG emissions, although technology dynamics would differ. Market-based measures use price to influence both consumer and thus industry behavior, leaving to consumers or to industries the choice of what vehicle to buy. In addition, price penalties applied to fuel or GHG emissions also make lower-GHG-emitting fuels and the underlying production processes more competitive and influence the way vehicles are operated. These interventions usually are also economically more efficient than regulatory measures because they apply the same marginal cost across all sources of emissions, within and outside the transport sector.

Despite these advantages of a market-based approach, regulations have been the basis of nearly all environmental legislation to date, in part because they tend to hide the cost and thus meet the least public resistance, and in part because they appear to give a greater guarantee of emissions reduction than does a price-based system. This pattern also applies to transportation, as evidenced by the vehicle fuel economy regulations in the United States, Europe, Japan, and elsewhere. In the future, regulations could gain increasing attractiveness because of the apparent resistance of transport emissions to price penalties and the sheer difficulty of putting an effective price-based measure in place.

A related choice is whether to emphasize a single policy approach—for example, a price penalty on CO_2 emissions or a set of regulations such as tailpipe emission standards—or to apply some portfolio of measures. Although either a pure price or a pure regulatory strategy could reduce emissions to almost any desired level, such an approach is unlikely in practice. One reason is that policies of both types are already in place, such as taxes on petroleum fuels, subsidies for alternative fuels or hybrid-electric vehicles, fuel economy regulations for LDVs, and so forth. Also, a single-policy approach could be less effective than a carefully coordinated set of measures: no single policy approach influences equally well all stages of a technology life-cycle, from vehicle concept to use on the road or in the air. Indeed, in some situations more than one policy measure would be needed just to compensate for the undesired consequences of some other intervention. For example, regulatory measures that exclusively aim to increase the fuel efficiency of new automobiles also reduce the marginal costs of driving, a side effect that could be mitigated by an increase in fuel tax.

Perhaps the most fundamental choice is whether emissions control for passenger transport is to be approached as a component of a national strategy, applying across all sectors, or handled as a unique problem. Passenger travel could be treated as an economic sector equivalent to others, having the same price penalty on CO_2 emissions applied to it as elsewhere in the economy. Such an equating of marginal costs across CO_2 emission sources would be most easily accomplished by means of a national carbon tax or cap-and-trade system. Implicitly, consumer and industry behavior when choosing among different possible features of personal mobility would not be viewed as fundamentally different from that in other aspects of modern economic life. Transportation fuels would simply be a component in the design of any governmental emissions trading or tax system. Combinations of policies might be applied, but an effort would be made to calibrate regulatory constraints so that a roughly equal carbon price could be applied across all sources of CO_2 emissions.[8] Following such a comprehensive price-based approach to limiting CO_2 emissions, the initial impact on transport emissions would likely be modest under a range of CO_2 emission penalties, as shown above. Lower-cost opportunities for mitigating CO_2 emissions would be realized in other sectors of the economy.

Alternatively, personal transport could be treated as a special case, with targets for emission reductions remaining independent of policy intervention elsewhere in the economy and with little concern shown for equality of treatment across sectors. Arguments for this approach usually make two assumptions. First, that consumer and industry behavior in this sector are unique—and that the low price elasticity of motor and aircraft fuel is an example of this uniqueness. Second, that controls are possible on passenger transport, but perhaps not elsewhere in the economy—and the value of marginal gains from reductions in this sector exceeds any costs suffered because of its differential treatment. This latter argument emerges when several policy concerns are being pursued, such as reducing GHG emissions while mitigating oil dependence and the associated push for a drastic increase in synthetic oil products made from GHG-intensive feedstocks and processes. In fact, the EU has pursued that approach by excluding surface transportation from the ETS and separately mandating fuel economy targets for LDVs. California is headed down the same path, proposing regulation of the carbon content of motor fuels.

Whatever mitigation strategy they advance, governments also can influence the development of technology through research and development (R&D) policies. Points of focus are improvements in propulsion and energy storage systems, reducing the driving resistance of road vehicles through even more aggressive weight reduction and downsizing, and exploring radical changes in aircraft design. Given the uncertainty of progress in each of these technology clusters, R&D investments should be distributed among promising alternative options. Research on battery technology is critical, but competing technologies such as the production and storage of hydrogen should be supported as well. Because biomass-derived synthetic oil products can be readily used in automobiles and aircraft, R&D investments in these fuels could have an especially high payoff.

Given this wide range of policy choice, and lack of clarity as to how governments will move, the road and air vehicle industries face challenging uncertainties in their planning and investment. Some certainties can nonetheless be identified. Perhaps most fundamental: the GHG emission problem is not going to fade away. Rather, as we show in chapter 1, the relative importance of transportation-related GHG emissions is likely to continue to increase, simply as a result of the sector shifts in the economy toward services which in turn involve transportation as a major energy consumer and emitter of greenhouse gases. Second, our analysis has shown that investments in more fuel-efficient mainstream vehicle technologies and low-CO_2 fuels will have to be the major part of a GHG mitigation solution, at least over the next several decades. R&D investments in fuel-saving technology would also minimize industry risk with regard to any government policy approach: reductions in road and air vehicle fuel consumption cut the emissions of CO_2, increase the relative impact of alternative fuel resources, and mitigate possible policy conflicts between reducing GHG emissions and reducing oil dependence.

The opportunities for consumer-initiated reductions in emissions, outside price pressures or regulatory restraint, are substantial as well. However, realizing these and other opportunities will require reversing the trends that have shaped our transportation system over at least the past fifty years. More environmentally conscious behavior could be spurred by information and education campaigns, stressing the impact of consumer choices. Indeed consumers could already reduce GHG emissions at small personal cost by lowering highway speed, monitoring tire pressure, and cutting down on aggressive driving. Comprehensive GHG mitigation strategies will need the joint effort of all three: government, industry, and consumers.

Even though the formulation of polices for the passenger transport sector presents daunting challenges, we believe the message of this book is clear. Substantial opportunities for reducing passenger travel GHG emissions do exist. However, their exploitation requires successfully confronting several problems, including, perhaps most important, the implementation of policies that shape consumer and airline preferences among vehicle attributes and influence total travel demand. Policies must also be implemented to develop technologies that can ease the now difficult trade-off between certain desired characteristics of personal transport and GHG emissions.

Also, given the long time lags in the system—technological, economic, political, and behavioral—it is crucial that increased efforts to implement policies that reduce GHG emissions be undertaken now. Otherwise, the seemingly inexorable growth of passenger travel will overwhelm any later attempt to meet atmospheric concentration targets. We all want to continue to enjoy unfettered personal mobility while protecting the planet. Unfortunately, there are too many of "us," and we are becoming too wealthy, for that to be feasible without changes in traveler behavior and improvements in technology, somehow spurred by government policy.

Notes

Chapter 1

1. Tarr, J., C. McShane, 1997. The Centrality of the Horse to the Nineteenth-Century American City, in Mohl R. (ed.), *The Making of Urban America*, SR Publishers, New York, 105–130.

2. Thompson, F.M.L., 1970. *Victorian England: The Horse-Drawn Society*, an Inaugural Lecture, Bedford College, London.

3. See, e.g., Nakićenović N., 1986. The Automobile Road to Technological Change, *Technological Forecasting and Social Change*, 29: 309–340.

4. The importance of transport modes is defined here in terms of the provision of passenger kilometers.

5. National Museum of American History, *America on the Move*; http://americanhistory.si.edu/ONTHEMOVE/exhibition/.

6. According to surveys, the share of the U.S. van, minivan, pickup truck, and sport-utility vehicle fleet used for personal transport has increased from about one-quarter in 1963 to more than three-quarters in 2002. U.S. Census Bureau, various years. *Vehicle Inventory and Use Survey* (formerly *Truck Inventory and Use Survey*), Washington, DC.

7. McAlinden, S.P., K. Hill, B. Swiecki, 2003. *Economic Contribution of the Automotive Industry to the U.S. Economy—An Update*, a study prepared for the Alliance of Automobile Manufacturers, Center for Automotive Research, Ann Arbor, MI.

8. U.S. Environmental Protection Agency, 2005. *National Emissions Inventory—Air Pollutant Emissions Trends Data and Estimation Procedures*, 1970–2002 *Average Annual Emissions, All Criteria Pollutants*, Washington, DC.

9. Another area where the benefits of technical improvements have been overwhelmed by traffic growth is vehicle occupant safety. Partly as a reaction to vehicle safety regulations, the automobile industry has made significant progress in safety engineering. In the 1920s, there were 11 traffic fatalities per 100 million vehicle kilometers traveled (VKT) within the United States (18 fatalities per 100

million vehicle miles traveled); this rate has declined to about 2 fatalities per 100 million VKT in 1970 and to about 0.6 per 100 million VKT (some 3 to 1 fatality per 100 million vehicle miles traveled, respectively). However, in absolute terms, the number of traffic fatalities has increased from about 22,000 in the mid-1920s to nearly 55,000 in 1970. After 1970, the fatality rate gradually declined to 46,000 in 2005, that is, more than twice the 1920 level, and the introduction of active safety devices, such as driver intervention systems, is likely to lead to a further gradual reduction in absolute levels. Yet, reducing the absolute number of fatalities to a level below that of the mid-1920s will remain challenging. U.S. Census Bureau, 1975. *Historical Statistics of the United States, Colonial Times to 1970*, U.S. Department of Commerce, Washington, DC. U.S. Census Bureau, 2008. *Statistical Abstract of the United States*, U.S. Department of Commerce, Washington, DC.

10. Some streamliners reached top speeds in excess of 180 km/h (112 mph).

11. As suggested by travel surveys from the United States and European countries, about two-thirds of total PKT (by all modes) involve trip distances shorter than 100 kilometers (62 miles). Thus, long-distance travel (all PKT with trip distances of 100 kilometers and more) account for the remaining third of total PKT.

12. International Civil Aviation Organization, ICAOdata; www.icaodata.com/.

13. Lee J.J., S.P. Lukachko, I.A. Waitz, A. Schäfer, 2001. Historical and Future Trends in Aircraft Performance, Cost, and Emissions, *Annual Review of Energy and the Environment 2001*, 26: 167–200.

14. U.S. Environmental Protection Agency, 2005. *National Emissions Inventory—Air Pollutant Emissions Trends Data and Estimation Procedures, 1970–2002 Average Annual Emissions, All Criteria Pollutants*, Washington, DC.

15. Sequeira, C.J., 2008. *An Assessment of the Health Implications of Aviation Emissions Regulations*, S.M. Aeronautics and Astronautics, Massachusetts Institute of Technology, Cambridge.

16. Neufville R.L. de, 2005. The Future of Secondary Airports—Nodes of a Parallel Air Transport Network? *Cahiers Scientifiques du Transport* 47: 11–38.

17. During the 1930s, there were 27 passenger fatalities per 100 million aircraft kilometers (43 passenger fatalities per 100 million aircraft miles). Not counting the fatalities associated with the September 11 terrorist attacks, this number declined to 0.3 passenger fatalities per 100 million aircraft kilometers during the 2000–2005 period (0.5 passenger fatalities per 100 million aircraft miles), a result mainly of improvements in aircraft technology and air traffic control. Yet, the approximately two orders of magnitude growth in aircraft revenue kilometers has maintained the original number of fatalities at a level of about 20–25 per year. U.S. Census Bureau, 1975. *Historical Statistics of the United States, Colonial Times to 1970*, U.S. Department of Commerce, Washington, DC. U.S. National Transportation Safety Board, *Aviation Accident Statistics*, Washington, DC.

18. Among all major environmental and societal impacts, only aircraft noise has declined in absolute terms. In 1971, legislation of the International Civil Aviation

Organization, a United Nations body, recommended standards that led to the first noise legislation for jet aircraft. Subsequent legislation, which extended these noise standards to all types of aircraft, has become tighter over time. Thus, significant reductions in aircraft noise have been achieved through mainly technological changes at the aircraft level. Additional reductions were achieved through land-use planning and management around airports, operational aircraft procedures, and aircraft operating restrictions. Since 1970, the number of people exposed to aircraft noise has declined by about 95 percent, although the number of aircraft operations has increased by 125 percent. See Waitz, I.A., J. Townsend, J. Cutcher-Gershenfeld, E. Greitzer, J. Kerrebrock, 2004. *Aviation and the Environment*, report to the United States Congress, Partnership for AiR Transportation Noise and Emissions Reduction, Massachusetts Institute of Technology, Cambridge. The increase in the number of aircraft operations was concluded from Air Transport Association of America, 1950–2007. *Economic Report* (formerly *Air Transport—Facts and Figures*), Washington, DC.

19. International Civil Aviation Organization (ICAO), 2007. *ICAO Environmental Report 2007*, Environmental Unit of the ICAO, Montreal.

20. International Energy Agency (IEA), 2007. *Energy Balances of OECD and Non-OECD Countries*, IEA/OECD, Paris.

21. British Petroleum, 2007. *BP Statistical Review of World Energy*; www.bp.com/statisticalreview/.

22. The updated global warming potential is 25 for methane and 298 for nitrous oxide. Since the old numbers are still used by the UN Framework Convention on Climate Change, we continue to use the old numbers in the text. Intergovernmental Panel on Climate Change, 2007. Working Group 1, *Fourth Assessment Report (Technical Summary)*; http://ipcc-wg1.ucar.edu/wg1/wg1-report.html.

23. Penner, J.E., D.H. Lister, D.J. Griggs, D.J. Dokken, M. McFarland, eds., 1999. *Aviation and the Global Atmosphere*, Cambridge University Press, Cambridge, UK.

24. Intergovernmental Panel on Climate Change, 2007. *Climate Change 2007: Synthesis Report*, Summary for Policy Makers, Cambridge University Press, New York.

25. Schäfer A., 2005. Structural Change in Energy Use, *Energy Policy*, 33(4): 429–437.

26. See, e.g., Lovins, A.B., E.K. Datta, O.-E. Bustnes, J.G. Koomey, N.J. Glasgow, 2004. *Winning the Oil Endgame—Innovation for Profits, Jobs, and Security*, Rocky Mountain Institute, Snowmass, CO.

Chapter 2

1. The basic data estimation methods applied here are a further development of those described in Schäfer, A., 1998. The Global Demand for Motorized Mobility, *Transportation Research A*, 32(6): 455–477.

2. *North America* mainly consists of Canada, the United States, Puerto Rico; *Pacific OECD* mainly consists of Australia, Japan, New Zealand; *Western Europe* mainly consists of European Community, Norway, Switzerland, Turkey; *Eastern Europe* mainly consists of Bulgaria, Hungary, Czech and Slovak Republics, former Yugoslavia, Poland, Romania; *Former Soviet Union* mainly consists of Azerbaijan, Baltic countries, Belarus, Kazakhstan, Russia, Ukraine, Uzbekistan; *Latin America and the Caribbean* mainly consists of Argentina, Colombia, Brazil, Chile, Mexico, Uruguay, Venezuela; *Middle East/North Africa* mainly consists of Algeria, Gulf States, Egypt, Iran, Morocco, Saudi Arabia, Sudan; *Sub-Saharan Africa* mainly consists of Ethiopia, Kenya, Niger, Nigeria, Senegal, South Africa, Tanzania, Zaire, Zimbabwe; *Centrally Planned Asia* mainly consists of Cambodia, China, Hong Kong, Mongolia, Vietnam; *South Asia* mainly consists of Bangladesh, India, Pakistan, Sri Lanka; *Other Pacific Asia* mainly consists of Indonesia, Malaysia, Philippines, Singapore, South Korea, Taiwan, Thailand. Nakićenović, N., J. Alcamo, G. Davis, B. de Vries, J. Fenhann, S. Gaffin, K. Gregory, A. Grübler, T. Yong Jung, T. Kram, E. Lebre La Rovere, L. Michaelis, S. Mori, T. Morita, W. Pepper, H. Pitcher, L. Price, K. Riahi, A. Roehrl, H.-H. Rogner, A. Sankovski, M. Schlesinger, P. Shukla, S. Smith, R. Swart, S. van Rooijen, N. Victor, Z. Dadi, 2000. *Special Report on Emission Scenarios*, Intergovernmental Panel on Climate Change, Cambridge University Press.

3. We use gross domestic product (GDP) as a proxy for income. GDP per capita data is purchasing power parity adjusted and derived from Penn World Table 6.2. Heston, A., R. Summers, B. Aten, 2006. *Penn World Table*, Version 6.2, Center for International Comparisons of Production, Income, and Prices at the University of Pennsylvania, Philadelphia, PA. The gaps in that database were filled through scaling macroeconomic data from Maddison, A., 2001. *The World Economy—A Millennial Perspective*, OECD, Paris.

4. U.S. Bureau of Statistics, 1901. *Statistical Abstract of the United States*, Washington, DC. Conversion of 1882 U.S. dollars (1882) into 2000 U.S. dollars from Economic History Services; www.eh.net/hmit/compare/.

5. Maddison, A., 2001. *The World Economy—A Millennial Perspective*, OECD, Paris.

6. Unless otherwise noted, all monetary figures are in 2000 U.S. dollars.

7. *A Short History of the Model T Automobile*, http://media.ford.com/article_display.cfm?article_id=860. Note that such returns from changes in production scale would not have been possible without Ford's ingenious redesign of the motor vehicles of his time. Liberating these economies of scale required a combination of three factors: interchangeability of parts, simplicity of design, and easy attachment of parts to the vehicle on the assembly line. Womack, J.P., D.T. Jones, D. Roos, 1990. *The Machine That Changed the World*, Harper Perennial, New York.

8. According to data by the American Society of Mechanical Engineers, the respective current prices were U.S. $850 in 1908 and U.S. $380 in 1927 (http://anniversary.asme.org/2005landmarks2.shtml). We use the GDP deflator from

the Economic History Services to calculate the year 2000 inflation-adjusted costs (www.eh.net/hmit/compare/).

9. Lean production combines the advantages of craft and mass production—that is, teams of multiskilled workers at all levels of an organization and the use of highly flexible, increasingly automated machines to produce volumes of products in enormous variety—while avoiding their disadvantages. Overall, lean production "uses less of everything compared with mass production—half the human effort in the factory, half the manufacturing space, half the investment in tools, half the engineering hours to develop a new product in half the time. Also, it requires keeping far less than half the needed inventory on site, results in fewer defects, and produces a greater and ever growing variety of products." See, Womack, J.P., D.T. Jones, D. Roos, 1990. *The Machine That Changed the World*, Harper Perennial, New York.

10. Morrison, S.A., C. Winston, 1997. *Market-Based Solutions to Air Service Problems for Medium-Sized Communities*, Hearing before the Subcommittee on Aviation, Committee on Transportation and Infrastructure, U.S. House of Representatives, Washington, DC.

11. U.S. Department of Transportation, *Government Transportation Financial Statistics 2003*; www.bts.gov/.

12. A closer look at the expenditure-revenue balances with respect to the transportation infrastructure reveals that direct subsidies were largest for highway transport ($24 billion), followed by transit ($20 billion), waterways ($7 billion), and air travel ($3 billion), with incomplete data available for rail transport. Although air travel received the smallest apparent direct subsidy, comparison to the other modes must be conducted with great care given the large indirect subsidies that go into this sector in the form of relatively low fuel taxes (see chapter 7 for more detail).

13. Zahavi defined the travel time budget for groups of urban travelers, consisting of people making at least one trip on the survey day using a motorized mode. Studies have shown that the travel time budget *per traveler* is typically higher at lower incomes. The poor face more constraints on their choice of living locations and transport modes and thus find it more difficult to stay within the travel time budget. Since the share of travelers to total population is lower in low-income societies, the average per person travel time is similar to that of other, high-income societies. See Zahavi, Y., 1981. *The UMOT–Urban Interactions*, DOT-RSPA-DPB 10/7, U.S. Department of Transportation, Washington, DC; Roth, G.J., Y. Zahavi, 1981. Travel Time Budgets in Developing Countries, *Transportation Research A*, 15(1): 87–95.

14. Ausubel, H., A. Grübler, 1995. Working Less and Living Longer: Long-Term Trends in Working Time and Time Budgets, *Technological Forecasting and Social Change*, 50: 113–131.

15. Mokhtarian, P.L., C. Chen, 2004. TTB or Not TTB, That Is the Question: A Review and Analysis of the Empirical Literature on Travel Time (and Money) Budgets, *Transportation Research A*, 38(9–10): 643–675.

16. Marchetti, C., 1994. Anthropological Invariants in Travel Behavior, *Technological Forecasting and Social Change*, 47:75–88.

17. The entire range of the travel time/speed/distance relationship can still be observed today when travel patterns in different parts of the world are compared. In very low-income rural areas of developing countries, nearly all trips undertaken are less than five kilometers (around three miles)—the distance that can be covered on foot within the roughly one-hour travel time budget. With rising income, people can afford faster modes of transport and thus increase the distance of their daily interactions. In the high-income, automobile-dominated industrialized countries, more than 95 percent of all trips occur within a trip length of 50 kilometers (some 30 miles)—the distance an automobile travels under mixed traffic conditions within 1–1.2 hours. With a shift toward still faster modes of transportation, that range of daily interaction may grow further in the future. See Schäfer, A., 2000. Regularities in Travel Demand: An International Perspective, *Journal of Transportation and Statistics*, 3(3): 1–32.

18. The key challenge, perhaps, is to come to understand how different individual behaviors, described by a skewed distribution, can produce rough stability at aggregate levels. During the transition from individual to aggregate levels, the extremes of time expenditures must be compensated somehow; that is, the more time-consuming travel of the comparatively few extreme travelers must be balanced by somewhat less time-consuming travel by a larger number of travelers.

19. Hu, P.S., J.R. Young, 1999. *Summary of Travel Trends*, 1995 Nationwide Personal Transportation Survey, Federal Highway Administration, U.S. Department of Transportation, Washington, DC.

20. Kloas, J., U. Kunert, H. Kuhfeld, 1993. *Vergleichende Auswertungen von Haushaltsbefragungen zum Personennahverkehr (KONTIV 1976, 1982, 1989)*, Gutachten im Auftrage des Bundesministers für Verkehr, Deutsches Institut für Wirtschaftsforschung, Berlin.

21. U.S. Department of Transportation, 2003. *National Survey of Distracted and Drowsy Driving Attitudes and Behaviors: 2002*, National Highway Traffic Safety Administration, Washington, DC.

22. For an excellent summary of the travel time budget dispute, see the foreword to the Personal Travel Budgets special issue by Kirby, H., 1981. *Transportation Research A*, 15: 1–6.

23. See, e.g., Zahavi, Y., 1981. *The UMOT–Urban Interactions*, DOT-RSPA-DPB 10/7, U.S. Department of Transportation, Washington, DC.

24. Schäfer, A., D.G. Victor, 2000. The Future Mobility of the World Population, *Transportation Research A*, 34(3): 171–205.

25. Echenique, M., 2007. Mobility and Income, *Environment and Planning A*, 39(8): 1783–1789.

26. GDP per capita numbers expressed in a common currency using market exchange rates (MER) are not necessarily an indicative measure of relative wealth, especially when comparing countries from the industrialized and developing

world, since prices for goods and services can differ greatly. Purchasing power parity (PPP) adjustments are a way of accounting for such price differences. In low-income countries, PPP-adjusted GDP per capita levels are relatively higher than the MER conversion would indicate.

27. In a few world regions, such as Sub-Saharan Africa and Middle East & North Africa, lasting per capita GDP losses caused, over the long term, a decrease in aggregate traffic volume, although the effect was delayed. This delay can be explained by the fact that existing transport systems, for which investments already had been made, continue to be used throughout their useful lifetime. Only when they can't be replaced—because of GDP losses—will the effect become recordable.

28. Schäfer, A., 2000. Regularities in Travel Demand: An International Perspective, *Journal of Transportation and Statistics*, 3(3): 1–32.

29. We derived the door-to-door speed of aircraft in two independent ways, which lead to very similar results. The first approach divides the weighted average trip distance from domestic and international travel by the flight time plus four hours for airport access, egress, and transfers. The second approach divides the same average trip distance by a survey-based expression of aircraft speed as a function of passenger trip distance plus three hours for airport access and egress. Air Transport Association of America, 1950–2007. *Economic Report* (formerly *Air Transport Facts and Figures*), Washington, DC. Jamin S., A. Schäfer, M.E. Ben-Akiva, I.A. Waitz, 2004. Aviation Emissions and Abatement Policies in the United States: A City Pair Analysis, *Transportation Research D*, 9(4): 294–314.

30. The general form of the regression equation can be derived from the following identity: PKT/cap = GDP/cap · TE/GDP · PKT/TE with TE being travel expenditures and TE/PKT the travel costs (P) to the consumer. Assuming TE/GDP (the travel money budget) to be constant and substituted by the parameter χ, introducing the income elasticity (α) and the price elasticity (β) leads to PKT/cap $= \chi \cdot (\mathrm{GDP/cap})^{\alpha} \cdot (\mathrm{P})^{\beta}$. Because long-term historical data of travel costs (P) are available only for very few countries, we exclude this factor from our projections. Thus, χ includes the averages of travel-money budget and price of travel in a particular region. Imposing the target point condition leads to an estimate of χ. To understand the implications of dropping P, we estimated the coefficients of this general form for North America, using the development of travel costs shown in figure 2.1. To reproduce the projected PKT/cap trajectory of the regression equation used here, which omits P, the 2050 average travel costs in the general form of the regression equation would need to be about one-third below those in 2005 (SRES-B1) or slightly less than half of those in 2005 (EPPA-Ref). Most of this decline in average travel costs would come as a byproduct from the rising share of the comparatively low-cost air travel (see figure 2.1).

31. For example, the United Nations Population Division predicts the world population to rise from 6.1 billion in 2000 to between 7.8 billion (low variant) to 11.9 billion (constant fertility variant) by 2050. The 2050 level of the medium variant (employed here) is 9.2 billion. United Nations, 2004. *World Population*

Prospects: The 2004 Revision Population Database, United Nations Population Division, New York; http://esa.un.org/unpp/.

32. Nakićenović, N., J. Alcamo, G. Davis, B. de Vries, J. Fenhann, S. Gaffin, K. Gregory, A. Grübler, T. Yong Jung, T. Kram, E. Lebre La Rovere, L. Michaelis, S. Mori, T. Morita, W. Pepper, H. Pitcher, L. Price, K. Riahi, A. Roehrl, H.-H. Rogner, A. Sankovski, M. Schlesinger, P. Shukla, S. Smith, R. Swart, S. van Rooijen, N. Victor, Z. Dadi, 2000. *Special Report on Emission Scenarios*, Intergovernmental Panel on Climate Change, Cambridge University Press.

33. One standard deviation or 68 percent of the projections were found to be between 3.9 and 5.6 times the 1990 level. Nakićenović, N., J. Alcamo, G. Davis, B. de Vries, J. Fenhann, S. Gaffin, K. Gregory, A. Grübler, T. Yong Jung, T. Kram, E. Lebre La Rovere, L. Michaelis, S. Mori, T. Morita, W. Pepper, H. Pitcher, L. Price, K. Riahi, A. Roehrl, H.-H. Rogner, A. Sankovski, M. Schlesinger, P. Shukla, S. Smith, R. Swart, S. van Rooijen, N. Victor, Z. Dadi, 2000. *Special Report on Emissions Scenarios*, Intergovernmental Panel on Climate Change, Cambridge University Press.

34. Paltsev, S., J.M. Reilly, H.D. Jacoby, R.S. Eckaus, J. McFarland, M. Sarofim, M. Asadoorian, M. Babiker, 2005. *The MIT Emissions Prediction and Policy Analysis (EPPA) Model: Version 4*. MIT Joint Program on the Science and Policy of Global Change, Report 125, Massachusetts Institute of Technology, Cambridge; http://web.mit.edu/globalchange/.

35. Data source: Heston, A., R. Summers, B. Aten, 2006. *Penn World Table*, Version 6.2, Center for International Comparisons of Production, Income, and Prices at the University of Pennsylvania, Philadelphia, PA. Since this dataset covers macroeconomic data only until 2004, the 2005 data were estimated from International Monetary Fund, 2007. *International Financial Statistics Yearbook*, Washington, DC.

36. United Nations, 2004. *World Population Prospects: The 2004 Revision Population Database*, United Nations Population Division, New York; http://esa.un.org/unpp/.

37. Assuming an average economic growth rate of 2.1 percent/yr, North Americans, the most mobile society, would reach the target point in only 2105.

38. Another hidden trend is the aggregation of buses and ordinary railways into low-speed transportation modes. Within that transportation market, a systematic trend toward lower cost and more flexible buses can be observed, with the exception of the Pacific OECD region, which is dominated by high-population-density Japan. There, ordinary railways have maintained a market share of 70 percent of the total low-speed traffic volume since 1970.

39. The two balancing equations correspond to the "conservation" of distance and travel time, that is, the sum of all travel times by mode plus nonmotorized modes must equal 1.2 hours per person per day and the sum of all PKT by mode must equal the projected total per person PKT. See Schäfer, A., D.G. Victor, 2000. The Future Mobility of the World Population, *Transportation Research A*, 34(3): 171–205.

40. To maintain a travel-time budget of 1.2 hours per person per day, the value of time coefficient in the utility function of the multinomial choice model needs to be iterated. The smallest increase in the value of time coefficient of 9 percent occurs in the case of lower economic growth rates (SRES-B1), with gradually increasing speeds per mode. In contrast, the largest increase in the value of time coefficient of 42 percent occurs in the case of higher economic growth rates (EPPA-Ref) in combination with constant 2005 speeds per mode.

41. For an introduction to diffusion theory, with application to transportation, see Grübler, A. 1990. *The Rise and Fall of Infrastructures*, Heidelberg, Physica-Verlag.

42. Given a projected 52–66 percent share of high-speed transportation in 2050 for the economic growth rates of the SRES-B1 scenario and the EPPA reference run, a 10 percent increase in LDV speed would reduce these shares to 46–62 percent. (As we describe in chapter 3, however, such an increase would be unlikely, given the substitution of air travel for long-distance automobile travel and the associated rising relative importance of lower-speed urban travel). In contrast, a 10 percent increase in the speed of high-speed transportation modes would reduce the share of this mode to only 50–64 percent. In comparison, a 10 percent increase in the travel-time budget would result in a 43–59 percent share of high-speed transportation.

43. U.S. Department of Labor, 2000. *Telework: The New Workplace of the 21st Century*, Washington, DC.

44. Schrank, D., T. Lomax, 2005. *The 2005 Urban Mobility Report*, Texas Transportation Institute, the Texas A&M University System.

45. *People's Daily*, 2001. The 96-km Beijing Fifth-Ring Road to Start Construction. Thursday, 06 December; http://english.people.com.cn/200112/06/eng20011206_86078.shtml.

46. Svercl, P.V., R.H. Asin, 1973. *Nationwide Personal Transportation Study: Home-to-Work Trips and Travel*, Report No. 8, Federal Highway Administration, U.S. Department of Transportation, Washington, DC. Hu, P.S., T.R. Reuscher, 2004. *Summary of Travel Trends*, 2001 National Household Travel Survey, Federal Highway Administration, U.S. Department of Transportation, Washington, DC.

47. Airbus, 2007. *Flying by Nature—Global Market Forecast 2007–2026*, Airbus S.A.S., 31707 Blagnac Cedex, France. Boeing, 2007. *Current Market Outlook 2007*, Boeing Commercial Airplanes, Market Analysis, Seattle, WA.

48. Using a travel time budget of 1.5 and 1.2 hours per person per day, our projected growth rates in *global* air traffic PKT are 4.6–5.2 percent per year if using the economic growth rates of EPPA-Ref and 4.9–5.8 for the SRES-B1 scenario over forty-five years. These ranges compare well with the twenty-year industry forecasts of 4.9 percent per year (Airbus) and 5.0 percent per year (Boeing).

49. One thousand feet correspond to nearly 305 meters.

50. Neufville, R.L. de, A.R. Odoni, 2003. *Airport Systems: Planning, Design, and Management*, New York, McGraw-Hill.

51. Joint Planning and Development Office, 2004. *Next Generation Air Transport System—Integrated Plan*, December; www.jpdo.aero/integrated_plan.html. Bonnefoy, P.A., R.J. Hansman, 2007. Potential Impacts of Very Light Jets on the National Airspace System, *Journal of Aircraft*, 44(4): 1318–1326.

52. The separation requirements are in the form of a matrix giving minimum distance or time spacings required for each combination of "leader" and "follower" weight category aircraft and are highest when a small aircraft is following a much larger one.

53. Hileman, J.I., Z.S. Spakovszky, M. Drela, M. Sargeant, 2007. Airframe Design for Silent Aircraft, 45th AIAA Aerospace Sciences Meeting and Exhibit, Jan 8–11, 2007, Reno, Nevada, AIAA Paper No. 2007-453.

54. High-speed rail could substitute flights within selected high-density corridors, freeing up capacity at some of the busiest airports in the United States. In 2005, 36 percent of all air passengers traveled between cities less than 800 kilometers (500 miles) apart. Assuming a mean rail speed of 300 kilometers per hour (nearly 180 miles per hour), that distance could be managed in less than three hours. Obviously only a fraction of these city pairs could be served efficiently with high-speed ground transportation. U.S. Department of Transportation, 2008. *Air Carriers (Form 41 Traffic), U.S. Carriers: T-100 Domestic Segment*, Bureau of Transportation Statistics, Washington, DC.

55. See Air Transport Association of America, 1950–2007. *Economic Report* (formerly *Air Transport Facts and Figures*), Washington, DC.

56. Joint Planning and Development Office, 2006. *NGATS 2005 Progress Report to the Next Generation Air Transportation Integrated Plan*; www.jpdo.gov/library/2005_Progress_Report.pdf.

57. Quentin, F., 2007. *ACARE: The European Technology Platform for Aeronautics*, Seminar of the Industrial Leaders of European Technology Platforms, Brussels, 12 December; ftp://ftp.cordis.europa.eu/pub/technology-platforms/docs/acare_en.pdf.

58. U.S. Census Bureau, various years. *Statistical Abstract of the United States*, U.S. Department of Commerce, Washington, DC. U.S. Bureau of Transportation Statistics, 1997. *1995 American Travel Survey*, BTS/ATS95-U.S., U.S. Department of Transportation, Washington, DC.

59. A further growth in mean trip distance would allow passengers to travel at high cruise speed over a larger proportion of their trip. Thus, the average door-to-door travel speed would rise even when current aircraft speeds remain unchanged. Airline data show that by 1980, aircraft gate-to-gate speeds had saturated at about 660 kilometers per hour (410 miles per hour) for domestic U.S. flights and at about 800 kilometers per hour (500 miles per hour) for international flights (see chapter 3). Since then, the average passenger trip distance continued to increase—by 30 percent—and reached 1,800 kilometers (1,120 miles) in 2005. According to our calculations, during the same twenty-five-year period, the average air travel speed increased by about 17 percent. A continuation of this long-term historical (roughly linear) increase in average passenger trip distance to

2,500 kilometers (1,550 miles) in 2050 would result in a further increase in mean passenger door-to-door travel speed by about 20 percent, all other factors being equal. More direct-flight routings would also increase the passengers' mean door-to-door speed. In a hypothetical air transport network, in which all origin-destination flights exclusively consist of direct connections, the average door-to-door speed would be increased by roughly 17 percent. (We derived that value from current average air transport characteristics, by conservatively eliminating one hour for flight transfers). Thus, a combination of a continuously growing passenger trip distance and an air traffic network in which half of the current hub-and-spoke flights become direct flights, would increase mean door-to-door speed by about 30 percent. A further opportunity for reducing air travel time lies in increased use of secondary and tertiary airports. Our estimates suggest that reducing total ground time by thirty minutes to one hour would increase the mean door-to-door travel speed by another 8–16 percent. Yet, we assume only half of that potential can be exploited. (Data from the 1995 American Travel survey suggests that the median distance to and from the airport is nearly 30 kilometers [18 miles]. Due to the typically asymmetric trip distance distribution, the mean distance to and from the airport is likely to be significantly longer. This suggests average airport access and egress times of at least one hour each. Using current average air transport characteristics, reducing the travel time to and from the airport by thirty minutes results in an 8 percent increase in door-to-door speed.) Thus, the combination of a growth in mean trip distance, more direct routings, and the increased use of secondary and tertiary airports would result in an increase in mean speed by $1.200 \cdot 1.085 \cdot 1.060 = 1.38$ or nearly 40 percent.

60. U.S. Department of Labor, 2000. *Telework: The New Workplace of the 21st Century*, Washington, DC.

61. Mokhtarian, P.L., 2004. Reducing Road Congestion: A Reality Check—A Comment, *Transport Policy*, 11: 183–184.

62. *Scientific American*, 1919. Declining Supply of Motor Fuel, March 8, p. 220.

63. According to extraction data from the Historical Statistics of the United States, about 5 billion barrels of oil had been extracted since the first successful drilling for oil by Drake in 1859. U.S. Census Bureau, 1975. *Historical Statistics of the United States, Colonial Times to 1970*, U.S. Department of Commerce, Washington, DC.

64. See, e.g., Campbell, C.J., J.H. Laherrère, 1998. The End of Cheap Oil, *Scientific American*, March: 78–83.

65. Hubbert, M.K., 1956. Nuclear Energy and the Fossil Fuels, presented before the Spring Meeting of the Southern District Division of Production, American Petroleum Institute, Plaza Hotel, San Antonio, TX, March 7–9; www.energybulletin.net/13630.html.

66. Deffeyes, K.S., 2001. *Hubbert's Peak—The Impending World Oil Shortage*, Princeton University Press.

67. Gautier, D.L., G.L. Dolton, K.I. Takahashi, K.L. Varnes, eds., 1996. *1995 National Assessment of United States Oil and Gas Resources—Results, Methodology, and Supporting Data*, U.S. Geological Survey Digital Data Series DDS-30, Release 2.

68. Adelman, M.A., M.C. Lynch, 1997. Fixed View of Resource Limits Creates Undue Pessimism, *Oil & Gas Journal*, April 7: 56–60. Despite Hubbert's correct prediction of peak oil, his projection of annual production levels was far off. U.S. oil production in 2005 was 2.5 billion barrels, two and three times Hubbert's projected levels.

69. Hubbert's projected world oil production peak in 2000 was based on estimated initial reserves of 1,250 billion barrels. The most recent assessment of world petroleum reserves by the U.S. Geological Survey suggests slightly more than 3,000 billion barrels, nearly two and a half times as large as the 1956 estimate by Hubbert, and this level is likely to continue to rise. Thus, not surprisingly, Hubbert's global projection turned out to be wrong. For statistics on oil production and consumption, see British Petroleum, 2007. *BP Statistical Review of World Energy*; www.bp.com/statisticalreview/.

70. See, e.g. Campbell, C.J., J.H. Laherrère, 1998. The End of Cheap Oil, *Scientific American*, March: 78–83.

71. Lynch, M.C., 2003. The New Pessimism about Petroleum Resources: Debunking the Hubbert Model (and Hubbert Modelers), *Minerals and Energy*, 18(1): 21–32. According to Lynch's analysis, many forecasters extrapolate questionably arranged datasets through simple curve-fitting techniques and create causality from correlations.

72. British Petroleum, 2007. *BP Statistical Review of World Energy*; www.bp.com/statisticalreview/.

73. In the United States alone, these strategic petroleum reserves hold 727 million barrels. They would be sufficient to keep U.S. road transportation moving at current levels for only about three months.

Chapter 3

1. Field tests suggest that differences in fuel consumption due to driving at higher engine efficiencies through higher gear ratios and anticipatory driving that reduces the amount of braking and reacceleration can be up to 5 percent in urban driving and 25 percent under highway conditions, averaging about 6 percent. However, average consumer data is not available. According to a 2001 survey by the U.S. Department of Transportation, 27 percent of all tested automobiles and 32 percent of vans, pickups, and sport-utility vehicles had at least one tire underinflated, that is, a tire pressure of at least 25 percent below the recommended value; 6 percent of light trucks and 3 percent of cars checked had all four tires underinflated. Extrapolating this distribution to the entire U.S. LDV fleet suggests that fuel savings of 1–2 percent would be possible given properly inflated tires in the entire fleet.

The rapid pace of motorization has gradually changed the environment in which vehicles are operated. Rising levels of traffic congestion and the associated increase in stop-and-go traffic reduce engine efficiencies and thus increase vehicle fuel consumption. According to the 2004 *Urban Mobility Report* by the Texas Transportation Institute (Schrank and Lomax 2005), urban traffic congestion alone wasted 22 billion liters (nearly 6 billion gallons) of fuel in the United States in 2002. That additional energy use corresponds to 5 percent of the total fuel used by the U.S. LDV fleet and increases LDV energy intensity at the same rate compared to an uncongested road network.

2. To allow consistent comparisons, fuel consumption and emission characteristics of vehicles are compared on the basis of a driving cycle, that is, a standardized vehicle speed profile over a given amount of time. The U.S. combined driving cycle consists of an urban component (Federal Test Procedure-75 [FTP-75]) and a highway component.

3. U.S. Environmental Protection Agency, 2007. *Light-Duty Automotive Technology and Fuel Economy Trends: 1975 through 2007*, Compliance and Innovative Strategies Division and Transportation and Climate Division, Office of Transportation and Air Quality, Washington, DC.

4. U.S. Census Bureau, 1980, 2004. *Vehicle Inventory and Use Survey*, U.S. Department of Commerce, Washington, DC.

5. Ross, M., T. Wenzel, 2002. *An Analysis of Traffic Deaths by Vehicle Type and Model*, American Council for an Energy-Efficient Economy, Washington DC. This analysis also indicates that the risk of traffic death to drivers of sport-utility vehicles is slightly higher compared to midsize and large automobiles, suggesting choosing a sport-utility vehicle because of safety reasons is not well grounded.

6. Choo, S., P.L. Mokhtarian, 2004. What Type of Vehicle Do People Drive? The Role of Attitude and Lifestyle in Influencing Vehicle Type Choice, *Transportation Research A*, 38(3): 201–222.

7. Ward's Communications, 2007. *Ward's Automotive Yearbook 2006*, Detroit, MI.

8. See, e.g., Burke, A., E. Abeles, B. Chen, 2004. *The Response of the Auto Industry and Consumers to Changes in the Exhaust Emission and Fuel Economy Standards (1975–2003): A Historical Review of Changes in Technology, Prices, and Sales of Various Classes of Vehicles*, Institute of Transportation Studies, University of California, Davis.

9. U.S. Census Bureau, 1958. *Statistical Abstract of the United States*, U.S. Department of Commerce, Washington, DC.

10. Ward's Communications, 2005. *Ward's Automotive Yearbook 2004*, Detroit, MI. Unless otherwise noted, all monetary figures are in 2000 U.S. dollars.

11. For example, in 2003, the strong performance of the finance division has more than compensated the losses from the Ford automotive sector, resulting in an overall positive economic balance for that company. Ford Motor Com-

pany, 2003. *Annual Report*; www.ford.com/en/company/investorInformation/companyReports/annualReports/default.htm.

12. Ward's Communications, 2005. *Ward's Automotive Yearbook 2004*, Detroit, MI.

13. Mannering, F., C. Winston, W. Starkey, 2002. An Exploratory Analysis of Automobile Leasing by U.S. Households, *Journal of Urban Economics*, 52(1): 154–176.

14. A new vehicle operating in the United States averages about 24,000 kilometers (roughly 15,000 miles) per year. This annual distance declines sharply with rising vehicle age. The average vehicle fleet covers about 16,000 kilometers (roughly 10,000 miles). Hu, P.S., T.R. Reuscher, 2004. *Summary of Travel Trends*, 2001 National Household Travel Survey, Federal Highway Administration, U.S. Department of Transportation, Washington, DC. The cost figures were derived from Davis, S.C., 2000. *Transportation Energy Data Book*, Edition 20, Oak Ridge National Laboratory, Oak Ridge, TN; American Automobile Association, 2005. *Your Driving Costs in 2005*, Heathrow, FL. U.S. Department of Transportation, 2005. *Highway Statistics 2006*, Federal Highway Administration, Washington, DC.

15. *CNN*, 2004. Car Buyers Still Shun Fuel Efficiency, Surveys Find Even $2-a-Gallon Gasoline Isn't Likely to Cause Big Change in Car-Buying Habits, 19 October; http://money.cnn.com/2004/10/18/pf/autos/fuel_economy. The September 2004 *Automotive Attitudes Survey* from HarrisInteractive and Kelley Blue Book Marketing Research suggested that most preferred vehicle characteristics are quality/reliability, followed by vehicle price, safety, cost of ownership, looks and styling, the ability to retain value, with fuel economy being the least important criteria. HarrisInteractive Market Research, 2004. *The AutoVibes Report, Automotive Attitudes Summary*, September 2004 results.

16. See, e.g., Ministère de l'Écologie, du Développement et de l'Aménagement Durables, 2006. *Les Comptes des Transports en 2007*, Paris, France. Department for Transport, 2005. *Transport Statistics Great Britain—2006 Edition*, London.

17. Bundesverkehrsministerium, 2005/2006. *Verkehr in Zahlen*, Deutscher Verkehrsverlag Hamburg. European Automobile Manufacturers' Association (ACEA), *New Passenger Car Registrations in Western Europe, Breakdown by Specifications: Average Power, Historical Series: 1990–2005*; www.acea.be/.

18. European Automobile Manufacturers' Association, *New Passenger Car Registrations in Western Europe, Breakdown by Specifications: Average Power, Historical Series: 1990–2005*; www.acea.be/.

19. *Christian Science Monitor*, 2005. Hot New Style for Stodgy Old Detroit, Art Spinella quoted, 24 January; http://csmonitor.com/2005/0124/p13s01-wmgn.html.

20. A major source of inconsistency of measuring vehicle occupancy rates is the survey coverage of long-distance travel. Since many travel surveys typically concentrate on daily short-distance travel, and since occupancy rates are lower for that type of travel, they underestimate average occupancy rates.

21. Lave, C., 1991. *Things Won't Get a Lot Worse: The Future of U.S. Traffic Congestion*, Working Paper 33, University of California Transportation Center, University of California, Berkeley.

22. Trips for personal matters include shopping, health care, and religious services.

23. Oak Ridge National Laboratory, Online Analysis Tool of the 2001 U.S. National Household Travel Survey; http://nhts.ornl.gov/2001.

24. As will be discussed in more detail in chapter 4, a decline in vehicle mass by 50 kilograms (110 lbs) or roughly 2–3 percent of vehicle weight results in a reduction in fuel use by 1–1.5 percent.

25. U.S. Federal Highway Administration, various years. *Highway Statistics*, U.S. Department of Transportation, Washington, DC.

26. Lubowski, R.N., M. Vesterby, S. Bucholtz, A. Baez, M.J. Roberts, 2006. *Major Uses of Land in the United States, 2002*, Economic Information Bulletin (EIB-14), U.S. Department of Agriculture, Washington, DC. Wooten, H.H., K. Gertel, W.C. Pendleton, 1962. *Major Uses of Land and Water in the United States, Summary for 1959*, Farm Economics Division, Economic Research Service, U.S. Department of Agriculture, Washington, DC.

27. Research Triangle Institute, 1997. *User's Guide for the Public Use Data Tape*, 1995 Nationwide Personal Transportation Survey, Federal Highway Administration, U.S. Department of Transportation, Publication No. FHWA-PL-98-002.

28. The price elasticity of demand is defined as the percentage change in demand divided by the percentage change in price. Integrating that expression results in $Q_2/Q_1 = (P_2/P_1)^{\varepsilon}$, with Q the demand, P the price, and ε the price elasticity.

29. See, e.g., Goodwin, P.B., 1992. A Review of New Demand Elasticities with Special Reference to Short and Long Run Effects of Price Changes, *Journal of Transport Economics and Policy*, May: 155–169. For a more recent analysis, see Goodwin, P.B., J. Dargay, M. Hanly, 2004. Elasticities of Road Traffic and Fuel Consumption with Respect to Price and Income: A Review, *Transport Reviews*, 24(3): 275–292.

30. J.E. Hughes et al. find a short-run price elasticity of gasoline demand of −0.034 to −0.077. Hughes, J.E., C.R. Knitiel, D. Sperling, 2008. Evidence of a Shift in the Short-Run Price Elasticity of Gasoline Demand, *Energy Journal*, 29(1): 113–134.

31. See, e.g., Small, K.A., K. Van Dender, 2007. Fuel Efficiency and Motor Vehicle Travel: The Declining Rebound Effect, *Energy Journal*, 28(1): 25–51.

32. The basic identity for LDV energy intensity corresponds to $E/PKT = \Sigma(E_i/VKT_i \cdot VKT_i/PKT_i \cdot PKT_i/PKT)$ with subscript i representing automobiles and light trucks.

33. Air Transport Association of America, 2008. *Annual Earnings: U.S. Airlines*, data series from 1938 to 2006; www.airlines.org/economics/finance/Annual+U.S.+Financial+Results.htm.

34. If passenger load factors were equal, such point-to-point operation would result in lower fuel use for stage lengths below approximately 2,000 kilometers (about 1,200 miles) because there would be fewer energy-intensive takeoffs and climbs. Our own analysis at MIT of the impact of direct routings compared to hub-and-spoke logistics suggests direct routing would produce a total reduction in air travel energy use by about 10 percent for domestic flights within the United States. See, Jamin, S., A. Schäfer, M.E. Ben-Akiva, I.A. Waitz, 2004. Aviation Emissions and Abatement Policies in the United States: A City Pair Analysis, *Transportation Research D*, 9(4): 294–314.

35. Lee, J.J., S.P. Lukachko, I.A. Waitz, A. Schäfer, 2001. Historical and Future Trends in Aircraft Performance, Cost, and Emissions, *Annual Review of Energy and the Environment 2001*, 26: 167–200.

36. U.S. Department of Transportation, 2008. *Air Carrier Financial Reports (Form 41 Financial Data), Schedule P-52*, U.S. Bureau of Transportation Statistics; www.bts.gov/.

37. Belobaba, P., 2006. Personal communication at MIT.

38. Babikian, R., S.P. Lukachko, I.A. Waitz, 2002. Historical Fuel Efficiency Characteristics of Regional Aircraft from Technological, Operational, and Cost Perspectives, *Journal of Air Transport Management*, 8(6): 389–400.

39. Bonnefoy, P.A., R.J. Hansman, 2007. Potential Impacts of Very Light Jets on the National Airspace System, *Journal of Aircraft*, 44(4): 1318–1326.

40. Eclipse Aviation, *Eclipse 500 Performance*; www.eclipseaviation.com/eclipse_500/performance/.

41. Assume a fleet of eight thousand large commercial carriers (LCC) and eight thousand VLJs. Even if using an average utilization of 0.6 for both LCC and VLJs, an average ramp-to-ramp speed of 660 km/h (LCC) and 550 km/h (VLJs), and an average seat capacity of 127 (LCC) and 6 (VLJs), a passenger load factor of 80 percent (LCC) and 50 percent (VLJs), total PKT would result in 2,820 billion for LCCs and 70 billion for VLJs, or some 2 percent of the total. Assuming an average energy intensity of 1.5–2.0 MJ/pkm for LCC and 3.2–5.1 MJ/pkm for VLJs (for shorter and longer stage lengths, respectively), energy use results in 4.2–5.6 PJ for LCC and 0.2–0.4 PJ for VLJs, or 4–8 percent of the total.

42. Babikian, R., S.P. Lukachko, I.A. Waitz, 2002. Historical Fuel Efficiency Characteristics of Regional Aircraft from Technological, Operational, and Cost Perspectives, *Journal of Air Transport Management*, 8(6): 389–400.

43. Greener by Design Science and Technology Sub-Group, 2005. *Air Travel—Greener by Design, Mitigating the Environmental Impact of Aviation: Opportunities and Priorities*, Royal Aeronautical Society, London.

44. Morrison, S.A., 1984. An Economic Analysis of Aircraft Design, *Journal of Transport, Economics and Policy*, 18(2): 123–143.

45. Since the denominator is aircraft kilometers along a great circle route and not actual kilometers traveled, poor routing results as an increase in energy usage.

46. The price elasticity ε is defined as $\varepsilon = dQ/Q/dP/P$ with Q being the quantity of air travel demand and P the ticket price. Integration of that expression leads to $Q_2 = Q_1 \cdot (P_2/P_1)^{\varepsilon}$. Thus, an X percent increase in P2/P1 leads to a $(1 + X)^{\varepsilon}$ percent increase in Q_2/Q_1.

47. Gillen, D., W. Morrison, C. Stuart, 2004. *Air Travel Demand Elasticities: Concepts, Issues and Measurement*, School of Business and Economics, Wilfrid Laurier University, Waterloo, Canada.

48. This estimate is based on an airfare elasticity for business travel of −0.71 (the average of long-haul international, long-haul domestic, and short-haul business), for leisure travel of −1.22 (the average of long-haul international, long-haul domestic, and short-haul domestic leisure), and for personal business travel of −0.96 (the mean value of business and leisure travel).

49. The combined effect of a shift toward larger vehicles and an increase in the share of urban driving results in $1.1 \cdot 1.1 = 1.21$, that is, a 21 percent increase in LDV energy intensity.

50. The combined effect of a shift toward larger vehicles and a reduction in the average vehicle occupancy rate from 2.5 to 2.0 pkm/vkm $= 1.1 \cdot 2.5/2.0 = 1.38$, that is, a 38 percent increase in LDV energy intensity for the SRES-B1 economic growth rates. Should the average vehicle occupancy rate remain unchanged, as a result of lower economic growth in the developing world as suggested by the EPPA-Ref economic growth rates, the combined effect corresponds to only the shift toward larger vehicles, that is, a 10 percent increase in LDV energy intensity.

51. The economic growth rates of the EPPA-Ref case result in a 4 percent increase in world passenger travel energy intensity. Because economic growth in the developing world is higher in the SRES-B1 scenario, the motor vehicle fleet grows more rapidly and household size declines more strongly, as do vehicle occupancy rates. As a result, energy intensity rises by 18 percent in this scenario.

52. 1 exajoule = 10^{18} joules.

Chapter 4

1. For this example we used an automobile with a mass of 1,500 kilograms (3,300 lbs), a cross-sectional area of 2.2 m^2, an aerodynamic drag coefficient of 0.3, and a tire rolling-resistance coefficient of 0.01.

2. Unless otherwise noted, all monetary figures are in 2000 U.S. dollars.

3. As an example, see the penetration of modern diesel engines in Europe. It took twenty-five years to increase their market share from 10 percent in all new automobiles sold to 50 percent.

4. Greene, D.L., J. DeCicco, 2000. Engineering-Economic Analyses of Automotive Fuel Economy Potential in the United States, *Annual Review of Energy and the Environment*, 25: 477–535.

5. Numbers are derived from the U.S. Partnership for a New Generation of Vehicles program.

6. A 10 percent increase in drivetrain efficiency leads to a new energy use of $1/1.1 = 0.91$ of the original level, that is, 9 percent below the original level.

7. See Lovins, A.B., E.K. Datta, O.-E. Bustnes, J.G. Koomey, N.J. Glasgow, 2005. *Winning the Oil Endgame—Innovation for Profits, Jobs, and Security*, Rocky Mountain Institute, Snowmass, CO.

8. As the name suggests, four strokes are required for a complete cycle. In the first stroke, the piston inhales a fuel-air mixture with constant proportions into a cylinder. After compression (second stroke), the fuel-air mixture is combusted with a spark; the burned gases then expand, delivering power to the crankshaft (third stroke); the torque generated forces the crankshaft to rotate, which then, through a transmission, turns the wheels. During the fourth piston stroke, the burnt gases in the cylinder are exhausted.

9. The compression ratio of an internal combustion engine is defined as the maximum to minimum cylinder volume, determined by the position of the piston at the beginning and end of the compression stroke.

10. The power-to-volume ratio is defined as the maximum engine power output divided by the volume displaced by the pistons during their intake strokes. One horsepower (hp) equals about 0.75 kW.

11. The number of oxygen molecules, liberated by the reduction of nitrogen oxides to nitrogen, exactly matches the required number for oxidizing unburned hydrocarbons and carbon monoxide into water vapor and carbon dioxide.

12. In part-load operation, an engine produces significantly less than maximum power. That condition corresponds to most driving.

13. A turbocharger compresses the intake air before the air enters the cylinder to increase the power from a given displacement engine. A compressor is mounted on the same shaft with a turbine, which in turn is driven by the engine exhaust.

14. Weiss, M.A., J.B. Heywood, A. Schäfer, V.K. Natarajan, 2003. *A Comparative Assessment of Advanced Fuel Cell Vehicles*, MIT Laboratory for Energy and the Environment 2003-001 RP, January, Massachusetts Institute of Technology, Cambridge. Weiss, M.A., J.B. Heywood, E.M. Drake, A. Schäfer, F. AuYeung, 2000. *On the Road in 2020: A Well-to-Wheels Assessment on New Passenger Car Technologies*, MIT Energy Laboratory, October, Massachusetts Institute of Technology, Cambridge. Kasseris, E., J.B. Heywood, 2007. *Comparative Analysis of Automotive Powertrain Choices for the Next 25 Years, Society of Automotive Engineers*, SAE Paper 2007-01-1605. National Research Council, 2002. *Effectiveness and Impact of Corporate Average Fuel Economy (CAFE) Standards*, Committee on the Effectiveness and Impact of Corporate Average Fuel Economy (CAFE) Standards, National Academy Press, Washington, DC. Austin, T.C., R.G. Dulla, T.R. Carlson, 1997. *Automotive Fuel Economy Improvement Potential Using Cost-Effective Design Changes*, Draft #3, prepared for the American Automobile Manufacturers Association, Sierra Research, Inc., Sacramento, CA.

Austin, T.C., R.G. Dulla, T.R. Carlson, 1999. *Alternative and Future Technologies for Reducing Greenhouse Gas Emissions from Road Vehicles*, prepared for the Transportation Table Subgroup on Road Vehicle Technology and Fuels, Sierra Research, Inc., Sacramento, CA.

15. A naturally aspirated engine draws air directly from the atmosphere rather than through a turbocharger. Note that we deliberately use the term energy, not fuel. Since a liter of diesel fuel contains 10–12 percent more energy than the same volume of gasoline, a diesel-fueled vehicle with a 20–25 percent higher energy efficiency experiences a 32–40 percent higher fuel efficiency and thus consumes 24–30 percent less fuel per distance traveled.

16. European Automobile Manufacturers' Association, *New Passenger Car Registrations in Western Europe*; www.acea.be/.

17. Japan Automobile Manufacturers Association, *Motor Vehicle Statistics*; www.jama.org/.

18. Sadoway, D.R., 2005. Advanced Batteries for Automotive Applications, presentation at Advanced Transportation Workshop, Global Climate and Energy Project, Stanford University, October 11–12.

19. While a battery's energy density determines the range and cost of an electric vehicle, a battery's power density, its ability to deliver a high level of electric power, determines a vehicle's acceleration capability. High values in both of these performance characteristics are desirable, if not necessary, for a vehicle to have a long range and high acceleration capability; in reality, there is a trade-off between the two. Unfortunately, current battery technology is deficient in both areas. This is an especially important matter for all-electric vehicles, since they require batteries having a high energy density if they are to have an extended travel range. In contrast, current hybrid systems require high-power-density batteries. Duvall, M., D. Taylor, M. Wehrey, N. Pinsky, W. Warf, F. Kahlhammer, L. Browning, 2004. *Advanced Batteries for Electric Vehicles, A Technology and Cost-Effectiveness Assessment for Battery Electric, Power Assist Hybrid Electric Vehicles, and Plug-in Hybrid Electric Vehicles*, Electric Power Research Institute, Palo Alto, CA.

20. A recent study by the Pacific Northwest Laboratory concluded that 73 percent of the LDV fleet (consisting of automobiles, pickup trucks, sport-utility vehicles, and vans) could be substituted by plug-in hybrid-electric vehicles driving an average of 53 kilometers per day in an electricity-only mode using the existing electricity-generation infrastructure. Similar results would apply to battery-only electric vehicles, if (conservatively) assuming similar levels of electricity consumption per VKT, ranging from 0.16 kWh/vkm for automobiles to 0.29 kWh/vkm for full-size SUVs. For comparison, the two-seater GM EV1 had an average electricity consumption of 0.10 kWh/vkm. Kintner-Meyer, M., K. Schneider, R. Pratt, 2007. *Impacts Assessment of Plug-in Hybrid Vehicles on Electric Utilities and Regional U.S. Power Grids*, Pacific Northwest National Laboratory, Richland, WA.

21. To allow consistent comparisons, fuel consumption and emission characteristics of vehicles are compared on the basis of a driving cycle, that is, a standard-

ized vehicle speed profile over a given amount of time. The U.S. Federal Test Procedure consists of an urban driving cycle (FTP-75) and a highway driving cycle.

22. Kintner-Meyer, M., K. Schneider, R. Pratt, 2007. *Impacts Assessment of Plug-in Hybrid Vehicles on Electric Utilities and Regional U.S. Power Grids*, Pacific Northwest National Laboratory, Richland, WA.

23. See, e.g., Weiss, M.A., J.B. Heywood, A. Schäfer, V.K. Natarajan, 2003. *A Comparative Assessment of Advanced Fuel Cell Vehicles*, MIT Laboratory for Energy and the Environment 2003-001 RP, January 2003, Massachusetts Institute of Technology, Cambridge.

24. Ward's Communications, various years. *Ward's Automotive Yearbook*, Detroit, MI.

25. While this trend toward lighter materials is uniform across the automobile industry, the specific characteristics of the trend within individual countries have differed. Today's German automobiles, for example, have slightly lower weight fractions in iron and steel (60 percent) than the world average, similar shares for aluminum (8 percent), but about twice the quantities of plastics (15 percent), which have become the second most important material in German cars. Verband Kunststofferzeugende Indutrie, no date. *Kunststoff im Automobil—Einsatz und Verwertung*, Frankfurt/Main; www.vke.de/de/infomaterial/download/. See also the material balances of various vehicle companies.

26. American Iron and Steel Institute, *ULSAB Engineering Report: Executive Summary* and ULSAC Project Results; www.autosteel.org//AM/.

27. Although aluminum is only one-third as dense as steel, it is less stiff, which requires larger geometries, stronger wall thickness, or a combination of both, thus offsetting part of its weight advantage.

28. Among them, the large luxury Audi A8 has a 41 percent lower empty weight than the same car produced using steel components, while the compact Audi A2 has a 15–28 percent lower vehicle weight—depending on engine size and the extent of aluminum use—compared with other automobiles of similar size. To achieve these weight reductions, Audi had to exploit secondary weight savings as well. Otherwise, the reduction in vehicle weight because of having an aluminum body alone would be only 12–13 percent. (The body-in-white accounts for about 25 percent of total vehicle weight. If reducing its weight by half through an aluminum body, the total reduction in vehicle weight is $0.25 \cdot 0.50 = 0.125$ or 12.5 percent.)

29. The density of carbon fiber–reinforced plastics is one-third less than that of aluminum. Because of their inherently superior strength, carbon fiber composites can translate their entire weight advantage into a reduction of a finished vehicle's mass.

30. Field, F.R., J.P. Clark, 1997. A Practical Road to Lightweight Cars, *Technology Review*, January: 28–36. Kelkar, A., R. Roth, J. Clark, 2001. Automobile Bodies: Can Aluminum be an Economical Alternative to Steel? *Journal of Metals*, 53(8): 28–32.

31. We derived that value from the statistical relationship between automobile fuel consumption and weight in figure 4.2: Fuel consumption = 0.0057 · mass + 1.7589. According to a U.S. Environmental Protection Agency analysis, the engine power–to–vehicle weight ratio is similar for the average new midsize, compact, and subcompact vehicles sold. U.S. Environmental Protection Agency, 2007. *Light-Duty Automotive Technology and Fuel Economy Trends: 1975 through 2007*, Compliance and Innovative Strategies Division and Transportation and Climate Division, Office of Transportation and Air Quality, Washington, DC.

32. As was shown in figure 3.2 and table 3.2, European automobiles have lower power-to-weight ratios than their U.S. counterparts, which also contributes to their lower fuel consumption.

33. Since a tire's rolling-resistance coefficient declines with increasing tire radii, the coefficients of sport-utility vehicles and pickup trucks should be lower compared with automobile tires, all other factors being equal. However, the design for off-road capability significantly compromises the rolling-resistance coefficient, which overall is about 50 percent higher.

34. Austin, T.C., R.G. Dulla, T.R. Carlson, 1997. *Automotive Fuel Economy Improvement Potential Using Cost-Effective Design Changes*, Draft #3, prepared for the American Automobile Manufacturers Association, Sierra Research, Inc., Sacramento, CA. Austin, T.C., R.G. Dulla, T.R. Carlson, 1999. *Alternative and Future Technologies for Reducing Greenhouse Gas Emissions from Road Vehicles*, prepared for the Transportation Table Subgroup on Road Vehicle Technology and Fuels, Sierra Research, Inc., Sacramento, CA.

35. The vehicle selected from the CAFE study represents the average new midsize automobile from 1999; that by DeCicco, An, and Ross assumes a model year 2000 Ford Taurus SE; the MIT study's reference car is a year 2001 Toyota Camry–like midsize automobile, while the Sierra report assumes the sales-weighted 1998 U.S. automobile.

36. The CAFE study does not rank technologies according to cost effectiveness, and thus is more conservative at lower reduction levels. It also does not consider reductions in vehicle weight.

37. Since 2006, the maximum sulfur content is 15 parts per million (ppmv) for diesel fuel, and the average sulfur content is 30 ppmv for gasoline in the United States. By contrast, the sulfur content in gasoline and diesel is currently 500 ppm in China, down from 800 and 2,000 respectively.

38. Benmaamar, M., 2003. *Urban Transport Services in Sub-Saharan Africa, Improving Vehicle Operations*, Sub-Saharan Africa Transport Policy Program, The World Bank and Economic Commission for Africa, SSATP Working Paper No. 75.

39. Sperling, D., E. Clausen, 2002. The Developing World's Motorization Challenge. *Issues in Science and Technology*, Fall 2002.

Chapter 5

1. Penner, J.E., D.H. Lister, D.J. Griggs, D.J. Dokken, M. McFarland, eds., 1999. *Aviation and the Global Atmosphere*, Cambridge University Press, Cambridge, UK. Marais, K., S.P. Lukachko, M. Jun, A. Mahashabde, I.A. Waitz, 2007. Assessing the Impact of Aviation on Climate, *Meteorologische Zeitschrift*.

2. The Breguet equation, rearranged for energy intensity (E/PKT), is

$$\frac{E}{PKT} = \frac{Q \cdot SFC}{PAX \cdot V \cdot (L/D)} \cdot \frac{W_F}{\ln\left[\frac{W_0}{W_0 - W_F}\right]}$$

with SFC being engine specific fuel consumption, PAX the number of passengers, V the aircraft speed, L/D the lift-to-drag ratio, Q the jet fuel's lower heating value, W_F the fuel weight before takeoff, and W_0 the aircraft weight at takeoff.

3. An engine's bypass ratio is defined as the air mass flow rate of air that passes through the engine but outside the combustor divided by the air mass moving through the core, which is typically composed of the high-pressure compressor, combustor, and high-pressure turbine. Thus, higher bypass ratios increase an engine's diameter.

4. St. Peter, J, 1999. *The History of Aircraft Gas Turbine Engine Development in the United States: A Tradition of Excellence*, International Gas Turbine Institute of the American Society of Mechanical Engineers, Atlanta, GA.

5. Babikian, R., S.P. Lukachko, I.A. Waitz, 2002. Historical Fuel Efficiency Characteristics of Regional Aircraft from Technological, Operational, and Cost Perspectives. *Journal of Air Transport Management*, 8(6): 38–400.

6. Ogden, C.L., C.D. Fryar, M.D. Carroll, K.M. Flegal, 2004. *Mean Body Weight, Height, and Body Mass Index, United States 1960–2002*. U.S. Department of Health and Human Services, Centers for Disease Control and Prevention, National Center for Health Statistics, Hyattsville, MD.

7. Lee, J.J., S.P. Lukachko, I.A. Waitz, A. Schäfer, 2001. Historical and Future Trends in Aircraft Performance, Cost, and Emissions, *Annual Review of Energy and the Environment 2001*, 26: 167–200.

8. The propulsive power corresponds to the product of thrust and flight velocity.

9. Lee, J.J., S.P. Lukachko, I.A. Waitz, A. Schäfer, 2001. Historical and Future Trends in Aircraft Performance, Cost, and Emissions, *Annual Review of Energy and the Environment 2001*, 26: 167–200.

10. Greener by Design Science and Technology Sub-Group, 2005. *Air Travel—Greener by Design, Mitigating the Environmental Impact of Aviation: Opportunities and Priorities*, Royal Aeronautical Society, London.

11. Liebeck, R.H, 2004. Design of the Blended Wing Body Subsonic Transport, *AIAA Journal of Aircraft*, 41(1): 10–25.

12. Hileman, J., Z. Spakovszky, M. Drela, M. Sargeant, 2007. Airframe Design for 'Silent Aircraft,' *AIAA Journal of Aircraft*, Paper 2007-0453.

13. Lee, J.J., S.P. Lukachko, I.A. Waitz, A. Schäfer, 2001. Historical and Future Trends in Aircraft Performance, Cost, and Emissions, *Annual Review of Energy and the Environment 2001*, 26: 167–200.

14. The total weight of a flight entertainment system with video screens at every seat ranges from around 840 kilograms (1,850 lbs) on a B767-300 to 1,360 kilograms (3,000 lbs) on a B747-400. See Evans D., 2003. Safety: Safety vs. Entertainment, *Avionics Magazine*, February 1; www.aviationtoday.com/av/categories/military/717.html.

15. Lee, J.J., S.P. Lukachko, I.A. Waitz, A. Schäfer, 2001. Historical and Future Trends in Aircraft Performance, Cost, and Emissions, *Annual Review of Energy and the Environment 2001*, 26: 167–200.

16. Mozdzanowska, A., R. Weibel, E. Lester, R.J. Hansman, 2007. The Dynamics of Air Transportation System Transition, 7th USA/Europe Air Traffic Management R&D Seminar, Barcelona, July.

17. U.S. Federal Aviation Agency, 2005. *RVSM Status Worldwide*; www.faa.gov/about/office_org/headquarters_offices/ato/service_units/enroute/rvsm/. Jelinek, F., S. Carlier, J. Smith, A. Quesne, 2002. *The EUR RVSM Implementation Project Environmental Benefit Analysis*, EEC/ENV/2002/008, October.

18. The percentage deviation from optimal routing is typically greatest for short-distance flights. This is because the airspace at airport vicinities is typically densest, thus imposing the single largest excess distance. A comparatively large excess distance divided by a comparatively short stage length results in a large deviation.

19. Kettunen, T., J.-C. Hustache, I. Fuller, D. Howell, J. Bonn, D. Knorr, 2005. *Flight Efficiency Studies in Europe and the United States*, presented at 6th USA/Europe Air Traffic Management R&D Seminar, Baltimore, MD, June.

20. Joint Planning and Development Office, 2007. *Next Generation Air Transportation System: Concept of Operations (ConOps)*, Version 2.0, 13 June.

21. See, e.g., *BBC News*, 2007. Extra Fuel Burnt in Air Fee Dodge, 3 December; http://news.bbc.co.uk/1/hi/england/7124021.stm.

22. Clarke, J.-P., A. Chidthaisong, P. Ciais, P.M. Cox, R.E. Dickinson, D. Hauglustaine, C. Heinze, El. Holland, D. Jacob, U. Lohmann, S. Ramachandran, P.L. da Silva Dias, S.C. Wofsy, X. Zhang, 2006. *Development, Design, and Flight Test Evaluation of a Continuous Descent Approach Procedure for Night-time Operation at Louisville International Airport*, PARTNER Report PARTNER-COE-2005-02; www.partner.aero/. Reynolds, T.G., L. Ren, J.-P.B. Clarke, 2007. Advanced Noise Abatement Approach Activities at Nottingham East Midlands Airport, UK, 7th USA/Europe Air Traffic Management R&D Seminar, Barcelona, Spain, July.

23. Green, J.E., 2002. Greener by Design—The Technology Challenge, *Aeronautical Journal*, 106(1056): 57–103.

24. Jamin, S., A. Schäfer, M.E. Ben-Akiva, I.A. Waitz, 2004. Aviation Emissions and Abatement Policies in the United States: A City Pair Analysis, *Transportation Research D*, 9(4): 294–314.

25. A thorough estimate of the combined effect of the independent projections of the determinants of aircraft energy use would require sophisticated aircraft design software. In the absence of such a tool, the Breguet range equation (described in this chapter's note 2) can only estimate the first-order impacts of technology changes. We inherently assume that a design optimization would eliminate any technology trade-off that may occur from extrapolating the performance characteristics independently.

26. Lee, J.J., S.P. Lukachko, I.A. Waitz, A. Schäfer, 2001. Historical and Future Trends in Aircraft Performance, Cost, and Emissions, *Annual Review of Energy and the Environment 2001*, 26: 167–200. Unless otherwise noted, all monetary figures are in 2000 U.S. dollars.

Chapter 6

1. For more information on aircraft designed for cryogenic fuels, see the Lockheed studies from the 1970s and 1980s. Brewer, G.D., 1990. *Hydrogen Aircraft Technology*, CRC Press.

2. Unless otherwise noted, all monetary figures are in 2000 U.S. dollars.

3. Marchetti, C., N. Nakićenović, 1979. *The Dynamics of Energy Systems and the Logistic Substitution Model*, RR-79-13, International Institute for Applied Systems Analysis, Laxenburg, Austria.

4. The policies included guaranteed ethanol fuel prices, a tax reduction for alcohol-fueled vehicles, subsidized loans for ethanol producers to increase capacity, compulsory sales of ethanol fuel at retail stations, and other measures.

5. Moreira, J.R., J. Goldemberg, 1999. The Alcohol Program, *Energy Policy*, 27: 229–245.

6. McFarlane, S., 2008. Brazil Ethanol Consumption Overtakes Gasoline, *Dow Jones Newswires*; www.cattlenetwork.com/International_Content.asp?contentid=216018.

7. Moreira, J.R., J. Goldemberg, 1999. The Alcohol Program, *Energy Policy*, 27: 229–245; supplemented by our own estimates.

8. Following the first oil crisis, the federal government promoted ethanol as a gasoline substitute. The 1978 Energy Policy Act partly exempts ethanol from federal excise tax on gasoline; in some states, ethanol is further supported through additional tax relief. Although these subsidies improve ethanol economics, the main reason for its rise is the phase-out of methyl tertiary-butyl ether (MBTE). MBTE, first introduced in 1979, is a substitute for lead additives, which were used as an octane booster. Since MBTE was already in place, refiners also used it to satisfy the 1990 amendment of the 1970 Clean Air Act. That piece of legislation requires a minimum oxygen content in gasoline of 2 percent (by weight) in areas where air quality does not meet federal standards. Since oxygenated hydrocarbon fuels undergo a more complete combustion in vehicle engines and thus reduce the amount of urban air pollutants, they proved highly successful. However,

MBTE turned out to be a potential human carcinogen and—since it is not readily biodegradable—it contaminated groundwater. Following the ban of MBTE in several states, ethanol, blended at the 10 percent level (E10), has become the oxygen component of choice, also widely known as "gasohol."

9. Renewable Fuels Association, *Industry Statistics*; www.ethanolrfa.org/industry/statistics/.

10. European Union, 2003. *Directive 2003/30/EC of the European Parliament and of the Council of 8 May 2003 on the Promotion of the Use of Biofuels or Other Renewable Fuels for Transport*; http://ec.europa.eu/energy/res/legislation/doc/biofuels/en_final.pdf.

11. In contrast to these goals, fuel production has been quite modest. In 2006, Europe produced 1.3 million tons of ethanol (European Biomass Industry Association; www.eubia.org/). Europe is also the world's largest producer of biodiesel, reaching nearly 5 million tons in 2006 (European Biodiesel Board; www.ebb-eu.org/).

12. *BBC News*, 2007. EU Ministers Agree Biofuel Target, 15 February; http://news.bbc.co.uk/2/hi/europe/6365985.stm.

13. *BBC News*, 2008. EU Rethinks Biofuels Guidelines, 14 January; http://news.bbc.co.uk/2/hi/europe/7186380.stm. Rosenthal, E., 2008. Europe, Cutting Biofuel Subsidies, Redirects Aid to Stress Greenest Options, *New York Times*, 22 January; www.nytimes.com/2008/01/22/business/worldbusiness/22biofuels.html. Godoy, J., 2008. *Subsidy Loss Threatens German Bio-Fuel Industry*, Enerpub, Energy Publisher, 7 January; www.energypublisher.com/article.asp?id=13410.

14. Light oil has a standard gravity of lower than 870 kg/m^3, medium oil up to 920 kg/m^3, and heavy oil above 920 kg/m^3. Extra-heavy oils have a standard gravity above 1,000 kg/m^3.

15. These numbers describing the fuel-cycle energy use of gasoline and diesel fuel are averages from the literature. See Schäfer, A., J.B. Heywood, M.A. Weiss, 2006. Future Fuel Cell and Internal Combustion Engine Automobile Technologies: A 25 Year Lifecycle and Fleet Impact Assessment, *Energy—The International Journal*, 31(12): 1728–1751.

16. The difference in life-cycle emissions between diesel and jet fuel is only a few percent. See Wong, H.M., 2008. *Life-cycle Assessment of Greenhouse Gas Emissions from Alternative Jet Fuels*, S.M. Aeronautics and Astronautics, Massachusetts Institute of Technology, Cambridge.

17. Since the global warming impact of methane is twenty-one times that of CO_2, the release of 0.05 gram of methane per MJ of extracted crude oil corresponds to $0.05 \cdot 21 = 1.05$ gram CO_2 per MJ of extracted crude oil. By comparison, we assume the release of 0.1 gram of methane per MJ of extracted coal and 0.2 gram of methane per MJ of methane (or 1 percent of the original amount of methane extracted, transmitted, and distributed). The main sources underlying these figures are U.S. Environmental Protection Agency, 2007. *Inventory of U.S.*

GHG Emissions and Sinks: 1990–2005, U.S. EPA #430-R-07-002, Washington, DC, and DeLucchi, M.A., 2003. *A Lifecycle Emissions Model (LEM): Lifecycle Emissions from Transportation Fuels, Motor Vehicles, Transportation Modes, Electricity Use, Heating and Cooking Fuels, and Materials—Documentation of Data and Methods*, UCD-ITS-RR-03-17, Institute of Transportation Studies, University of California, Davis.

18. Meyer, R.F., E.D. Attanasi, 2003. *Heavy Oil and Natural Bitumen—Strategic Petroleum Resources*, USGS Fact Sheet 70-03, U.S. Geological Survey; http://pubs.usgs.gov/fs/fs070-03/fs070-03.html.

19. These oil sands originate from crude oil deposits, which have migrated away from the sedimentary rock under which they were trapped and penetrated through and saturated large areas of sandstone. Bacteria present in sandstone degraded hydrocarbons, starting with the lightest compounds.

20. The electric elements heat up parts of oil shale formations over three to four years to release shale oil into wells specifically drilled for that purpose. However, to extract a sufficiently large amount of shale oil, the amount of electricity consumption would be enormous. A RAND study estimated that electricity requirements for a plant producing 100,000 barrels of shale oil per day would have power requirements of 1.2 gigawatts, which corresponds to two to three large dedicated power plants. See, Bartis, J.T., T. LaTourette, L. Dixon, 2005. *Oil Shale Development in the United States, Prospects and Policy Issues*, RAND Corporation, Santa Monica, CA.

21. Assuming a similar concentration of organic material per ton of raw material.

22. National Energy Board, 2004. *Canada's Oil Sands: Opportunities and Challenges to 2015*; www.neb.gc.ca/. Bartis, J.T., T. LaTourette, L. Dixon, 2005. *Oil Shale Development in the United States, Prospects and Policy Issues*, RAND Corporation, Santa Monica, CA.

23. In 2005, about 46 percent of the oil sands were surface mined, while the remaining 44 percent were recovered in situ through steam injection. The latter method leads to an energy use for extracting and upgrading the bitumen of about 0.5 MJ per MJ of synthetic crude.

24. The fuel-cycle energy use and GHG emissions of extra-heavy oil are between those of products from conventional crude and oil sands. Assuming an energy input of 0.12 MJ for heavy oil extraction and upgrading, the fuel-cycle energy use results in 0.24 MJ per MJ_{Diesel} or per $MJ_{Jet\ Fuel}$ and 0.30 MJ per $MJ_{Gasoline}$. If the energy input for upgrading is derived from oil products, fuel-cycle GHG emissions result in 17.6 gCO_2-eq per MJ_{Diesel} or per $MJ_{Jet\ Fuel}$ and 21.6 gCO_2-eq per $MJ_{Gasoline}$.

25. Arro, H., A. Prikk, T. Pihu, 2006. Calculation of CO_2 Emission from CFB Boilers of Oil Shale Power Plants, *Oil Shale*, 23(4): 356–365.

26. U.S. Department of Energy, 1983. *Energy Technology Characterizations Handbook: Environmental Pollution and Control Factors*, Third Edition, DOE/EP-0093, Washington, DC.

27. National Energy Board, 2004. *Canada's Oil Sands: Opportunities and Challenges to 2015*; www.neb.gc.ca/.

28. Bartis, J.T., T. LaTourette, L. Dixon, 2005. *Oil Shale Development in the United States, Prospects and Policy Issues*, RAND Corporation, Santa Monica, CA.

29. UK Department of Trade and Industry (DTI), 1999. *Technology Status Report: Coal Liquefaction, Technology Status Report 010*, DTI, London.

30. Gibson, P., 2007. *Coal to Liquids at Sasol*, Kentucky Energy Security Summit, 11 October, Lexington, KY.

31. British Petroleum, 2007. *BP Statistical Review of World Energy*; www.bp.com/statisticalreview/.

32. Laherrere, J., 2004. *Natural Gas Future Supply*, presented at the International Energy Workshop at the International Institute for Applied Systems Analysis, Laxenburg, Austria, June 22–24. British Petroleum, 2007. *BP Statistical Review of World Energy*; www.bp.com/statisticalreview/.

33. The six criteria pollutants include carbon monoxide, lead, nitrogen dioxide, sulfur dioxide, particulate matter, and ground-level ozone.

34. UK Department of Trade and Industry (DTI), 1999. *Technology Status Report: Coal Liquefaction*, Technology Status Report 010, DTI, London. Literature estimates for GTL conversion efficiencies range from a conservative 53 percent to a more aggressive 66 percent. The 53 percent estimate is based on the assumption of a truly remote plant with no use of any of the byproducts arising during Fischer-Tropsch fuel production. Weiss, M.A., J.B. Heywood, E.M. Drake, A. Schäfer, F. AuYeung, 2000. *On the Road in 2020—A Lifecycle Analysis of New Automobile Technologies*, Energy Laboratory Report MIT EL 00-003, Energy Laboratory, Massachusetts Institute of Technology, Cambridge. By contrast, other studies assume efficiencies as high as 66 percent. See, e.g., PriceWaterhouseCoopers, 2003. *Shell Middle Distillate Synthesis (SMDS), Update to a Lifecycle Approach to Assess the Environmental Inputs and Outputs, and Associated Environmental Impacts, of Production and Use of Distillates from a Complex Refinery and SMDS Route*; www.shell.com/.

35. These figures include methane leaks during natural gas extraction of 0.5 percent, or 0.1 g/MJ.

36. See, e.g., Weiss, M.A., Heywood J.B. Heywood, E.M. Drake, A. Schäfer, F. AuYeung, 2000. *On the Road in 2020—A Lifecycle Analysis of New Automobile Technologies*, Energy Laboratory Report MIT EL 00-003, Energy Laboratory, Massachusetts Institute of Technology, Cambridge. Bajura, R.A., E.M. Eyring, 2005. *Coal and Liquid Fuels*, GCEP Advanced Coal Workshop, March 15–16, Provo, Utah. Ansolabehere, S., J. Beer, J. Deutch, D. Ellerman, J. Friedmann, H. Herzog, D. Jacoby, G. McRae, R. Lester, J. Moniz, E. Steinfeld, J. Katzer, 2007. *The Future of Coal: An Interdisciplinary MIT Study*, Massachusetts Institute of Technology, Cambridge.

37. Ansolabehere, S., J. Beer, J. Deutch, D. Ellerman, J. Friedmann, H. Herzog, D. Jacoby, G. McRae, R. Lester, J. Moniz, E. Steinfeld, J. Katzer, 2007. *The*

Future of Coal: An Interdisciplinary MIT Study, Massachusetts Institute of Technology, Cambridge.

38. The economic competitiveness of synthetic fuels also depends on the fuel quality. For example, since synthetic diesel fuel is very clean and has a higher combustion quality, it has a premium.

39. See, e.g., Greene N., 2004. *Growing Energy—How Biofuels Can Help End America's Oil Dependence*, National Resources Defense Council, New York City.

40. Fargione, J., J. Hill, D. Tilman, S. Polasky, P. Hawthorne, 2008. Land Clearing and the Biofuel Carbon Debt, *Science*, 29 February, 319(5867): 1235–1238. Searchinger, T., R. Heimlich, R.A. Houghton, F. Dong, A. Elobeid, J. Fabiosa, S. Tokgoz, D. Hayes, T.-H. Yu, 2008. Use of U.S. Croplands for Biofuels Increases Greenhouse Gases through Emissions from Land-Use Change, *Science*, 29 February, 319(5867): 1238–1240. These studies, however, do not take into account any increase in land productivity. For a study that also includes the increase in land productivity, see Wang, X., 2008. Impacts of Greenhouse Gas Mitigation Policies on Agricultural Land, Ph.D. Urban and Regional Studies, Massachusetts Institute of Technology, Cambridge.

41. $1 \text{ EJ} = 10^{18} \text{ J}$.

42. This range in the global biomass potential is based on our review of nearly twenty studies published since 1991.

43. Perlack, R.D., L.L. Wright, A.F. Turhollow, R.L. Graham, B.J. Stokes, D.C. Erbach, 2005. *Biomass as Feedstock for a Bioenergy and Bioproducts Industry: The Technical Feasibility of a Billion-Ton Annual Supply*, Oak Ridge National Laboratory, Oak Ridge, TN.

44. A similar study funded by the European Commission has identified agricultural and forestry residues of about half that amount for the European Union, Switzerland, Norway, and Ukraine combined. As this study shows, the collection and delivery costs for biomass residues rise strongly the more that potential is exploited. Wit, M.P. de, A.P.C. Faaij, 2008. *Biomass Resources Potential and Related Costs, Assessment of the EU-27, Switzerland, Norway and Ukraine*, Refuel Work Package 3 Final Report, Refuel, Netherlands.

45. Biodiesel can be blended with crude oil–derived diesel fuel up to 5 percent by volume without modifying vehicle components. Higher biofuel shares require material substitutions of power-train components to avoid malfunctioning or material damage. European Automobile Manufacturers' Association (ACEA), *ACEA Position on the Use of Bio-Diesel (FAME) and Synthetic Bio-Fuel in Compression Ignition Engines*; www.acea.be/images/uploads/070208_ACEA_FAME_BTL_final.pdf.

46. A survey by the U.S. Environmental Protection Agency shows that pure biodiesel offers reductions in unburned hydrocarbons by 67 percent, carbon monoxide by 48 percent, particulates by 47 percent, sulfates by 100 percent, at a simultaneous increase in nitrogen oxide by 10 percent compared with diesel from refined oil. National Biodiesel Board; www.biodiesel.org/.

47. *International Herald Tribune*, 2008, Virgin Atlantic Flies Jumbo Jet Powered by Biofuel, 24 February; www.iht.com/articles/2008/02/24/business/biofuel.php.

48. *Energy & Environment*, 2006. *Ethanol-Boosted Gasoline Engine Promises High Efficiency at Low Cost*, October. MIT Laboratory for Energy and the Environment, Massachusetts Institute of Technology, Cambridge.

49. The energy content of biodiesel is about 10 percent lower than that of diesel fuel, and the energy content of synthetic oil products is very similar to that of crude oil.

50. Feedstock costs are about 30 cents per gallon of ethanol, which is roughly one-third the costs of other feedstocks used in other countries. Shapouri, H., M. Salassi, 2006. *The Economic Feasibility of Ethanol Production from Sugar in the United States*, U.S. Department of Agriculture, July.

51. The fermentation process releases comparatively large amounts of CO_2, on a mass basis roughly identical to the produced ethanol. Ethanol is fermented from glucose sugar following $C_6H_{12}O_6 \rightarrow 2CO_2 + 2C_2H_5OH$. The molecular mass of the reaction products is 44 for CO_2 and 46 for ethanol (C_2H_5OH) and thus nearly identical. However, these emissions are absorbed by the next generation of plant matter.

52. On average, coproducts account for slightly more than half of a processing facility's outputs by weight in the United States. Shapouri, H., P. Gallagher, 2005. *USDA's 2002 Ethanol Cost-of-Production Survey*, U.S. Department of Agriculture, Washington, DC.

53. For details, see Shapouri, H., J. Duffield, M. Wang, 2002. *The 2001 Net Energy Balance of Corn-Ethanol*, U.S. Department of Agriculture, Office of the Chief Economist, Washington, DC.

54. Roig-Franzia, M., 2007. A Culinary and Cultural Staple in Crisis: Mexico Grapples With Soaring Prices for Corn—and Tortillas, *WashingtonPost.com*, 27 January; www.washingtonpost.com/wp-dyn/content/article/2007/01/26/AR2007012601896_pf.html.

55. *BBC News*, 2008. Why Are Wheat Prices Rising? 26 February; http://news.bbc.co.uk/1/hi/business/7264653.stm#subject.

56. For processes that generate marketable coproducts, such as corn- or sugar beet–derived ethanol or rapeseed-based diesel fuel, the amount of life-cycle GHG emissions the biofuels produce depends on how the coproducts are taken into account. For example, if a processing plant produces a given amount of ethanol—and also coproducts—how should the energy input into the facility and the released GHG emissions be apportioned? A team of researchers from the U.S. Department of Agriculture and Argonne National Laboratory has attempted to answer these questions by outlining the energy use implications of four different ways of coproduct accounting. The team's analysis suggests that attributing total corn-to-ethanol conversion energy use by recording the mass flows of all products assigns a larger energy share to the coproducts. This results in an estimate that shows the lowest levels of energy use for ethanol production.

The associated GHG emissions would thus also be lowest. In contrast, if energy use for producing the coproducts were based on the energy known to have been expended in comparable dedicated processes (to produce similar products), the energy use assigned to ethanol production would be 67 percent higher; and estimated GHG emissions also would be highest. The life-cycle GHG emissions in table 6.3 are based on this latter approach, leading to more conservative estimates. Shapouri, H., J.A. Duffield, M. Wang, 2002. *The Energy Balance of Corn Ethanol: An Update*, U.S. Department of Agriculture, Washington, DC.

57. Iogen Corporation; www.iogen.ca/.

58. Choren Industries; www.choren.com/.

59. Ng, H., M. Biruduganti, K. Stork, 2005. *Comparing the Performance of Sundiesel and Conventional Diesel in a Light-Duty Vehicle and Heavy-Duty Engine*, SAE Technical Paper Series, 2005-01-3776.

60. Leigh Haag, A., 2007. Algae Bloom Again, *Nature*, 447: 520–521.

61. Note also that the marginal gains in land from an increase in fuel-energy yield decline. A doubling in fuel-energy yield from 50 to 100 GJ/ha reduces the land area by one-half while another doubling to 200 GJ/ha reduces the land area by only one-quarter.

62. Lubowski, R.N., M. Vesterby, S. Bucholtz, A. Baez, M.J. Roberts, 2006. *Major Uses of Land in the United States, 2002*, Economic Information Bulletin (EIB-14), U.S. Department of Agriculture, Washington, DC.

63. The projected 50 percent increase in corn yield through 2050 is consistent with a recent study conducted by the Oak Ridge National Laboratory. Perlack, R.D., L.L. Wright, A.F. Turhollow, R.L. Graham, B.J. Stokes, D.C. Erbach, 2005. *Biomass as Feedstock for a Bioenergy and Bioproducts Industry: The Technical Feasibility of a Billion-Ton Annual Supply*, Oak Ridge National Laboratory, Oak Ridge, TN.

64. Depending on the type of plant, the theoretical maximum photosynthetic efficiency, that is, the energy stored in the plant divided by the solar energy input, is between 3.3 percent and 6.7 percent. For a thorough discussion on photosynthetic efficiency and biomass resource potentials, see Hall, D.O., F. Rosillo-Calle, R.H. Williams, J. Woods, 1993. Biomass for Energy: Supply Prospects, in Johansson, T.B., H. Kelly, A.K.N. Reddy, R.H. Williams, L. Burnham, 2003 (eds.). *Renewable Energy—Sources for Fuels and Electricity*, Island Press, Washington, DC.

65. Nakićenović, N., A. Grübler, A. McDonald, 1998. *Global Energy Perspectives*, Cambridge University Press, Cambridge, UK.

66. International Association for Natural Gas Vehicles; www.iangv.org/.

67. The relative importance of the hydrocarbon molecules varies, but methane typically accounts for 80–95 percent by volume, with minor shares of heavier gaseous components, such as ethane, propane, and butane. Other molecules with a still smaller share include nitrogen, oxygen, and carbon dioxide.

68. The total leakage of the example given in the text results from (1 – 0.995) · (1 – 0.9975) · (1 – 0.9975) = 0.9900, that is, 99 percent of the original amount of fuel is left at the end of the fuel cycle. A concise literature review of methane leaks is given in DeLucchi, M.A., 2003. *Appendix E: Methane Emissions from Natural Gas Production, Oil Production, Coal Mining, and Other Sources*, UCD-ITS-RR-03-17E, Institute of Transportation Studies, University of California, Davis.

69. According to table 6.1, 1 liter of natural gas contains only one-third of the energy in the same volume of gasoline, even if compressed to a pressure of 250 bar. To store the equivalent of 56 liters (15 gallons) of gasoline onboard the vehicle, the extra mass of a lightweight plastic composite fuel tank compared with a full gasoline tank corresponds to about 70 kilograms (some 150 lbs) or 4.3 percent extra weight based on an average automobile weight of 1,580 kilograms (3,480 lbs) as shown in table 3.2. Since every 10 percent change in vehicle weight leads to a roughly 6 percent change in fuel use, the 4.3 percent extra weight translates into a 2.6 percent increase in fuel consumption.

70. In addition, it is questionable whether CNG vehicles are allowed to use non-ventilated tunnels. Permission for using tunnels and underpasses is typically granted by local authorities.

71. Although hydrogen is not a GHG, hydrogen leaks would increase the amount of hydroxyl radicals in the atmosphere, which in turn would increase the lifetime of methane and other gases. Since methane is a GHG, hydrogen leaks indirectly contribute to the greenhouse effect.

72. The life-cycle performance of hydrogen fuel cell vehicles is also superior to the direct use in internal combustion engines. Successful tests with hydrogen-fueled engines go back to the 1920s, and internal combustion engine vehicles with liquid hydrogen storage were first tested in 1971. For a summary of hydrogen use in internal combustion engines, see, e.g., Peschka, W., 1992. *Liquid Hydrogen—Fuel of the Future*, Springer Verlag Wien, New York.

73. Note that the fuel cycle energy use of the U.S. electricity mix strongly depends on how the primary energy equivalent of nuclear energy and renewable electricity is taken into account.

74. First proposed in the United States by General Atomics in the 1970s, thermochemical hydrogen production is based on the sulfur-iodine cycle. That cycle consists of three chemical reactions, in the course of which the sulfur and iodine compounds are recovered and reused. The net reaction is the split of water into hydrogen and oxygen.

75. Due to hydrogen's low boiling temperature of only 20 K (–253°C), liquefaction is extremely energy intensive, with electricity requirements of one-third hydrogen's heating value. Since no fuel tank can be insulated completely, heat conduction leads to boil-off losses as high as a few percent per day.

76. The U.S. Department of Energy's targets for onboard hydrogen storage reflect these difficulties. The 2010 target of hydrogen accounting for 6 percent of the total system weight is only half the hydrogen storage capacity of gasoline,

where hydrogen accounts for 12–13 percent of the total weight of the hydrogen and carbon atoms. (Of course, the carbon in gasoline is a source of energy too.) The hydrogen target increases to 9 percent of the total system weight by 2015. However, these technological targets are combined with ambitious cost projections. The joint objective is even more difficult to achieve. Also, associated with using gaseous hydrogen are all the inconveniences of handling fuel discussed earlier for natural gas. However, hydrogen's nearly one order of magnitude lower physical density (at standard temperature and pressure) makes these problems even more challenging. U.S. Department of Energy, 2006, *Energy Efficiency and Renewable Energy, Hydrogen Fuel Cells & Infrastructure Program*; www.eere.energy.gov/hydrogenandfuelcells/storage/storage_challenges.html. Burke, A.F., M. Gardiner, 2005. *Hydrogen Storage Options: Technologies and Comparisons for Light-Duty Vehicle Applications*, UCD-ITS-RR-05-01, Institute of Transportation Studies, University of California, Davis.

77. In 2005, LDVs consumed about 15.4 EJ for passenger travel, which corresponds per year to a power of 485 GW. To estimate the electricity input into the electrolysis plant, that capacity needs to be divided by plant efficiency of 65 percent. Additional electricity is required for transmission and compression, totaling in an electric power demand of nearly 900 GW_{el}.

78. For a description of the vehicle cycle, see Weiss, M.A., J.B. Heywood, E.M. Drake, A. Schäfer, F. AuYeung, 2000. *On the Road in 2020: A Well-to-Wheels Assessment on New Passenger Car Technologies*, MIT Energy Laboratory, October, Massachusetts Institute of Technology, Cambridge.

79. As with table 3.3, all vehicle use–related energy, fuel use, and CO_2 emission figures are based on the U.S. combined driving cycle and thus underreport the real values by about 20 percent.

80. Ansolabehere, S., J. Beer, J. Deutch, D. Ellerman, J. Friedmann, H. Herzog, D. Jacoby, G. McRae, R. Lester, J. Moniz, E. Steinfeld, J. Katzer, 2007. *The Future of Coal: An Interdisciplinary MIT Study*, Massachusetts Institute of Technology, Cambridge.

81. To ensure internal consistency with the assumed oil price of $50 per barrel, the lower-end supply costs of $6/GJ was used for natural gas (see notes to table 6.4) plus the costs for gas compression of about $4/GJ. The supply costs of cellulosic ethanol is the average of the range indicated in table 6.3, that is, $15/GJ.

82. In the context of alternative fuels, the gasoline break-even costs again correspond to the minimum gasoline costs to make the alternative vehicle–alternative fuel combination cost effective, provided that the respective alternative fuel can be supplied at these costs. As can be seen from tables 6.2 through 6.4, all the projected fuel costs in figure 4.3 are below $1 per liter ($3.8 per gallon).

83. As can be seen from figure 3.5, the average aircraft speed of 690 kilometers per hour (430 miles per hour) is the average of U.S. domestic and international flights. The average speed of automobiles is derived from the various U.S. Nationwide Personal Transportation Surveys and the 2001 National Household Travel Survey.

Chapter 7

1. A summary of this work is provided by Greene, D.L., O.D. Patterson, M. Singh, J. Li, 2005. Feebates, Rebates and Gas-Guzzler Taxes: A Study of Incentives for Increased Fuel Economy, *Energy Policy*, 33: 757–775.

2. Unless otherwise noted, all monetary figures are in 2000 U.S. dollars.

3. These numbers are based on an average VKT of 20,000 kilometers (12,430 miles) per year for fifteen years.

4. Sometimes rebates are used to directly oppose incentives brought about by higher gasoline prices. For example, when in mid-2008 U.S. gasoline prices were in the neighborhood of $4 per gallon, General Motors offered a rebate program for purchases of Dodge, Jeep, and Chrysler vehicles. All gas expenditures above $2.99 per gallon would be refunded during the first three years of ownership.

5. Following conversation with the airline industry, the amortization period for an increase in cost to gain fuel economy is around seven years in this sector. For retrofits there is evidence that the payback period is shorter, an example being the limited use of "winglets," wing tip extensions that reduce lift-induced drag, typically recovering fuel costs within two to three years (*Airline Fleet & Network Management*, 2006. Flying Further for Less: Blended Winglets and Their Benefits, Aviation Industry Press, London, March/April: 12–14).

6. For example, the price elasticity of the demand for household energy has been estimated to be approximately twice that of gasoline over both the short and long run. See Poyer, D.A., M. Williams, 1993. Residential Energy Demand: Additional Empirical Evidence by Minority Household Type, *Energy Economics*, (15)2: 93–100.

7. Espey, J.A., M. Espey, 2004. Turning on the Lights: A Meta-Analysis of Residential Electricity Demand Elasticities, *Journal of Agricultural and Applied Economics*, 36(1): 65–81.

8. See, e.g., Wachs, M., 2003. *Improving Efficiency and Equity in Transportation Finance*, Center on Urban and Metropolitan Policy, The Brookings Institution, Washington, DC.

9. Commission of the European Communities, 2008. *Directive of the European Parliament and of the Council amending Directive 203/87/EC so as to include aviation activities in the scheme for greenhouse gas emissions allowances within the Community* [COM(2008) 548 final], Brussels, 18 September.

10. For analysis of interaction of emissions prices with prior distortions, see Paltsev, S., H. Jacoby, J. Reilly, L. Viguier, M. Babiker, 2005. Modeling the Transport Sector: The Role of Existing Fuel Taxes in Climate Policy, in Loulou R., J.-P. Waaub, G. Zaccour (eds.), *Energy and Environment*, Springer-Verlag, New York: 211–238.

11. National Research Council, 2002. *Effectiveness and Impact of Corporate Average Fuel Economy (CAFE) Standards*, Committee on the Effectiveness and Impact of Corporate Average Fuel Economy (CAFE) Standards, National

Academy Press, Washington, DC. The study did not include a penalty for national security, nor did it address a concern, expressed in a study by the U.S. Council on Foreign Relations, about the costs to U.S. national interest of actions by terrorist organizations or regimes unfavorable to U.S. foreign policy objectives that are supported by crude oil revenues (U.S. Council on Foreign Relations, 2006. *National Security Consequences of U.S. Oil Dependency*, New York).

12. The study related security costs to the level of military expenditures (Leiby, P., D.W. Jones, T.R. Curlee, R. Lee, 1997. *Oil Imports: An Assessment of Benefits and Costs*, U.S. Oak Ridge National Laboratory, Oak Ridge, TN). A study by the U.S. Department of Transportation of regulations to reduce oil imports did not credit any benefit to reduced military expenditure (U.S. Department of Transportation, 2006. *Average Fuel Economy Standards for Light Trucks, Model Years 2008–2011*, 49 CFR Parts 523, 533, and 537, RIN 2127-AJ61, National Highway Traffic Safety Administration, Washington, DC).

13. McCubbin, D.R., M.A. DeLucchi, 1999. The Health Costs of Motor-Vehicle-Related Air Pollution, *Journal of Transport Economics and Policy*, 33(3): 253–286. We scaled the monetized health impact of the various pollutants with updated emissions data from the U.S. Environmental Protection Agency's *Trends Report* and the most recent increase in VKT from the U.S. Department of Transportation's *Highway Statistics*. The health impact today is about 40 percent lower than in 1990, the reference year of the McCubbin and DeLucchi study. With further turnover of the vehicle fleet and the prospect of increasingly tight emission standards, the health costs associated with motor vehicle emissions are likely to continue their historical decline.

The average external noise cost can be derived from the published marginal costs by road type, weighted by the 2004 VKT per road type, as reported by the U.S. *Highway Statistics*, and inflated to 2000 U.S. dollars. DeLucchi, M., S.-L. Hsu, 1996. *The External Damage Cost of Direct Noise From Motor Vehicles*, Report 14 in the series The Annualized Social Cost of Motor-Vehicle Use in the United States, based on 1990–1991 data, Institute of Transportation Studies, University of California, Davis.

14. For an analysis of this effect, see Greene, D.L., 1992. Vehicle Use and Fuel Economy: How Big Is the Rebound Effect? *Energy Journal*, 13(1): 117–143.

15. Bandivandekar, A., J.B. Heywood, 2004. *Coordinated Policy Measures for Reducing the Fuel Consumption of the U.S. Light-Duty Vehicle Fleet*, MIT Laboratory for Energy and the Environment, Report LFEE 2004-001RP, Massachusetts Institute of Technology, Cambridge. Also see Greene, D.L., A. Schäfer, 2003. *Reducing Greenhouse Gas Emissions from U.S. Transportation*, Pew Center for Global Climate Change, Washington DC.

16. In addition, public infrastructure development and land-use controls can influence both VKT and choice of transport mode. These measures could have substantial influence on CO_2 emissions from countries in early stages of developing modern transport infrastructure. For the United States, Europe, and Japan, on the other hand, modern infrastructure already exists and it is unlikely that land use and urban policies can significantly influence the growth patterns described in chapter 1, at least not for many decades.

17. An introduction to cap-and-trade systems is provided by Ellerman, A.D., P. Joskow, D. Harrison, 2003. *Emissions Trading: Experience, Lessons, and Considerations for Greenhouse Gases*, Pew Center on Global Climate Change, Washington, DC.

18. See, e.g., Goodwin, P.B., 1992. A Review of New Demand Elasticities with Special Reference to Short and Long Run Effects of Price Changes, *Journal of Transport Economics and Policy*, May: 155–169. Hughes, J.E., C.R. Knitiel, D. Sperling, 2008. Evidence of a Shift in the Short-Run Price Elasticity of Gasoline Demand, *Energy Journal*, 29(1): 113–134.

19. Gillen, D., W. Morrison, C. Stuart, 2004. *Air Travel Demand Elasticities: Concepts, Issues and Measurement*, School of Business and Economics, Wilfrid Laurier University, Waterloo, Canada.

20. If the short-term elasticity proves to have fallen as estimated by Hughes, Knittel, and Sperling (op. cit), the long-term elasticity will have fallen somewhat as well from the time when most estimates in the literature were made.

21. Schäfer A., H.D. Jacoby, 2005. Technology Detail in a Multi-Sector CGE Model: Transport under Climate Policy, *Energy Economics*, 27(1): 1–24.

22. The ultimate references for these technologies are Austin, T.C., R.G. Dulla, T.R. Carlson, 1997. *Automotive Fuel Economy Improvement Potential Using Cost-Effective Design Changes*, Draft #3, prepared for the American Automobile Manufacturers Association, Sierra Research, Inc., Sacramento, CA. Austin, T.C., R.G. Dulla, T.R. Carlson, 1999. *Alternative and Future Technologies for Reducing Greenhouse Gas Emissions from Road Vehicles*, prepared for the Transportation Table Subgroup on Road Vehicle Technology and Fuels, Sierra Research, Inc., Sacramento, CA. We aggregated individual technologies from these analyses into sensible technology packages for 2005, 2010, and 2020.

23. While the differences in fuel consumption and CO_2 emissions between the average new automobile sold in 1995 and 2005 are less than 5 percent, vehicle weight and engine power have continued to increase. However, the vehicle data we employ allows for some increase in weight and engine power. In addition, and more generally, such changes are not relevant for the type of conclusions we intend to draw from that analysis.

24. Hathaway, P., 2000. *The Effects of the Fuel 'Protest' on Road Traffic*, Transport Statistics Road Traffic Division, DETR, UK.

25. Still another proposal is to apply the cost of auto insurance, which rivals fuel expense in annual terms, as a charge per gallon of fuel in a "pay at the pump" system. The idea likely fails on the ground of administrative cost given the complexities of auto coverage and state regulation.

26. Wit, R., B.H. Boon, A. van Velzen, M. Cames, O. Deuber, D.S. Lee, 2005. *Giving Wings to Emissions Trading—Inclusion of Aviation under the European Emission Trading System: Design and Impacts*, report for the European Commission, DG Environment, No. ENV.C.2/ETU/2004/0074r.

27. Poterba, J., 1991. Is the Gasoline Tax Regressive? in D. Bradford (ed.) *Tax Policy and the Economy 5*, MIT Press, Cambridge, MA.

28. The LCB results from 240,000 vkm · 0.09 L/vkm · 0.64 kgC/L = 13,824 kgC, or about 14 tC (some 50 tCO_2).

29. The average lifetime of emitted CO_2 in the atmosphere is on the order of a hundred years, while the emissions over the fifteen-year life of a typical vehicle are centered around its middle years. Thus it is not a bad approximation, so far as the climate system is concerned, to assume that all its emissions occur at the time of purchase, to be penalized at the CO_2 price at that time.

30. A simpler but more approximate approach would tie the initial price penalty to a vehicle characteristic correlated with fuel efficiency, such as vehicle weight or cylinder volume. A fanciful alternative that could give the same incentive to shift among attributes in favor of efficiency would be a subsidy to all vehicles, starting at some high level of consumption per mile (perhaps that of the worst guzzler) and providing a rebate that increases linearly with each decrement of L/vkm. Subsidies are applied in the United States, but they are limited to hybrid-drive vehicles.

31. Fuel consumption values are taken from the U.S. Department of Energy's www.fueleconomy.gov/.

32. Greene, D.L., O.D. Patterson, M. Singh, J. Li, 2005. Feebates, Rebates and Gas-Guzzler Taxes: A Study of Incentives for Increased Fuel Economy, *Energy Policy*, 33: 757–775. These results are based on the assumption that consumers amortize their investments in fuel-saving technologies over three years.

33. National Resource Council, 2001. *Evaluating Vehicle Emissions Inspection and Maintenance Programs*, National Academy Press, Washington, DC.

34. These and the following estimates in this section are derived from the 2001 National Household Travel Survey data. http://nhts.ornl.gov/.

35. For aircraft fleet characteristics, BackAviation Solutions, 2007. *Fleet Worldwide Commercial Aircraft, Ownership & Transactions Database*, Version 4.9, June.

36. We are not aware of any study examining the airline response to various retirement incentives, so we can only attempt a qualitative estimate. The residual value of thirty-year-old or older aircraft is likely to be very small, although under favorable market conditions (i.e., manageable fuel and maintenance costs), an old amortized aircraft can continue to generate profits. Yet, according to figure 3.5, maintenance and fuel costs accounted for about two-thirds of direct operating expenditures plus investments of airlines in 2005, while the associated depreciation and amortization expenditures accounted for less than 10 percent. This suggests that an early retirement scheme coupled with low interest rate loans for the purchase of new aircraft could lead to retirement of most old aircraft, provided airlines have access to the capital market. The difference in cost structure compared with the automobile, where depreciation accounts for about half of the ownership and operating costs, coupled with the profit-driven environment in which passenger aircraft operate, might enable lower CO_2 reduction costs than in the automobile sector.

37. Federal and state subsidies cover every stage of biofuel's life-cycle: the building of biofuel processing plants, the growing of feedstocks, the production of the fuels and blending with oil products, the distribution and storage of biofuels, and the provision of incentives to the vehicle industry to produce "flexible fuel" vehicles that can operate with biofuel-oil blends of up to 85 percent biofuel. These subsidies can take many forms, including cash allowances, loan guarantees, reduced tax rates, government-provided loans and below-market-rate insurance, tariffs on competing products, and purchase mandates (such as provided by the Renewable Fuels Standard of 2005, which requires 7.5 billion gallons of renewable fuels to be produced a year by 2012). See Koplow, D., 2006. *Biofuels—At What Cost? Government Support for Ethanol and Biodiesel in the United States*, prepared for the Global Subsidies Initiative (GSI) of the International Institute for Sustainable Development (IISD), Geneva, Switzerland.

38. This percentage is about the average of the findings of several studies, although there is wide variation. See Farrell, A.E., R.J. Plevin, B.T. Turner, A.D. Jones, M. O'Hare, D. Kammen, 2006. Ethanol Can Contribute to Energy and Environmental Goals, *Science*, 311: 506–508.

39. An average vehicle lifetime of fifteen years means that half the automobile fleet of a given vintage is retired after that period. It would take another fifteen years to retire nearly the entire vehicle fleet of the 2020 vintage.

40. An, F., A. Sauer, 2004. *Comparison of Passenger Vehicle Fuel Economy and Greenhouse Gas Emission Standards Around the World*, Pew Center on Global Climate Change, Washington, DC.

41. Other technological improvements include those made to vehicle tires and air conditioners. See European Commission, 2007. Reducing CO_2 Emissions on Light-Duty Vehicles, *Europa*; http://ec.europa.eu/environment/co2/co2_home.htm.

42. Austin, D., T. Dinan, 2005. Clearing the Air: The Costs and Consequences of Higher CAFE Standards and Increased Gasoline Taxes, *Journal of Environmental Economics and Management*, 50(3): 562–582.

43. Small, K., K. Van Dender, 2005. *The Effect of Improved Fuel Economy on Vehicle Miles Traveled: Estimating the Rebound Effect Using U.S. State Data, 1966–2001*, Department of Economics, University of California, Irvine.

44. Austin, D., T. Dinan, 2005. Clearing the Air: The Costs and Consequences of Higher CAFE Standards and Increased Gasoline Taxes, *Journal of Environmental Economics and Management*, 50(3): 562–582. Also see Lutter, R., T. Kravitz, 2003. *Do Regulations Requiring Light Trucks to Be More Fuel Efficient Make Economic Sense*, AEI–Brookings Joint Center for Regulatory Studies, Regulatory Analysis 03-2. On the other hand, it is argued by Gerard and Lave that the magnitude of the rebound effect is an artifact of the failure of the United States to incorporate the cost of congestion, accidents, and so forth into the price of fuel. If these externalities were properly priced, it is countered, the rebound effect would largely disappear (Gerard, D., L. Lave, 2004. *CAFE Increases: Missing the Elephant in the Room*, AEI–Brookings Joint Center for Regulatory

Studies, Regulatory Analysis, Related Publication 04-11; www.aei-brookings.org/).

45. Farrell, A., D. Sperling, 2007. *A Low-Carbon Fuel Standard for California, Part 1: Technical Analysis*, California Energy Commission; www.energy.ca.gov/low_carbon_fuel_standard/.

46. See the annual reports by the mentioned automobile manufacturers.

47. Boeing, various years, *Annual Report*; www.boeing.com/.

48. Sperling, D., 2001. Public-Private Technology R&D Partnerships: Lessons from U.S. Partnership for a new Generation of Vehicles, *Transport Policy*, 8: 247–256.

49. National Research Council, 2001. *Review of the Research Program of the Partnership for a New Generation of Vehicles*, Seventh Report, Standing Committee to Review the Research Program of the Partnership for a New Generation of Vehicles, National Academy Press, Washington, DC.

50. In automotive research, this support typically has been far smaller than government help in the United States. Several small-scale automobile-related research programs have been funded by the European Union and individual member countries. Similarly, in Japan, a large number of government-initiated research programs have focused on fuel efficiency and the reduction of tailpipe emissions, and on electric propulsion systems. See Åhman, M., 2004. *Government Policy and Environmental Innovation in the Automobile Sector in Japan*, Department of Environmental and Energy Systems Analysis, Lund University, Report No. 53, January.

51. Aerospace Industry Association, 2004. *Aerospace Facts and Figures*, Arlington, VA.

52. Weber, H.J., A.J. Gellman, G.W. Hamlin, 2005. *Study of European Government Support to Civil Aeronautics R&D*, prepared for the National Aeronautics and Space Administration under Contract No. NNH04CC62C, TECOP International, Inc.

53. Sperling, D., 2001. Public-Private Technology R&D Partnerships: Lessons from U.S. Partnership for a New Generation of Vehicles, *Transport Policy*, 8: 247–256.

54. Subcommittee on Aviation, 2005. *Hearing on the U.S. Jet Transport Industry: Global Market Factors Affecting U.S. Producers*, U.S. House of Representatives, 25 May, Washington, DC; www.house.gov/transportation/aviation/05-25-05/05-25-05memo.html.

55. *BusinessWeek Online*, 2005. Boeing vs. Airbus: Time to Escalate, 21 March; www.businessweek.com/bwdaily/dnflash/mar2005/nf20050321_4418_db046.htm.

56. See, e.g., European Commission, 2004. *EU–U.S. Agreement on Large Civil Aircraft 1992: Key Facts and Figures*, Press Release, October 6.

57. National Research Council, 2001. *Review of the Research Program of the Partnership for a New Generation of Vehicles*, Seventh Report, Standing Com-

mittee to Review the Research Program of the Partnership for a New Generation of Vehicles, National Academy Press, Washington, DC.

58. The full range of these arguments can be found in the public comments on the discount rate in setting CAFE standards for light trucks. U.S. Department of Transportation, 2006. *Average Fuel Economy Standards for Light Trucks, Model Years 2008–2011*, 49 CFR Parts 523, 533, and 537, RIN 2127-AJ61, National Highway Traffic Safety Administration, Washington, DC. The rate finally chosen was 7 percent.

59. Researchers at MIT and the University of Michigan have suggested that this inefficiency could be corrected by linking the two systems through a regime of emissions trading: Ellerman, A.D., H.D. Jacoby, M. Zimmerman, 2006. *Bringing Transportation into a Cap-and-Trade Regime*, MIT Joint Program on the Science and Policy of Global Change, Report No. 136, Massachusetts Institute of Technology, Cambridge.

Chapter 8

1. As during the past fifty-five years, the increase in life-cycle GHG emissions would largely result from the projected growth in travel demand. We kept GGE/E constant. The increase in the global average E/PKT would be 12 percent (EPPA-Ref) or 25 percent (SRES-B1). This growth is smaller than the increase in LDV energy intensity alone, because of a rise in the relative importance of less energy-intensive aircraft and of (still) less energy-intensive travel in the developing world.

2. In this estimate, we assume the energy intensity of the North American LDV fleet to decline by 50 percent. Because of the smaller average vehicle size and the resulting smaller potential in reducing energy use, we assume the energy intensity in the LDV fleet of Western Europe, the Pacific OECD, and the reforming economies of the Former Soviet Union and Eastern Europe to decline by 40 percent. Due to infrastructure constraints, we assume a 30 percent reduction in LDV energy intensity within the developing world.

3. The reversal of the bar heights for the world and the United States is a consequence of the different regional economic growth rates in the underlying scenarios. The U.S. economic growth rate and thus travel demand, energy use, and GHG emissions are higher in the EPPA-Ref case than in the SRES-B1 scenario. In contrast, the significantly higher economic growth rates in the developing world in the SRES-B1 scenario cause global travel demand, energy use, and GHG emissions to grow more strongly than in the EPPA-Ref case.

4. Clarke, L.E, J.A. Edmonds, H.D. Jacoby, H.M. Pitcher, J.M. Reilly, R.G. Richels, 2007. *Scenarios of Greenhouse Gas Emissions and Atmospheric Concentrations*, U.S. Climate Change Science Program, Washington, DC.

5. Sokolov, A.P., C.A. Schlosser, S. Dutkiewicz, S. Paltsev, D.W. Kicklighter, H.D. Jacoby, R.G. Prinn, C.E. Forest, J. Reilly, C. Wang, B. Felzer, M.C. Sarofim, J. Scott, P.H. Stone, J.M. Melillo, J. Cohen, 2005. *The MIT Integrated*

Global System Model (IGSM) Version 2: Model Description and Baseline Evaluation, MIT Joint Program on the Science and Policy of Global Change, Report No. 124, Massachusetts Institute of Technology, Cambridge. The EPPA model component of this analysis framework was used in developing the sample analysis of emissions control measures in chapter 7.

6. Unless otherwise noted, all monetary figures are in 2000 U.S. dollars.

7. Not accounted for here is the fact that universal participation in emissions reductions would lower world oil prices (a response considered in the CCSP study), which would lower the incentive to emissions reduction. Countering this bias is the fact that the estimates in table 8.1 do not consider the potential savings from changes in vehicle attributes.

8. Even then, there would be some differences in emissions penalties among different sectors. Some fuels are already taxed at different levels depending on the fuel type, transport mode, and state. Also, a national price would apply to vehicles that are already subject to regulatory fuel economy standards, which differ by vehicle type. Economy-wide measures could thus amplify such divergence in consumer incentives to reduce emissions.

References

Adelman, M.A., M.C. Lynch, 1997. Fixed View of Resource Limits Creates Undue Pessimism, *Oil & Gas Journal*, April 7: 56–60.

Aerospace Industry Association, 2004. *Aerospace Facts and Figures*, Arlington, VA.

Åhman, M., 2004. *Government Policy and Environmental Innovation in the Automobile Sector in Japan*, Department of Environmental and Energy Systems Analysis, Lund University, Report No. 53, January.

Air Transport Association of America, 1950–2007. *Economic Report* (formerly *Air Transport Facts and Figures*), Washington, DC.

Air Transport Association of America, 2008. *Annual Earnings: U.S. Airlines*, data series from 1938 to 2006; www.airlines.org/economics/finance/Annual+U.S.+Financial+Results.htm.

Airbus, 2007. *Flying by Nature—Global Market Forecast 2007–2026*, Airbus S.A.S., 31707 Blagnac Cedex, France.

Airline Fleet & Network Management, 2006. Flying Further for Less: Blended Winglets and Their Benefits, Aviation Industry Press, London, March/April: 12–14.

American Automobile Association, 2005. *Your Driving Costs in 2005*, Heathrow, FL.

American Iron and Steel Institute, 2001. *ULSAB Engineering Report: Executive Summary and ULSAC Project Results*; www.autosteel.org//AM/.

American Public Transportation Association, various years. *Public Transportation Fact Book*, Washington, DC.

An, F., A. Sauer, 2004. *Comparison of Passenger Vehicle Fuel Economy and Greenhouse Gas Emission Standards Around the World*, Pew Center on Global Climate Change, Washington, DC.

Ansolabehere, S., J. Beer, J. Deutch, D. Ellerman, J. Friedmann, H. Herzog, D. Jacoby, G. McRae, R. Lester, J. Moniz, E. Steinfeld, J. Katzer, 2007. *The Future of Coal: An Interdisciplinary MIT Study*, Massachusetts Institute of Technology, Cambridge.

Arro, H., A. Prikk, T. Pihu, 2006. Calculation of CO_2 Emission from CFB Boilers of Oil Shale Power Plants, *Oil Shale*, 23(4): 356–365.

Austin, D., T. Dinan, 2005. Clearing the Air: The Costs and Consequences of Higher CAFE Standards and Increased Gasoline Taxes, *Journal of Environmental Economics and Management*, 50(3): 562–582.

Austin, T.C., R.G. Dulla, T.R. Carlson, 1997. *Automotive Fuel Economy Improvement Potential Using Cost-Effective Design Changes*, Draft #3, prepared for the American Automobile Manufacturers Association, Sierra Research, Inc., Sacramento, CA.

Austin, T.C., R.G. Dulla, T.R. Carlson, 1999. *Alternative and Future Technologies for Reducing Greenhouse Gas Emissions from Road Vehicles*, prepared for the Transportation Table Subgroup on Road Vehicle Technology and Fuels, Sierra Research, Inc., Sacramento, CA.

Ausubel, H., A. Grübler, 1995. Working Less and Living Longer: Long-Term Trends in Working Time and Time Budgets, *Technological Forecasting and Social Change*, 50: 113–131.

Babikian, R., S.P. Lukachko, I.A. Waitz, 2002. Historical Fuel Efficiency Characteristics of Regional Aircraft from Technological, Operational, and Cost Perspectives, *Journal of Air Transport Management*, 8(6): 389–400.

BackAviation Solutions, 2007. *Fleet Worldwide Commercial Aircraft, Ownership & Transactions Database*, Version 4.9, June.

Bajura, R.A., E.M. Eyring, 2005. *Coal and Liquid Fuels*, GCEP Advanced Coal Workshop, March 15–16, Provo, Utah.

Bandivadekar, A., K. Bodek, L. Cheah, C. Evans, T. Groode, J. Heywood, E. Kasseris, M. Kromer, M. Weiss, 2008. *On the Road in 2035: Reducing Transportation's Petroleum Consumption and GHG Emissions*, Report LFEE 2008-05RP, MIT Laboratory for Energy and the Environment Report, Massachusetts Institute of Technology, Cambridge.

Bandivandekar, A., J.B. Heywood, 2004. *Coordinated Policy Measures for Reducing the Fuel Consumption of the U.S. Light-Duty Vehicle Fleet*, MIT Laboratory for Energy and the Environment, Report LFEE 2004-001RP, Massachusetts Institute of Technology, Cambridge.

Bartis, J.T., T. LaTourette, L. Dixon, 2005. *Oil Shale Development in the United States, Prospects and Policy Issues*, RAND Corporation, Santa Monica, CA.

BBC News, 2007. EU Ministers Agree Biofuel Target, 15 February; http://news.bbc.co.uk/2/hi/europe/6365985.stm.

BBC News, 2007. Extra Fuel Burnt in Air Fee Dodge, 3 December; http://news.bbc.co.uk/1/hi/england/7124021.stm.

BBC News, 2008. EU Rethinks Biofuels Guidelines, 14 January; http://news.bbc.co.uk/2/hi/europe/7186380.stm.

BBC News, 2008. Why Are Wheat Prices Rising? 26 February; http://news.bbc.co.uk/1/hi/business/7264653.stm#subject.

Benmaamar, M., 2003. *Urban Transport Services in Sub-Saharan Africa, Improving Vehicle Operations*, Sub-Saharan Africa Transport Policy Program, SSATP Working Paper No. 75, The World Bank and Economic Commission for Africa.

Boeing, 2007. *Current Market Outlook 2007*, Boeing Commercial Airplanes, Market Analysis, Seattle, WA.

Boeing, various years. *Annual Report*; www.boeing.com/.

Bonnefoy, P.A., R.J. Hansman, 2007. Potential Impacts of Very Light Jets on the National Airspace System, *Journal of Aircraft*, 44(4): 1318–1326.

Brewer, G.D., 1990. *Hydrogen Aircraft Technology*, CRC Press.

British Petroleum, 2007. *BP Statistical Review of World Energy*; www.bp.com/statisticalreview/.

Bundesamt für Statistik, 1992. *Schweizerische Verkehrsstatistik 1990*, Bern.

Bundesverkehrsministerium, various years. *Verkehr in Zahlen*, Deutscher Verkehrsverlag Hamburg.

Burke, A., E. Abeles, B. Chen, 2004. *The Response of the Auto Industry and Consumers to Changes in the Exhaust Emission and Fuel Economy Standards (1975–2003): A Historical Review of Changes in Technology, Prices, and Sales of Various Classes of Vehicles*, Institute of Transportation Studies, University of California, Davis.

Burke, A.F., M. Gardiner, 2005. *Hydrogen Storage Options: Technologies and Comparisons for Light-Duty Vehicle Applications*, UCD-ITS-RR-05-01, Institute of Transportation Studies, University of California, Davis.

BusinessWeek Online, 2005. Boeing vs. Airbus: Time to Escalate, 21 March; www.businessweek.com/bwdaily/dnflash/mar2005/nf20050321_4418_db046.htm.

Campbell, C.J., J.H. Laherrère, 1998. The End of Cheap Oil, *Scientific American*, March: 78–83.

CBS News, 2005. Could You Save Gas Money By Driving A Model T? 16 August; http://cbs5.com/seenon/local_story_228191600.html.

Choo, S., P.L. Mokhtarian, 2004. What Type of Vehicle Do People Drive? The Role of Attitude and Lifestyle in Influencing Vehicle Type Choice, *Transportation Research A*, 38(3): 201–222.

Christian Science Monitor, 2005. Hot New Style for Stodgy Old Detroit, 24 January; http://csmonitor.com/2005/0124/p13s01-wmgn.html.

Clarke, J.-P., D. Bennett, K. Elmer, J. Firth, R. Hilb, N. Ho, S. Johnson, S. Lau, L. Ren, D. Senechal, N. Sizov, R. Slattery, K.-O. Tong, J. Walton, A. Willgruber, D. Williams, 2006. *Development, Design, and Flight Test Evaluation of a Continuous Descent Approach Procedure for Night-time Operation at Louisville International Airport*, PARTNER Report PARTNER-COE-2005-02; www.partner.aero/.

Clarke, L.E., J.A. Edmonds, H.D. Jacoby, H.M. Pitcher, J.M. Reilly, R.G. Richels, 2007. *Scenarios of Greenhouse Gas Emissions and Atmospheric Concentrations*, U.S. Climate Change Science Program, Washington, DC.

CNN, 2004. Car Buyers Still Shun Fuel Efficiency, Surveys Find Even $2-a-Gallon Gasoline Isn't Likely to Cause Big Change in Car-Buying Habits, 19 October; http://money.cnn.com/2004/10/18/pf/autos/fuel_economy.

Commission of the European Communities, 2008. *Directive of the European Parliament and of the Council amending Directive 203/87/EC so as to include aviation activities in the scheme for greenhouse gas emissions allowances within the Community* [COM(2008) 548 final]. Brussels, 18 September.

Commission of the European Communities, various years. *Monitoring of ACEA's Commitment on CO_2 Emission Reductions from Passenger Cars*, Joint Report of the European Automobile Manufacturers' Association and the Commission Services, Brussels.

Concawe, EUCAR, Joint Research Center, 2007. *Well-to-Wheels Analysis of Future Automotive Fuels and Powertrains in the European Context*, Version 2c; http://ies.jrc.ec.europa.eu/WTW.

Davis, S.C., S.W. Diegel, various years. *Transportation Energy Data Book*, Oak Ridge National Laboratory, Oak Ridge, TN.

De Carvalho Macedo, I., M.R.L. Verde Real, J.E.A. Ramos da Silva, 2004. *Assessment of Greenhouse Gas Emissions in the Production and Use of Ethanol in Brazil*, Government of the State of São Paulo, Brazil.

DeCicco J., F. An, M. Ross, 2001. *Technical Options for Improving the Fuel Economy of U.S. Cars and Light Trucks by 2010–2015*, American Council for an Energy-Efficient Economy, Washington, DC.

Deffeyes, K.S., 2001. *Hubbert's Peak—The Impending World Oil Shortage*, Princeton University Press.

DeLucchi, M.A., 2003. *Appendix E: Methane Emissions from Natural Gas Production, Oil Production, Coal Mining, and Other Sources*, UCD-ITS-RR-03-17E, Institute of Transportation Studies, University of California, Davis.

DeLucchi, M.A., 2003. *A Lifecycle Emissions Model (LEM): Lifecycle Emissions from Transportation Fuels, Motor Vehicles, Transportation Modes, Electricity Use, Heating and Cooking Fuels, and Materials—Documentation of Data and Methods*, UCD-ITS-RR-03-17, Institute of Transportation Studies, University of California, Davis.

DeLucchi, M., S.-L. Hsu, 1996. *The External Damage Cost of Direct Noise From Motor Vehicles*, Report 14 in the series The Annualized Social Cost of Motor-Vehicle Use in the United States, based on 1990–1991 data, Institute of Transportation Studies, University of California, Davis.

Denman, K.L., G. Brasseur, A. Chidthaisong, P. Ciais, P.M. Cox, R.E. Dickinson, D. Hauglustaine, C. Heinze, El. Holland, D. Jacob, U. Lohmann, S. Ramachandran, P.L. da Silva Dias, S.C. Wofsy, X. Zhang, 2007. Couplings Between Changes in the Climate System and Biogeochemistry, IPCC Fourth Assessment

Report, Working Group I Report, *Climate Change 2007—The Physical Science Basis*, Cambridge University Press, Cambridge, UK.

Department for Transport, various years. *Transport Statistics Great Britain*, London.

Duvall, M., D. Taylor, M. Wehrey, N. Pinsky, W. Warf, F. Kahlhammer, L. Browning, 2004. *Advanced Batteries for Electric Vehicles, A Technology and Cost-Effectiveness Assessment for Battery Electric, Power Assist Hybrid Electric Vehicles, and Plug-in Hybrid Electric Vehicles*, Electric Power Research Institute, Palo Alto, CA.

Dyni, J.R., 2005. *Geology and Resources of Some World Oil-Shale Deposits*, U.S. Geological Survey, VA.

Echenique, M., 2007. Mobility and Income, *Environment and Planning A*, 39(8): 1783–1789.

Eclipse Aviation, *Eclipse 500 Performance*; www.eclipseaviation.com/eclipse_500/performance/.

Ellerman, A.D., H.D. Jacoby, M. Zimmerman, 2006. *Bringing Transportation into a Cap-and-Trade Regime*, MIT Joint Program on the Science and Policy of Global Change, Report No. 136, Massachusetts Institute of Technology, Cambridge.

Ellerman, A.D., P. Joskow, D. Harrison, 2003. *Emissions Trading: Experience, Lessons, and Considerations for Greenhouse Gases*, Pew Center on Global Climate Change, Washington, DC.

Energy & Environment, 2006. *Ethanol-Boosted Gasoline Engine Promises High Efficiency at Low Cost*, October. MIT Laboratory for Energy and the Environment, Massachusetts Institute of Technology, Cambridge.

Eno Transportation Foundation, various years. *Transportation in America*, Washington, DC.

Espey, J.A., M. Espey, 2004. Turning on the Lights: A Meta-Analysis of Residential Electricity Demand Elasticities, *Journal of Agricultural and Applied Economics*, 36(1): 65–81.

European Automobile Manufacturers' Association (ACEA), *ACEA Position on the Use of Bio-Diesel (FAME) and Synthetic Bio-Fuel in Compression Ignition Engines*; www.acea.be/images/uploads/070208_ACEA_FAME_BTL_final.pdf.

European Automobile Manufacturers' Association, *New Passenger Car Registrations in Western Europe*; www.acea.be/

European Automobile Manufacturers' Association, *Trends in New Car Characteristics*; www.acea.be/.

European Automobile Manufacturers Association, *New Passenger Car Registrations in Western Europe, Breakdown by Specifications: Average Power, Historical Series: 1990–2005*; www.acea.be/.

European Commission, 2003. *Time Use of Different Stages of Life—Results from 13 European Countries*, Office for Official Publications of the European Communities, Luxembourg.

European Commission, 2004. EU–U.S. Agreement on Large Civil Aircraft 1992: Key Facts and Figures, Press Release, October 6.

European Commission, 2007. Reducing CO_2 Emissions on Light-Duty Vehicles, *Europa*; http://ec.europa.eu/environment/co2/co2_home.htm.

European Union, 2003. *Directive 2003/30/EC of the European Parliament and of the Council of 8 May 2003 on the Promotion of the Use of Biofuels or Other Renewable Fuels for Transport*; http://ec.europa.eu/energy/res/legislation/doc/biofuels/en_final.pdf.

Fargione, J., J. Hill, D. Tilman, S. Polasky, P. Hawthorne, 2008. Land Clearing and the Biofuel Carbon Debt, *Science*, 29 February, 319(5867): 1235–1238.

Farrell, A.E., R.J. Plevin, B.T. Turner, A.D. Jones, M. O'Hare, D. Kammen, 2006. Ethanol Can Contribute to Energy and Environmental Goals, *Science*, 311(5760): 506–508.

Farrell, A.E., D. Sperling, 2007. *A Low-Carbon Fuel Standard for California, Part 1: Technical Analysis*, California Energy Commission; www.energy.ca.gov/low_carbon_fuel_standard/.

Field, F.R., J.P. Clark, 1997. A Practical Road to Lightweight Cars, *Technology Review*, January: 28–36.

Ford Motor Company, 2003. *Annual Report*; www.ford.com/en/company/investorInformation/companyReports/annualReports/default.htm.

Gautier, D.L., G.L. Dolton, K.I. Takahashi, K.L. Varnes, eds., 1996. *1995 National Assessment of United States Oil and Gas Resources—Results, Methodology, and Supporting Data*, U.S. Geological Survey Digital Data Series DDS-30, Release 2.

Gerard, D., L. Lave, 2004. *CAFE Increases: Missing the Elephant in the Room*, AEI–Brookings Joint Center for Regulatory Studies, Regulatory Analysis, Related Publication 04-11; www.aei-brookings.org/.

Gibson, P., 2007. *Coal to Liquids at Sasol*, Kentucky Energy Security Summit, 11 October, Lexington, KY.

Gillen, D., W. Morrison, C. Stuart, 2004. *Air Travel Demand Elasticities: Concepts, Issues and Measurement*, School of Business and Economics, Wilfrid Laurier University, Waterloo, Canada.

Godoy, J., 2008. *Subsidy Loss Threatens German Bio-Fuel Industry*, Enerpub, Energy Publisher, 7 January; www.energypublisher.com/article.asp?id=13410.

Goodwin, P.B., 1992. A Review of New Demand Elasticities with Special Reference to Short and Long Run Effects of Price Changes, *Journal of Transport Economics and Policy*, May: 155–169.

Goodwin, P.B., J. Dargay, M. Hanly, 2004. Elasticities of Road Traffic and Fuel Consumption with Respect to Price and Income: A Review, *Transport Reviews*, 24(3): 275–292.

Green, J.E., 2002. Greener by Design—The Technology Challenge, *Aeronautical Journal*, 106(1056): 57–103.

Greene, D.L., 1992. Energy-Efficiency Improvement Potential of Commercial Aircraft. *Annual Review of Energy and the Environment*, 17: 537–573.

Greene, D.L., 1992. Vehicle Use and Fuel Economy: How Big Is the Rebound Effect? *Energy Journal*, 13(1): 117–143.

Greene, D.L., 1995. Commercial Air Transport Energy Use and Emissions: Is Technology Enough? Presented at 1995 Conference on Sustainable Transportation Energy Strategies, Pacific Grove, CA.

Greene, D.L., J. DeCicco, 2000. Engineering-Economic Analyses of Automotive Fuel Economy Potential in the United States, *Annual Review of Energy and the Environment*, 25: 477–535.

Greene, D.L., O.D. Patterson, M. Singh, J. Li, 2005. Feebates, Rebates and Gas-Guzzler Taxes: A Study of Incentives for Increased Fuel Economy, *Energy Policy*, 33: 757–775.

Greene, D.L., A. Schäfer, 2003. *Reducing Greenhouse Gas Emissions from U.S. Transportation*, Pew Center for Global Climate Change, Washington DC.

Greene, N., 2004. *Growing Energy—How Biofuels Can Help End America's Oil Dependence*, National Resources Defense Council, New York City.

Greener by Design Science and Technology Sub-Group, 2005. *Air Travel—Greener by Design, Mitigating the Environmental Impact of Aviation: Opportunities and Priorities*, Royal Aeronautical Society, London.

Grübler, A. 1990. *The Rise and Fall of Infrastructures*, Heidelberg, Physica-Verlag.

Hall, D.O., F. Rosillo-Calle, R.H. Williams, J. Woods, 1993. Biomass for Energy: Supply Prospects, in Johansson, T.B., H. Kelly, A.K.N. Reddy, R.H. Williams, L. Burnham, 2003 (eds.), *Renewable Energy—Sources for Fuels and Electricity*, pp. 593–651, Island Press, Washington, DC.

HarrisInteractive Market Research, 2004. *The AutoVibes Report, Automotive Attitudes Summary*, September 2004 results; www.harrisinteractive.com/.

Hathaway, P., 2000. *The Effects of the Fuel 'Protest' on Road Traffic*, Transport Statistics Road Traffic Division, DETR, UK.

Heston, A., R. Summers, B. Aten, 2006. *Penn World Table*, Version 6.2, Center for International Comparisons of Production, Income, and Prices at the University of Pennsylvania, Philadelphia.

Hileman, J., Z.S. Spakovszky, M. Drela, M. Sargeant, 2007. Airframe Design for 'Silent Aircraft,' *AIAA Journal of Aircraft*, Paper 2007-0453.

Hu, P.S., T.R. Reuscher, 2004. *Summary of Travel Trends*, 2001 National Household Travel Survey, Federal Highway Administration, U.S. Department of Transportation, Washington, DC.

Hu, P.S., J.R. Young, 1999. *Summary of Travel Trends*, 1995 Nationwide Personal Transportation Survey, Federal Highway Administration, U.S. Department of Transportation, Washington, DC.

Hubbert, M.K., 1956. Nuclear Energy and the Fossil Fuels, presented before the Spring Meeting of the Southern District Division of Production, American Petroleum Institute, Plaza Hotel, San Antonio, TX, March 7–9; www.energybulletin.net/13630.html.

Hughes, J.E., C.R. Knitiel, D. Sperling, 2008. Evidence of a Shift in the Short-Run Price Elasticity of Gasoline Demand, *Energy Journal*, 29(1): 113–134.

Institut National de la Statistique et des Études Économiques, various years. *Les comptes des transports*, Commission des Comptes des Transports de la Nation, Paris.

Institute of Energy Economics, 2007. *Handbook of Energy & Economic Statistics in Japan*, The Energy Data and Modeling Center, Tokyo.

Intergovernmental Panel on Climate Change, 2007. *Climate Change 2007: Synthesis Report*, Summary for Policy Makers, Cambridge University Press, UK.

Intergovernmental Panel on Climate Change, 2007. Working Group 1, *Fourth Assessment Report (Technical Summary)*; http://ipcc-wg1.ucar.edu/wg1/wg1-report.html.

International Civil Aviation Organization (ICAO), 2007. *ICAO Environmental Report 2007*, Environmental Unit of the ICAO, Montreal.

International Energy Agency (IEA), 2007. *Energy Balances of OECD and Non-OECD Countries*, IEA/OECD, Paris.

International Herald Tribune, 2008. Virgin Atlantic Flies Jumbo Jet Powered by Biofuel, 24 February; www.iht.com/articles/2008/02/24/business/biofuel.php.

International Monetary Fund, 2007. *International Financial Statistics Yearbook*, Washington, DC.

Jamin, S., A. Schäfer, M.E. Ben-Akiva, I.A. Waitz, 2004. Aviation Emissions and Abatement Policies in the United States: A City Pair Analysis, *Transportation Research D*, 9(4): 294–314.

Japan Automobile Manufacturers Association, *Motor Vehicle Statistics*; www.jama.org/.

Jelinek, F., S. Carlier, J. Smith, A. Quesne, 2002. *The EUR RVSM Implementation Project Environmental Benefit Analysis*, EEC/ENV/2002/008, October.

Joint Planning and Development Office, 2004. *Next Generation Air Transport System—Integrated Plan*, December; www.jpdo.aero/.

Joint Planning and Development Office, 2006. *NGATS 2005 Progress Report to the Next Generation Air Transportation Integrated Plan*; www.jpdo.gov/library/2005_Progress_Report.pdf.

Joint Planning and Development Office, 2007. *Next Generation Air Transportation System: Concept of Operations (ConOps)*, Version 2.0, 13 June.

Kasseris, E., J.B. Heywood, 2007. *Comparative Analysis of Automotive Powertrain Choices for the Next 25 Years*, Society of Automotive Engineers, SAE Paper 2007-01-1605.

Kelkar, A., R. Roth, J. Clark, 2001. Automobile Bodies: Can Aluminum Be an Economical Alternative to Steel? *Journal of Metals*, 53(8): 28–32.

Kettunen, T., J.-C. Hustache, I. Fuller, D. Howell, J. Bonn, D. Knorr, 2005. Flight Efficiency Studies in Europe and the United States, presented at 6th USA/Europe Air Traffic Management R&D Seminar, Baltimore, MD, June.

Kintner-Meyer, M., K. Schneider, R. Pratt, 2007. *Impacts Assessment of Plug-in Hybrid Vehicles on Electric Utilities and Regional U.S. Power Grids*, Pacific Northwest National Laboratory, Richland, WA.

Kirby, H., 1981. Foreword, *Transportation Research A*, 15: 1–6.

Kloas, J., U. Kunert, H. Kuhfeld, 1993. *Vergleichende Auswertungen von Haushaltsbefragungen zum Personennahverkehr (KONTIV 1976, 1982, 1989)*, Gutachten im Auftrage des Bundesministers für Verkehr, Deutsches Institut für Wirtschaftsforschung, Berlin.

Koplow, D., 2006. *Biofuels—At What Cost? Government Support for Ethanol and Biodiesel in the United States*, prepared for the Global Subsidies Initiative (GSI) of the International Institute for Sustainable Development (IISD), Geneva, Switzerland.

Laherrere, J., 2004. Natural Gas Future Supply, presented at the International Energy Workshop at the International Institute for Applied Systems Analysis, Laxenburg, Austria, June 22–24.

Lave, C., 1991. *Things Won't Get a Lot Worse: The Future of U.S. Traffic Congestion*, Working Paper 33, University of California Transportation Center, University of California, Berkeley.

Lee, J.J., S.P. Lukachko, I.A. Waitz, A. Schäfer, 2001. Historical and Future Trends in Aircraft Performance, Cost, and Emissions, *Annual Review of Energy and the Environment 2001*, 26: 167–200.

Leiby, P., D.W. Jones, T.R. Curlee, R. Lee, 1997. *Oil Imports: An Assessment of Benefits and Costs*, U.S. Oak Ridge National Laboratory, Oak Ridge, TN.

Leigh Haag, A., 2007. Algae Bloom Again, *Nature*, 447: 520–521.

Liebeck, R.H., 2004. Design of the Blended Wing Body Subsonic Transport, *AIAA Journal of Aircraft*, 41(1): 10–25.

Lovins, A.B., E.K. Datta, O.-E. Bustnes, J.G. Koomey, N.J. Glasgow, 2005. *Winning the Oil Endgame—Innovation for Profits, Jobs, and Security*, Rocky Mountain Institute, Snowmass, CO.

Lubowski, R.N., M. Vesterby, S. Bucholtz, A. Baez, M.J. Roberts, 2006. *Major Uses of Land in the United States, 2002*, Economic Information Bulletin (EIB-14), U.S. Department of Agriculture, Washington, DC.

Lutter, R., T. Kravitz, 2003. *Do Regulations Requiring Light Trucks to Be More Fuel Efficient Make Economic Sense*, AEI–Brookings Joint Center for Regulatory Studies, Regulatory Analysis 03-2.

Lynch, M.C., 2003. The New Pessimism about Petroleum Resources: Debunking the Hubbert Model (and Hubbert Modelers), *Minerals and Energy*, 18(1): 21–32.

Maddison, A., 2001. *The World Economy—A Millennial Perspective*, OECD, Paris.

Mannering, F., C. Winston, W. Starkey, 2002. An Exploratory Analysis of Automobile Leasing by U.S. Households, *Journal of Urban Economics*, 52(1): 154–176.

Marais, K., S.P. Lukachko, M. Jun, A. Mahashabde, I.A. Waitz, 2007. Assessing the Impact of Aviation on Climate, *Meteorologische Zeitschrift*, 17(2): 157–172.

Marchetti, C., 1994. Anthropological Invariants in Travel Behavior, *Technological Forecasting and Social Change*, 47: 75–88.

Marchetti, C., N. Nakićenović, 1979. *The Dynamics of Energy Systems and the Logistic Substitution Model*, RR-79-13, International Institute for Applied Systems Analysis, Laxenburg, Austria.

McAlinden, S.P., K. Hill, B. Swiecki, 2003. *Economic Contribution of the Automotive Industry to the U.S. Economy—An Update*, a study prepared for the Alliance of Automobile Manufacturers, Center for Automotive Research, Ann Arbor, MI.

McCubbin, D.R., M.A. DeLucchi, 1999. The Health Costs of Motor-Vehicle-Related Air Pollution, *Journal of Transport Economics and Policy*, 33(3): 253–286.

McFarlane, S., 2008. Brazil Ethanol Consumption Overtakes Gasoline, *Dow Jones Newswires*; www.cattlenetwork.com/International_Content.asp?contentid=216018.

Media.Ford.com, 2008. *A Short History of the Model T Automobile*; http://media.ford.com/article_display.cfm?article_id=860.

Meyer, R.F., E.D. Attanasi, 2003. *Heavy Oil and Natural Bitumen—Strategic Petroleum Resources*, USGS Fact Sheet 70-03, U.S. Geological Survey; http://pubs.usgs.gov/fs/fs070-03/fs070-03.html.

Meyer, R.F., E.D. Attanasi, P.A. Freeman, 2007. *Heavy Oil and Natural Bitumen Resources in Geological Basins of the World*, U.S. Geological Survey, VA.

Ministère de l'Écologie, du Développement et de l'Aménagement Durables, 2006. *Les Comptes des Transports en 2007*, Paris, France.

Model T Ford Club of America, *Model T Encyclopedia*; www.mtfca.com/encyclo/index.htm.

Mokhtarian, P.L., 2004. Reducing Road Congestion: A Reality Check—A Comment, *Transport Policy*, 11: 183–184.

Mokhtarian, P.L., C. Chen, 2004. TTB or Not TTB, That Is the Question: A Review and Analysis of the Empirical Literature on Travel Time (and Money) Budgets, *Transportation Research A*, 38(9–10): 643–675.

Moreira, J.R., J. Goldemberg, 1999. The Alcohol Program, *Energy Policy*, 27: 229–245.

Morrison, S.A., 1984. An Economic Analysis of Aircraft Design, *Journal of Transport, Economics and Policy*, 18(2): 123–143.

Morrison, S.A., C. Winston, 1997. *Market-Based Solutions to Air Service Problems for Medium-Sized Communities*, Hearing before the Subcommittee on Aviation, Committee on Transportation and Infrastructure, U.S. House of Representatives, Washington, DC.

Motor Vehicle Manufacturers Association, 1996. *World Motor Vehicle Data*, Detroit, MI.

Mozdzanowska, A., R. Weibel, E. Lester, R.J. Hansman, 2007. The Dynamics of Air Transportation System Transition, 7th USA/Europe Air Traffic Management R&D Seminar, Barcelona, July.

Nakićenović, N., 1986. The Automobile Road to Technological Change, *Technological Forecasting and Social Change*, 29: 309–340.

Nakićenović, N., J. Alcamo, G. Davis, B. de Vries, J. Fenhann, S. Gaffin, K. Gregory, A. Grübler, T. Yong Jung, T. Kram, E. Lebre La Rovere, L. Michaelis, S. Mori, T. Morita, W. Pepper, H. Pitcher, L. Price, K. Riahi, A. Roehrl, H.-H. Rogner, A. Sankovski, M. Schlesinger, P. Shukla, S. Smith, R. Swart, S. van Rooijen, N. Victor, Z. Dadi, 2000. *Special Report on Emission Scenarios*, Intergovernmental Panel on Climate Change, Cambridge University Press, UK.

Nakićenović, N., A. Grübler, A. McDonald, 1998. *Global Energy Perspectives*, Cambridge University Press, Cambridge, UK.

National Energy Board, 2004. *Canada's Oil Sands: Opportunities and Challenges to 2015*; www.neb.gc.ca/.

National Museum of American History, *America on the Move*; http://americanhistory.si.edu/ONTHEMOVE/exhibition/.

National Research Council, 2001. *Review of the Research Program of the Partnership for a New Generation of Vehicles*, Seventh Report, Standing Committee to Review the Research Program of the Partnership for a New Generation of Vehicles, National Academy Press, Washington, DC.

National Research Council, 2002. *Effectiveness and Impact of Corporate Average Fuel Economy (CAFE) Standards*, Committee on the Effectiveness and Impact of Corporate Average Fuel Economy (CAFE) Standards, National Academy Press, Washington, DC.

National Research Council, 2004. *The Hydrogen Economy: Opportunities, Costs, Barriers, and R&D Needs*, National Academy of Engineering, Board on Energy and Environmental Systems, Washington, DC.

National Research Council and Aeronautical and Space Engineering Board, 1992. *Aeronautical Technologies for the Twenty-First Century*, National Academy Press, Washington, DC.

National Resource Council, 2001. *Evaluating Vehicle Emissions Inspection and Maintenance Programs*, National Academy Press, Washington, DC.

Neufville, R.L. de, 2005. The Future of Secondary Airports—Nodes of a Parallel Air Transport Network? *Cahiers Scientifiques du Transport* 47: 11–38.

Neufville, R.L. de, A.R. Odoni, 2003. *Airport Systems: Planning, Design, and Management*, McGraw-Hill.

Ng, H., M. Biruduganti, K. Stork, 2005. *Comparing the Performance of Sundiesel and Conventional Diesel in a Light-Duty Vehicle and Heavy-Duty Engine*, SAE Technical Paper Series, 2005-01-3776.

Oak Ridge National Laboratory, 2004. Online Analysis Tool of the 2001 U.S. National Household Travel Survey, http://nhts.ornl.gov/publications.shtml.

Ogden, C.L., C.D. Fryar, M.D. Carroll, K.M. Flegal, 2004. *Mean Body Weight, Height, and Body Mass Index, United States 1960–2002*. U.S. Department of Health and Human Services, Centers for Disease Control and Prevention, National Center for Health Statistics, Hyattsville, MD.

Paltsev, S., H. Jacoby, J. Reilly, L. Viguier, M. Babiker, 2005. Modeling the Transport Sector: The Role of Existing Fuel Taxes in Climate Policy, in Loulou R., J.-P. Waaub, G. Zaccour (eds.), *Energy and Environment*, Springer-Verlag, New York: 211–238.

Paltsev, S., J.M. Reilly, H.D. Jacoby, R.S. Eckaus, J. McFarland, M. Sarofim, M. Asadoorian, M. Babiker, 2005. *The MIT Emissions Prediction and Policy Analysis (EPPA) Model: Version 4*. MIT Joint Program on the Science and Policy of Global Change, Report 125, Massachusetts Institute of Technology, Cambridge; http://web.mit.edu/globalchange/.

Penner, J.E., D.H. Lister, D.J. Griggs, D.J. Dokken, M. McFarland, eds., 1999. *Aviation and the Global Atmosphere*, Cambridge University Press, Cambridge, UK.

People's Daily, 2001. The 96-km Beijing Fifth-Ring Road to Start Construction. Thursday, 06 December; http://english.people.com.cn/200112/06/eng20011206_86078.shtml.

Perlack, R.D., L.L. Wright, A.F. Turhollow, R.L. Graham, B.J. Stokes, D.C. Erbach, 2005. *Biomass as Feedstock for a Bioenergy and Bioproducts Industry: The Technical Feasibility of a Billion-Ton Annual Supply*, Oak Ridge National Laboratory, Oak Ridge, TN.

Peschka, W., 1992. *Liquid Hydrogen—Fuel of the Future*, Springer Verlag Wien, New York.

Poterba, J., 1991. Is the Gasoline Tax Regressive? in D. Bradford (ed.), *Tax Policy and the Economy 5*, MIT Press, Cambridge, MA, 145–164.

Poyer, D.A., M. Williams, 1993. Residential Energy Demand: Additional Empirical Evidence by Minority Household Type, *Energy Economics*, (15)2: 93–100.

PriceWaterhouseCoopers, 2003. *Shell Middle Distillate Synthesis (SMDS), Update to a Lifecycle Approach to Assess the Environmental Inputs and Outputs, and Associated Environmental Impacts, of Production and Use of Distillates from a Complex Refinery and SMDS Route*; www.shell.com/.

Quentin, F., 2007. *ACARE: The European Technology Platform for Aeronautics*, Seminar of the Industrial Leaders of European Technology Platforms,

Brussels, 12 December; ftp://ftp.cordis.europa.eu/pub/technology-platforms/docs/acare_en.pdf.

Renewable Fuels Association, *Industry Statistics*; www.ethanolrfa.org/industry/statistics/.

Research Triangle Institute, 1997. *User's Guide for the Public Use Data Tape*, 1995 Nationwide Personal Transportation Survey, Federal Highway Administration, U.S. Department of Transportation, Publication No. FHWA-PL-98-002.

Reynolds, T.G., L. Ren, J.-P.B. Clarke, 2007. Advanced Noise Abatement Approach Activities at Nottingham East Midlands Airport, UK, 7th USA/Europe Air Traffic Management R&D Seminar, Barcelona, Spain, July.

Roig-Franzia, M., 2007. A Culinary and Cultural Staple in Crisis: Mexico Grapples With Soaring Prices for Corn—and Tortillas, *WashingtonPost.com*, 27 January; www.washingtonpost.com/wp-dyn/content/article/2007/01/26/AR2007012601896_pf.html.

Rosenthal, E., 2008. Europe, Cutting Biofuel Subsidies, Redirects Aid to Stress Greenest Options, *New York Times*, 22 January; www.nytimes.com/2008/01/22/business/worldbusiness/22biofuels.html.

Ross, M., T. Wenzel, 2002. *An Analysis of Traffic Deaths by Vehicle Type and Model*, American Council for an Energy-Efficient Economy, Washington DC.

Roth, G.J., Y. Zahavi, 1981. Travel Time Budgets in Developing Countries, *Transportation Research A*, 15(1): 87–95.

Sadoway, D.R., 2005. Advanced Batteries for Automotive Applications, presentation at Advanced Transportation Workshop, Global Climate and Energy Project, Stanford University, October 11–12.

Schäfer, A., 1998. The Global Demand for Motorized Mobility, *Transportation Research A*, 32(6): 455–477.

Schäfer, A., 2000. Regularities in Travel Demand: An International Perspective, *Journal of Transportation and Statistics*, 3(3): 1–32.

Schäfer, A., 2005. Structural Change in Energy Use, *Energy Policy*, 33(4): 429–437.

Schäfer, A., J.B. Heywood, M.A. Weiss, 2006. Future Fuel Cell and Internal Combustion Engine Automobile Technologies: A 25 Year Lifecycle and Fleet Impact Assessment, *Energy—The International Journal*, 31(12): 1728–1751.

Schäfer, A., H.D. Jacoby, 2005. Technology Detail in a Multi-Sector CGE Model: Transport under Climate Policy, *Energy Economics*, 27(1): 1–24.

Schäfer, A., D.G. Victor, 2000. The Future Mobility of the World Population, *Transportation Research A*, 34(3): 171–205.

Schrank, D., T. Lomax, 2005. *The 2005 Urban Mobility Report*, Texas Transportation Institute, The Texas A&M University System.

Scientific American, 1919. Declining Supply of Motor Fuel, March 8, p. 220.

Searchinger, T., R. Heimlich, R.A. Houghton, F. Dong, A. Elobeid, J. Fabiosa, S. Tokgoz, D. Hayes, T.-H. Yu, 2008. Use of U.S. Croplands for Biofuels Increases

Greenhouse Gases through Emissions from Land-Use Change, *Science*, 29 February, 319(5867): 1238–1240.

Sequeira, C.J., 2008. *An Assessment of the Health Implications of Aviation Emissions Regulations*, S.M. Aeronautics and Astronautics, Massachusetts Institute of Technology, Cambridge.

Shapouri, H., J. Duffield, M. Wang, 2002. *The 2001 Net Energy Balance of Corn-Ethanol*, U.S. Department of Agriculture, Office of the Chief Economist, Washington, DC.

Shapouri, H., J.A. Duffield, M. Wang, 2002. *The Energy Balance of Corn Ethanol: An Update*, U.S. Department of Agriculture, Washington, DC.

Shapouri, H., P. Gallagher, 2005. *USDA's 2002 Ethanol Cost-of-Production Survey*, U.S. Department of Agriculture, Washington, DC.

Shapouri, H., M. Salassi, 2006. *The Economic Feasibility of Ethanol Production from Sugar in the United States*, U.S. Department of Agriculture, July.

Small, K., K. Van Dender, 2005. *The Effect of Improved Fuel Economy on Vehicle Miles Traveled: Estimating the Rebound Effect Using U.S. State Data, 1966–2001*, Department of Economics, University of California, Irvine.

Small, K.A., K. Van Dender, 2007. Fuel Efficiency and Motor Vehicle Travel: The Declining Rebound Effect, *Energy Journal*, 28(1): 25–51.

Sokolov, A.P., C.A. Schlosser, S. Dutkiewicz, S. Paltsev, D.W. Kicklighter, H.D. Jacoby, R.G. Prinn, C.E. Forest, J. Reilly, C. Wang, B. Felzer, M.C. Sarofim, J. Scott, P.H. Stone, J.M. Melillo, J. Cohen, 2005. *The MIT Integrated Global System Model (IGSM) Version 2: Model Description and Baseline Evaluation*, MIT Joint Program on the Science and Policy of Global Change, Report No. 124, Massachusetts Institute of Technology, Cambridge.

Sperling, D., 2001. Public-Private Technology R&D Partnerships: Lessons from U.S. Partnership for a new Generation of Vehicles, *Transport Policy*, 8: 247–256.

Sperling, D., E. Clausen, 2002. The Developing World's Motorization Challenge. *Issues in Science and Technology*, Fall; www.issues.org/19.1/sperling.htm.

St. Peter, J., 1999. *The History of Aircraft Gas Turbine Engine Development in the United States: A Tradition of Excellence*, International Gas Turbine Institute of the American Society of Mechanical Engineers, Atlanta, GA.

Subcommittee on Aviation, 2005. *Hearing on the U.S. Jet Transport Industry: Global Market Factors Affecting U.S. Producers*, U.S. House of Representatives, 25 May, Washington, DC; www.house.gov/.

Svercl, P.V., R.H. Asin, 1973. *Nationwide Personal Transportation Study: Home-to-Work Trips and Travel*, Report No. 8, Federal Highway Administration, U.S. Department of Transportation, Washington, DC.

Szalai, A., P.E. Converse, P. Feldheim, K.E. Scheuch, P.J. Stone, 1972. *The Use of Time: Daily Activities of Urban and Suburban Populations in 12 Countries*, Mouton, The Hague.

Tarr, J., C. McShane, 1997. The Centrality of the Horse to the Nineteenth-Century American City, in Mohl R. (ed.), *The Making of Urban America*, SR Publishers, New York, 105–130.

Thompson, F.M.L., 1970. *Victorian England: The Horse-Drawn Society*, an Inaugural Lecture, Bedford College, London.

Tijmensen, M.J.A., A.P.C. Faaij, C.N. Hamelinck, M.R.M. van Hardeveld, 2002. Exploration of the Possibilities for Production of Fischer-Tropsch Liquids and Power via Biomass Gasification, *Biomass & Bioenergy*, 23: 129–152.

UK Department of Trade and Industry (DTI), 1999. *Technology Status Report: Coal Liquefaction*, Technology Status Report 010, DTI, London.

UN Framework Convention on Climate Change (UNFCCC), 2006–2007. *Greenhouse Gas Inventory Data—Detailed Data by Party*; http://unfccc.int/ghg_data/ghg_data_unfccc/ghg_profiles/items/3954.php.

United Nations, 2004. *World Population Prospects: The 2004 Revision Population Database*, United Nations Population Division, New York.

U.S. Bureau of Economic Analysis, 2008. *National Income and Product Accounts*, Washington, DC; www.bea.gov/.

U.S. Bureau of Statistics, 1901. *Statistical Abstract of the United States*, Washington, DC.

U.S. Bureau of Transportation Statistics, 1997. *1995 American Travel Survey*, BTS/ATS95-U.S., U.S. Department of Transportation, Washington, DC.

U.S. Bureau of Transportation Statistics, various years. *National Transportation Statistics*, U.S. Department of Transportation, Washington, DC.

U.S. Census Bureau, 1975. *Historical Statistics of the United States, Colonial Times to 1970*, U.S. Department of Commerce, Washington, DC.

U.S. Census Bureau, various years. *Statistical Abstract of the United States*, U.S. Department of Commerce, Washington, DC.

U.S. Census Bureau, various years. *Vehicle Inventory and Use Survey* (formerly *Truck Inventory and Use Survey*), U.S. Department of Commerce, Washington, DC.

U.S. Council on Foreign Relations, 2006. *National Security Consequences of U.S. Oil Dependency*, New York.

U.S. Department of Energy, 1983. *Energy Technology Characterizations Handbook: Environmental Pollution and Control Factors*, Third Edition, DOE/EP-0093, Washington, DC.

U.S. Department of Energy, 2006. *Annual Energy Outlook 2006—With Projections to 2030*, U.S. Department of Energy, Washington, DC.

U.S. Department of Energy, 2007. *Annual Energy Outlook 2007—With Projections to 2030*, U.S. Department of Energy, Washington, DC.

U.S. Department of Energy, 2006. *Energy Efficiency and Renewable Energy, Hydrogen Fuel Cells & Infrastructure Program*; www.eere.energy.gov/hydrogenandfuelcells/storage/storage_challenges.html.

U.S. Department of Labor, 2000. *Telework: The New Workplace of the 21st Century*, Washington, DC.

U.S. Department of Transportation, 2008. *Air Carrier Financial Reports (Form 41 Financial Data), Schedule P-52*, U.S. Bureau of Transportation Statistics, Washington, DC; www.bts.gov/.

U.S. Department of Transportation, 2008. *Air Carrier Summary Data (Form 41 and 298C Summary Data), T2: U.S. Air Carrier Traffic and Capacity Statistics by Aircraft Type*, Bureau of Transportation Statistics, Washington, DC; www.bts .gov/.

U.S. Department of Transportation, 2008. *Air Carriers (Form 41 Traffic), U.S. Carriers: T-100 Domestic Segment*, Bureau of Transportation Statistics, Washington, DC; www.bts.gov/.

U.S. Department of Transportation, 2004. *Government Transportation Financial Statistics 2003*; Bureau of Transportation Statistics, Washington, DC.

U.S. Department of Transportation, *Historical Passenger Car Fleet Average Characteristics*, U.S. National Highway Traffic Safety Administration, Washington, DC; http://www.nhtsa.gov/CARS/rules/CAFE/HistoricalCarFleet.htm.

U.S. Department of Transportation, 2003. *National Survey of Distracted and Drowsy Driving Attitudes and Behaviors: 2002*, National Highway Traffic Safety Administration, Washington, DC.

U.S. Department of Transportation, 2006. *Average Fuel Economy Standards for Light Trucks, Model Years 2008–2011*, 49 CFR Parts 523, 533, and 537, RIN 2127-AJ61, National Highway Traffic Safety Administration, Washington, DC.

U.S. Department of Transportation, various years. *Highway Statistics*, Federal Highway Administration, Washington, DC.

U.S. Energy Information Administration, *2005 Primer on Gasoline Prices*; Washington, DC.

U.S. Energy Information Administration, 2007. *Annual Energy Review 2006*, U.S. Department of Energy, Washington, DC.

U.S. Environmental Protection Agency, 2005. *National Emissions Inventory—Air Pollutant Emissions Trends Data and Estimation Procedures, 1970–2002 Average Annual Emissions, All Criteria Pollutants*, Washington, DC.

U.S. Environmental Protection Agency, 2007. *Inventory of U.S. GHG Emissions and Sinks: 1990–2005*, U.S. EPA #430-R-07-002, Washington, DC.

U.S. Environmental Protection Agency, 2007. *Light-Duty Automotive Technology and Fuel Economy Trends: 1975 through 2007*, Compliance and Innovative Strategies Division and Transportation and Climate Division, Office of Transportation and Air Quality, Washington, DC.

U.S. Federal Aviation Agency, 2005. *RVSM Status Worldwide*; www.faa.gov/about/office_org/headquarters_offices/ato/service_units/enroute/rvsm/.

U.S. National Transportation Safety Board, 2008. *Aviation Accident Statistics*, Washington, DC.

Verband Kunststofferzeugende Indutrie, no date. *Kunststoff im Automobil—Einsatz und Verwertung*, Frankfurt/Main; www.vke.de/de/infomaterial/download/.

Wachs, M., 2003. *Improving Efficiency and Equity in Transportation Finance*, Center on Urban and Metropolitan Policy, The Brookings Institution, Washington, DC.

Waitz, I.A., J. Townsend, J. Cutcher-Gershenfeld, E. Greitzer, J. Kerrebrock, 2004. *Aviation and the Environment*, report to the United States Congress, Partnership for AiR Transportation Noise and Emissions Reduction, Massachusetts Institute of Technology, Cambridge.

Ward's Communications, various years. *Ward's Automotive Yearbook*, Detroit, MI.

Weber, H.J., A.J. Gellman, G.W. Hamlin, 2005. *Study of European Government Support to Civil Aeronautics R&D*, prepared for the National Aeronautics and Space Administration under Contract No. NNH04CC62C, TECOP International, Inc.

Weiss, M.A., J.B. Heywood, E.M. Drake, A. Schäfer, F. AuYeung, 2000. *On the Road in 2020—A Lifecycle Analysis of New Automobile Technologies*, Energy Laboratory Report MIT EL 00-003, Energy Laboratory, Massachusetts Institute of Technology, Cambridge.

Weiss, M.A., J.B. Heywood, A. Schäfer, V.K. Natarajan, 2003. *A Comparative Assessment of Advanced Fuel Cell Vehicles*, MIT Laboratory for Energy and the Environment 2003-001 RP, January, Massachusetts Institute of Technology, Cambridge.

Wit, M.P. de, A.P.C. Faaij, 2008. *Biomass Resources Potential and Related Costs, Assessment of the EU-27, Switzerland, Norway and Ukraine*, Refuel Work Package 3 Final Report, Refuel, Netherlands.

Wit, R.C.N., B.H. Boon, A. van Velzen, M. Cames, O. Deuber, D.S. Lee, 2005. *Giving Wings to Emissions Trading—Inclusion of Aviation under the European Emission Trading System: Design and Impacts*, report for the European Commission, DG Environment, No. ENV.C.2/ETU/2004/0074r.

Womack, J.P., D.T. Jones, D. Roos, 1990. *The Machine That Changed the World*, Harper Perennial, New York.

Wooten, H.H., K. Gertel, W.C. Pendleton, 1962. *Major Uses of Land and Water in the United States, Summary for 1959*, Farm Economics Division, Economic Research Service, U.S. Department of Agriculture, Washington, DC.

Zahavi, Y., 1981. *The UMOT–Urban Interactions*, DOT-RSPA-DPB 10/7, U.S. Department of Transportation, Washington, DC.

Index